Organization of the Early
Vertebrate Embryo

NATO ASI Series

Advanced Science Institutes Series

A series presenting the results of activities sponsored by the NATO Science Committee, which aims at the dissemination of advanced scientific and technological knowledge, with a view to strengthening links between scientific communities.

The series is published by an international board of publishers in conjunction with the NATO Scientific Affairs Division

A	**Life Sciences**	Plenum Publishing Corporation
B	**Physics**	New York and London
C	**Mathematical and Physical Sciences**	Kluwer Academic Publishers Dordrecht, Boston, and London
D	**Behavioral and Social Sciences**	
E	**Applied Sciences**	
F	**Computer and Systems Sciences**	Springer-Verlag
G	**Ecological Sciences**	Berlin, Heidelberg, New York, London,
H	**Cell Biology**	Paris, Tokyo, Hong Kong, and Barcelona
I	**Global Environmental Change**	

PARTNERSHIP SUB-SERIES

1. **Disarmament Technologies**	Kluwer Academic Publishers
2. **Environment**	Springer-Verlag
3. **High Technology**	Kluwer Academic Publishers
4. **Science and Technology Policy**	Kluwer Academic Publishers
5. **Computer Networking**	Kluwer Academic Publishers

The Partnership Sub-Series incorporates activities undertaken in collaboration with NATO's Cooperation Partners, the countries of the CIS and Central and Eastern Europe, in Priority Areas of concern to those countries.

Recent Volumes in this Series:

Volume 277 — The Pineal Gland and Its Hormones: Fundamentals and Clinical Perspectives
edited by Franco Fraschini, Russel J. Reiter, and Bojidar Stankov

Volume 278 — Obesity Treatment: Establishing Goals, Improving Outcomes, and Reviewing the Research Agenda
edited by David B. Allison and F. Xavier Pi-Sunyer

Volume 279 — Organization of the Early Vertebrate Embryo
edited by Nikolas Zagris, Anne Marie Duprat, and Antony Durston

Series A: Life Sciences

Organization of the Early Vertebrate Embryo

Edited by

Nikolas Zagris

University of Patras
Patras, Greece

Anne Marie Duprat

CNRS/Paul Sabatier University
Toulouse, France

and

Antony Durston

Hubrecht Laboratory
Utrecht, The Netherlands

Plenum Press
New York and London
Published in cooperation with NATO Scientific Affairs Division

Proceedings of a NATO Advanced Study Institute on
Organization of the Early Vertebrate Embryo,
held September 16–26, 1994,
in Spetsai, Greece,

NATO-PCO-DATA BASE

The electronic index to the NATO ASI Series provides full bibliographical references (with keywords and/or abstracts) to about 50,000 contributions from international scientists published in all sections of the NATO ASI Series. Access to the NATO-PCO-DATA BASE is possible in two ways:

—via online FILE 128 (NATO-PCO-DATA BASE) hosted by ESRIN, Via Galileo Galilei, I-00044 Frascati, Italy

—via CD-ROM "NATO Science and Technology Disk" with user-friendly retrieval software in English, French, and German (©WTV GmbH and DATAWARE Technologies, Inc. 1989). The CD-ROM also contains the AGARD Aerospace Database.

The CD-ROM can be ordered through any member of the Board of Publishers or through NATO-PCO, Overijse, Belgium.

Library of Congress Cataloging-in-Publication Data

Organization of the early vertebrate embryo / edited by Nikolas
  Zagris, Anne Marie Duprat, and Antony Durston.
      p.   cm. -- (NATO ASI series. Series A. Life sciences ; v.
  279)
    "Proceedings of a NATO Advanced Study Institute on Organization of
  the Early Vertebrate Embryo, held September 16-26, 1994, in Spetsai,
  Greece"--T.p. verso.
    "Published in cooperation with NATO Scientific Affairs Division."
  Includes bibliographical references and index.
    ISBN 0-306-45132-8
    1. Embryology--Vertebrates--Congresses.   I. Zagris, Nikolas.
  II. Duprat, Anne Marie.   III. Durston, Antony.   IV. North Atlantic
  Treaty Organization.   Scientific Affairs Division.   V. NATO Advanced
  Study Institute on Organization of the Early Vertebrate Embryo (1994
  : Spetsai, Greece)   VI. Series.
  QL959.W575   1995
  596'.033--dc20   95-46100
                                                           CIP

ISBN 0-306-45132-8

© 1995 Plenum Press, New York
A Division of Plenum Publishing Corporation
233 Spring Street, New York, N. Y. 10013

10 9 8 7 6 5 4 3 2 1

Printed in the United States of America

Preface

This book is the product of a NATO Advanced Study Institute of the same name, held at the Anargyrios and Korgialenios School on the island of Spetsai, Greece, in September 1994. The institute considered the molecular mechanisms which generate the body plan during vertebrate embryogenesis. The main topics discussed included: commitment and imprinting during germ cell differentiation; hierarchies of inductive cell interactions; the molecular functioning of Spemann's organizer and formation of embryonic axes; the extracellular matrix and the cytoskeleton in relation to morphogenesis and cell migration; neurogenesis and patterning of the neuraxis; the regulation of pattern formation by Hox genes and other transcription factors.

This ASI was marked by a number of special features. An important one was that it brought together three different generations of embryologists: pioneers in classical embryology; scientists who are now leading the present molecular elucidation of vertebrate embryogenesis; and the promising younger ASI participants, some of whom are already making important contributions to this field. This aspect was very important in determining the character of the meeting. It exposed ambiguities in the classical embryological dogma and thus facilitated a subtle application of the recent molecular findings to classical problems. The second shining feature of this ASI was its evolutionary emphasis. The findings presented were obtained in four different vertebrate systems: mammals (the mouse), avians (the chicken), amphibians (Xenopus) and the teleost fishes (zebrafish). Several speakers explicitly used evolutionary comparisons between these different systems to identify universal mechanisms and to understand their functioning. In conclusion, this was a stimulating, interactive meeting with highlights which were too numerous to list in detail. The format of didactic lectures followed by tutorials stimulated extensive discussions in which the current transition in modern molecular embryology from an inventorial phase to a synthetic, model-building phase fortunately became very clear.

The ASI was, by all measures, very successful and enjoyable, and we would like to express our sincere thanks to all who made it so. We thank the Scientific Affairs Division of NATO, which provided the major portion of the funds. We are grateful to the director of the ASI Programme, Prof. L. Veiga da Cunha, for his keen interest throughout the planning stages of the ASI. Supplementary funds were supplied by FEBS and their assistance is gratefully acknowledged. The National Science Foundation (U.S.A.) funded the travel expenses of four students and we would like to express our sincere thanks. We would also like to thank the personnel of the Anargyrios and Korgialenios School for the assistance they provided. The island of Spetsai provided an ideal location for stimulating discussions and a friendly exchange of views. The success of any meeting is largely a reflection of the attitude and input of the participants, and we would like to thank everyone who participated for helping to make this a splendid meeting.

Nikolas Zagris
Anne Marie Duprat
Antony Durston

CONTENTS

PRIMORDIAL GERM CELLS IN MAMMALS

Anne McLaren

Wellcome/CRC Institute of Cancer and Developmental Biology
University of Cambridge
Tennis Court Road
Cambridge, CB2 1QR

INTRODUCTION

Germ cells, defined as cells all of whose surviving descendants give rise to gametes (eggs or sperm), are the most fascinating cells in the body. The germ cell lineage, or germline, provides the only biological link, the only route of genetic transmission, from one generation to the next. Unlike somatic cell lineages, the germline is potentially immortal. For reproduction, for heredity, for development and for evolution, germ cells are of fundamental importance.

In mammals, the first days or weeks after fertilisation are occupied with cleavage and implantation, and with the development of the extra-embryonic life-support systems required for survival and growth in the environment of the uterus. Establishment of the germline as well as of the somatic cell lineages is delayed at least until gastrulation is under way. In the mouse, the first germ cells (termed primordial germ cells from the time of their first appearance until their entry into the genital ridges) can be identified about seven days after the egg is fertilised, midway through the period of gastrulation, by their high alkaline phosphatase activity (Ginsburg *et al.* 1990). The value of alkaline phosphatase histochemistry for tracing the early development of the germ cell lineage in mice (though unfortunately, not in all mammals) was first appreciated forty years ago, by Chiquoine (1954).

Once the germline is established, there ensues a period of migration, taking the primordial germ cells from their initial extraembryonic location to the genital ridges, the site of the future gonads. Up to this time, the appearance and behaviour of germ cells in female and male embryos has been indistinguishable;

Organization of the Early Vertebrate Embryo
Edited by N. Zagris *et al.*, Plenum Press, New York, 1995

but once in the genital ridges, germ cells in female mouse embryos enter prophase of the first meiotic division, while at the same time in male embryos, germ cells arrest in the G1 stage of a mitotic cell cycle. After birth, the meiotic germ cells (oocytes) in females become enclosed in primordial ovarian follicles, and remain dormant for weeks or months (up to fifty years in women) until the follicle starts to grow. Growth and maturation of the oocyte follows and the resulting egg may finally be ovulated and fertilised. Thus the egg undergoes an immensely complex sequence of differentiation steps but (unlike all other cells in the body) it ends up where it began.

In this paper, I shall discuss the establishment of the germline, the migration of primordial germ cells and finally their fate in the early genital ridges.

Establishment of the germline

At 7 1/4 days *post coitum* (i.e. on the morning of the 8th day of embryonic development), a small cluster of cells staining strongly for alkaline phosphatase activity can be detected in the extraembryonic region of the mouse egg cylinder (Ginsburg *et al*. 1990). The cluster is just proximal to the posterior end of the primitive streak and to the developing amnion and is bounded on one side by visceral endoderm and elsewhere by extraembryonic mesoderm. This cluster of cells is the first appearance of the germ cell lineage. It has long been known from various lines of evidence (notably the transplantation experiments of Gardner and Rossant, 1979) that the ancestors of primordial germ cells are derived from epiblast (embryonic ectoderm) and not from primary endoderm. However, it is only recently that the cell lineage tracing studies of Lawson and Hage (1994) have located the precursor cell population. It lies in the proximal epiblast of the 6 1/2 day mouse egg cylinder, close to the extraembryonic ectoderm and distributed around the circumference of the egg cylinder.

For many years the questions was debated as to whether there is a determined germline in mice, i.e. some cytoplasmic determinant of germ cell development, present in the egg and segregated during cleavage so that any cell lacking it is incapable of giving rise to germ cells. If there is not a determined germ line, when and where does germline lineage restriction occur? The chimera studies of Kelly (1977) established that no segregation of germline determinants occurs during early cleavage. It now appears from the work of Lawson and Hage (1994) that germline lineage restriction does not occur until the progenitor cell population has left the epiblast and been translocated to the extraembryonic location in which primordial germ cells can first be identified.

Lawson and Hage injected a fluorescent lineage marker into single epiblast cells of 6 or 6 1/2 day *post coitum* mouse embryos and traced their descendants after 40 hours' development in culture. Primordial germ cells were identified by alkaline phosphatase staining. Injected cells in the proximal half of the epiblast contributed to extraembryonic mesoderm; a minority of these descendant clones included some primordial germ cells. The proportion of primordial germ cells in

2

such a clone was never more than 50%, and could be as low as 5%, so lineage restriction had clearly not taken place at the time of cell injection. From the proportion of primordial germ cells that were descended from the injected cell in each embryo, it was estimated that the founding population at the time of lineage restriction contained about 45 cells. Doubling times for the injected cells increased with the proportion of primordial germ cells in the descendant clone, suggesting that cell division Slowed down at the time of lineage restriction, to about the 16-17 hour doubling time characteristic of later primordial germ cells. The time of germline lineage restriction was estimated as just over 7 days *post coitum*, shortly before the time at which primordial germ cells can first be identified. This suggests that lineage restriction does not take place until the precursor cells have passed through the primitive streak into the extraembryonic region.

To what extent germline lineage restriction is a cell-autonomous event, and to what extent it is induced by the location of the cells at the time of restriction, cannot be ascertained until transplantation experiments have been carried out.

During much of gastrulation, primordial germ cells are positioned outside the embryo proper, in the extraembryonic region, not only in mammals, but also in birds and other Vertebrates. Such sequestration may be important to preserve the pluripotency of the germ cells during a stage of development when somatic cell lineages are being determined and cell fates laid down within the embryo. In particular, Monk *et al.* (1987) suggested that germline sequestration might allow the germ cells to escape the widespread tissue-specific DNA methylation that is occurring in the somatic cell lineages of the embryo at this time. Subsequent work (Grant *et al.* 1992; Kafri *et al.*, 1992) has confirmed that the germ cell lineage does indeed remain strikingly undermethylated during this period.

GERM CELL MIGRATION

As gastrulation draws to an end and the allantois starts to grow out, the posterior part of the mouse egg cylinder invaginates to form the hind gut. The cells in the primordial germ cell cluster spread out and are carried into the hind gut with the invaginating endoderm. Some active locomotion of the germ cells may be involved initially, but the major part of this translocation is passive, brought about by morphogenetic movements of the surrounding tissues.

From the hindgut endoderm the germ cells migrate dorsally up the mesentery, and then laterally round the coelomic angles and into the genital ridges. As they move from the hindgut into the mesentery and from the mesentery into the ridges, their adhesion to fibronectin, an extracellular matrix molecule present together with laminin along the migration pathway, decreases strikingly (ffrench-Constant *et al.* 1991). Soluble factors may play a directional role, since medium conditioned by genital ridge tissue has been shown to exert both a proliferative and a chemotropic effect on migratory germ cells in culture (Godin *et al.*, 1990). Once they have left the hind gut, the germ cells put out cell processes and contact one another to form an extensive network (Gomperts *et al.* 1994). Contraction of these processes as the germ cells enter the genital ridge presumably assists in assembling the cells into testicular or ovarian cords within the gonad.

Two much studied mouse mutants, Dominant white-spotting (*W*) and Steel (*Sl*) have been known for decades to reduce, sometimes to zero, the number of germ cells reaching the genital ridges. Both mutants also affect two other migratory cell types, melanoblasts and haemopoietic stem cells. With strong alleles, the homozygotes die of anaemia before birth; if they survive, they are sterile and have white coats, lacking germ cells and pigment cells. Although *W/W* and *Sl/Sl* mice have such similar phenotypes, the genes act very differently: *W* acts cell autonomously, while *Sl* acts through the tissue environment. Thus *W/W* mice dying of anaemia could be rescued with *Sl/Sl* haemopoietic stem cells, but a transfusion in the reverse direction would be doubly lethal.

The molecular basis for this intriguing complementarity became clear when the two genes were cloned and their products identified. *W* turns out to encode a cell-surface tyrosine kinase receptor molecule, c-kit, while the gene product of *Sl* is the ligand of c-kit, variously termed stem-cell factor (SCF), Steel factor, mast-cell growth factor and kit ligand (for review, see Witte, 1990).

Although breakdown of the SCF/c-kit signal transduction pathway was known to reduce the number of primordial germ cells reaching the genital ridges, it was not known how early the reduction in number occurred, nor how it was mediated. Was the founder population of normal size? Did primordial germ cells in *W/W* or *Sl/Sl* embryos die, or fail to proliferate, or just fail to migrate? Using a restriction fragment length polymorphism to identify homozygous mutant embryos in litters segregating for *W^e*, a strong allele at the *W* locus, we counted primordial germ cells and plotted their distribution in *W^e/W^e* embryos and litter-mate controls at different stages of development (Buehr *et al.*,1994).

At the beginning of germ cell migration, the number of primordial germ cells was about double that estimated for the founder population and did not depend on the genotype of the embryo. This suggests that, although *W* is known to be expressed early in the germ cell lineage (Manova and Bachvarova, 1991), its product is not required for the first cell division after lineage restriction. From that time forward, while germ cell numbers in litter mate controls increased at a steady rate, no further increase in numbers occurred in *W/W* embryos and after about 24 hours, numbers started to decline. We concluded that the SCF/c-kit signal transduction pathway was required for germ cell proliferation during the migratory period, and that in its absence the germ cells eventually die. SCF has been shown to promote survival of primordial germ cells *in vitro* by suppressing apoptosis (Pesce *et al.* ,1993).

Although the primordial germ cells in *W^e/W^e* embryos, as in litter-mate controls, were distributed along the length of the hind gut, the mutant germ cells failed to migrate normally from the ventral to the dorsal side of the hind gut, and were retarded in migration up the mesentery. They also tended to be clumped together, suggesting that the cell surface receptor defect was interfering with the normal interaction between the germ cells and the substrate on which they were migrating.

Germ cells isolated during the migratory period or after entry into the genital ridge have proved notoriously difficult to maintain *in vitro*. (De Felici, M., and McLaren, A., 1983; Wabik-Sliz, B., and McLaren, A., 1984). They will divide once or twice, but rarely survive more than a few days. On an appropriate feeder layer, the number of germ cells surviving is increased (Donovan *et al.*, 1986), probably because the feeder layer is supplying SCF (Dolci *et al.*, 1991), but the duration of survival remains short.

A very different result is obtained if an appropriate feeder layer is combined with a cocktail of growth factors (SCF, leukaemia inhibitory factor, and basic fibroblast growth factor). The germ cells proliferate, and on repeated subculture continue to proliferate indefinitely, i.e. they have become immortalised (Matsui *et al.*, 1992; Resnick *et al.*, 1992). Their phenotype changes: they come increasingly to resemble embryonic stem cells (ES cells) and, like ES cells, lose their dependence on growth factors. These embryonic germ cell derived lines, or EG cells as they are now termed, can be derived from embryos 8 1/2 days *post coitum*, at the onset of germ cell migration, and also (though less efficiently) from male genital ridges 11 1/2 or 12 1/2 but not 15 1/2 days *post coitum*.

If injected back into a blastocyst, EG cells like ES cells can contribute to all the cell lineages of the developing embryo, including the germ cell lineage (Labosky *et al.*, 1994; Stewart, 1994). Thus the early epiblast has the potential for giving rise to both somatic and germ cell lineages, or (under appropriate *in vitro* conditions) to immortalised ES cell lines which are themselves capable (under appropriate in vivo conditions) of regaining the early epiblast potential; while the germ cell lineage appears capable (under appropriate *in vitro* conditions) of achieving an ES-cell like condition which can again revert to the pluripotent state of the early epiblast.

Molecular analysis of ES cells and EG cells is likely to increase our understanding of cell potency and its restriction and to throw light on the special features characterising primordial germ cells.

GERM CELL SEX DETERMINATION

Primordial germ cells in mice enter the genital ridges 10-11 days *post coitum*. Their cell surface properties change, they cease to show migratory behaviour on explantation, but they continue for another three days or so to proliferate mitotically, with a doubling time of about 16 hours. *Sry*, the sex-determining gene on the Y chromosome, is expressed in the somatic supporting cell lineage of the genital ridge at 10-11 days *post coitum*; this results in the differentiation of Sertoli cells, and by 12 1/2 days *post coitum* the first testis cords can be seen in the genital ridges of male embryos (McLaren, 1991). By the following day, a difference between the germ cells of male and female embryos is for the first time apparent. In male embryos, the germ cells arrest in the G1 stage of the cell cycle, as T-prospermatogonia. They do not divide again until after birth, and none enter meiosis until at least a week after birth.. In female embryos, the germ cells enter first meiotic prophase; as oocytes, they pass through leptotene, zygotene and pachytene, before arresting in the diplotene stage of first meiotic prophase after birth.

In the normal female genital ridge, all germ cells are XX in chromosome constitution and enter meiosis. In the normal male genital ridge, all germ cells are XY in chromosome constitution and undergo mitotic arrest. Is it the germ cells' own chromosomes that determine their fate at this stage of development, or is it the tissue environment in which they find themselves?

To answer this question, it is necessary to examine the fate of XX germ cells in a male genital ridge, and XY germ cells in a female genital ridge. In the mouse, various genetic and embryological manipulations make this possible, and the answer is clear-cut. XX, XO and XY germ cells all undergo mitotic arrest and develop as T-prospermatogonia. Thus whether or not a germ cell enters meiosis before birth is unrelated to its own chromosome constitution. In contrast, the fate of germ cells after birth is crucially dependent on their sex chromosome constitution (Burgoyne, 1993).

If entry into meiosis before birth depends on the environment of the female rather than the male genital ridge, does the female environment induce meiosis or does the male inhibit it? Or both? The first clue came from the observation of Zamboni and his colleagues that growing oocytes could be found after birth in the adrenal glands of both female and male mice (Zamboni and Upadhyay, 1983). We confirmed (McLaren, 1984) that occasional germ cells, identified by their high alkaline phosphatase activity, were present in the adrenal promordia, which in mouse embryos are located close to the genital ridges, and that in male as well as female embryos these ectopic germ cells entered meiosis at the same time as did germ cells in the female genital ridge. In the intervening mesonephric region, occasional germ cells could also be found; in male embryos some of these were in meiosis but others were in mitotic arrest, as typical T-prospermatogonia. We concluded that a short-range diffusible substance emanating from the male genital ridge is capable of inhibiting entry of germ cells into meiosis, and that they then develop as T-prospermatogonia.

What is the nature of this hypothetical meiosis inhibitor, and what cells produce it? It does not depend on the formation of testis cords: the precursors of peritubular cells that are required for testis cord formation normally enter the genital ridge from the mesonephric region after 11 1/2 days *post coitum*, so if a male genital ridge is isolated at 11 1/2 days, no testis cords form, yet the germ cells develop as T-prospermatogonia, not as oocytes (Buehr *et al.* , 1993). Sertoli cells, however, have already differentiated, and are producing anti-Mullerian hormone (AMH : also known as Mullerian Inhibiting Substance). AMH might therefore seem a possible candidate for the meiosis inhibitor: but an experiment by Vigier *et al.* (1987) in which embryonic rat ovaries were cultured in the presence of purified AMH indicated that AMH killed meiotic germ cells but did not block entry into meiosis or induce differentiation of T-prospermatogonia.

Is there in addition a meiosis-inducing substance in the ovary? If so, it would need to be present also in the adrenals of male and female mouse embryos; but this is not inconceivable, since the adrenal and the genital ridge have a very similar embryological origin.

Recently, we have been disaggregating embryonic lung, a tissue of endodermal rather than mesodermal origin, adding primordial germ cells from 11 1/2-day male embryos, reaggregating and culturing. We have also disagrregated, reaggregated and cultured 11 1/2-day male genital ridges, a procedure that interferes with Sertoli cell differentiation. In both cases, the germ cells enter meiosis and develop as oocytes (McLaren and Southee, unpublished results). Since it seems unlikely that lung and male genital ridges produce a meiosis-inducing substance, we conclude that all germ cells, whatever their chromosome constitution, are programmed to enter meiosis before birth, but in or near the testis an inhibitory substance intervenes to shunt germ cells along the male pathway of development.

Does entry of germ cells into meiosis before birth necessarily determine their phenotypic sex, i.e. whether they follow the pathway of oogenesis or spermatogenesis? Apparently yes: germ cells that fail to enter meiosis before birth, if they survive at all will undergo spermatogenesis; those that enter meiosis, whether in the ovary, in the adrenal, in a lung reaggregate or in a reaggregated male genital ridge, develop as growing oocytes.

CONCLUSION

It is becoming increasingly clear that germ cells and somatic cells interact closely throughout development (McLaren, 1994). Whether or not the establishment of the germ cell lineage requires an inductive signal from the somatic cells, we know that germ cells are totally dependent during their migration period on growth factors and extracellular matrix emanating from their somatic environment. Once in the genital ridge, whether a germ cell undergoes oogenesis or spermatogenesis again depends on signals from the environment.

Some genes, both in mouse and human, are expressed differently according to whether the allele has passed through the paternal or the maternal germ line. Such genomic imprinting, which may depend either on DNA methylation or on changes in chromatin configuration, reflects a difference between the male and female environment during gametogenesis. Given the close contact between germ cells and either Sertoli cells or follicle cells, which in humans may last for many years, there are likely to be many subtle ways in which germ cells may be influenced by their somatic environment.

REFERENCES

Buehr, M., McLaren, A., Bartley, A. and Darling, S., 1994, Proliferation and migration of primordial germ cells in *We/We* mouse embryos. *Dev Dyn* 198: 182-9.

Burgoyne, P.S., 1993, Deletion mapping of the functions of the mouse Y chromosome. *in* "Sex chromosomes and sex-determining genes", K.C. Reed and J.A.M. Graves, eds., Harwood Academic Publishers, Chur, Switzerland, pp 357-372.

Chiquoine, A.D., 1954, The identification, origin and migration of the primordial germ cells in the mouse embryo, *Anat Rec* 118:135-146.

De Felici, M., and McLaren, A., 1983, *In vitro* culture of mouse primordial germ cells. *Exp. Cell Res*, 144:417-427.

Dolci, S., Williams, D.E., Ernst, M.K., Resnick, J.L., Brannan, C.I., Lock, L.F., Lyman, S.D., Boswell, H.S. and Donovan, P.J.,1991, Requirement for mast cell growth factor for primordial germ cell survival in culture. *Nature* 352:809-811.

Donovan, P.J., Stott, D., Cairns, L.A., Heasman, J., and Wylie, C.C., 1986, Migratory and post migratory germ cells behave differently in culture. *Cell* 44:831-838.

French-Constant, C., Hollingsworth, A., Heasman, J., and Wylie, C., 1991, Response to fibronectin of mouse primordial germ cells before, during and after migration, *Development* 113:1365-1373.

Gardner, R.L., and Rossant, J., 1979, Investigation of the fate of 4.5 day *post-coitum* mouse inner cell mass cells by blastocyst injection. *J. Embryol. exp. Morph*, 52:141-152.

Ginsburg, M., Snow, M.H.L., and McLaren, A., 1990, Primordial germ cells in the mouse embryo during gastrulation. *Development* 110: 521-528.

Godin, I., Wylie, C., and Heasman, J., 1990, Genital ridges exert long-range effects on mouse primordial germ cell numbers and direction of migration in culture. *Development* 108:357-63.

Gomperts, M., Garcia, C.M., Wylie, C., and Heasman, J.,1994, Interactions between primordial germ cells play a role in their migration in mouse embryos. *Development* 120:135-41.

Grant, M., Zucotti, M., and Monk M., 1992, Methylation of CpG sites of two X-linked genes coincides with X-inactivation in the female mouse embryo but not in the germ line,*Nature Genetics* 2:161-166.

Kafri, T., Ariel, M., Brandeis, M., et al., 1992, Developmental pattern of gene-specific DNA methylation in the mouse embryo and germline. *Genes & Dev*. 6:7805-7814.

Kelly, S.J.,1977, Studies of the developmental potential of 4- and 8-cell stage mouse blastomeres. *J. exp. Zool.*, 200:365-376.

Labosky, P.A., Barlow, D.P., and Hogan, B.L.M., 1994, Embryonic germlines and their derivation from mouse primordial germ cells. *in*: Germline Development, Wiley, Chichester , Ciba Found. Symp. 182, pp157-178.

Lawson, K.A., and Hage, W.J., 1994, Clonal analysis of the origin of primordial germ cells in the mouse. *in* "Germline Development", J. Marsh and J. Goode, eds., Ciba Fndn. Symp. 182:68-84.

Manova, K. and Bachvarova, R.F.,1991, Expression of *c-kit* encoded at the W locus of mice in developing embryonic germ cells and presumptive melanoblasts. *Dev Biol* 146: 312-24.

Matsui, Y., Zsebo, K., and Hogan, B.L.M., 1992, Derivation of pluripotential embryonic stem cells from murine primordial germ cells in culture. *Cell* 70:841-847.

McLaren, A., 1984, Meiosis and differentiation of mouse germ cells. *in*: 38th Symp.Soc.Exp.Biol. ,"Controlling Events in Meiosis", C.W. Evans and H.G. Dickinson, eds., Company of Biologists, Cambridge.

McLaren A., 1991, Sex determination in mammals. Oxford Rev. Reprod. Biol. 13:1-33.

McLaren, A., 1994, Germline and soma: interactions during early mouse development, *in*: Seminars in Developmental Biology, 5:182-189.

Monk, M., Boubelik, M., and Lehnert, S., 1987, Temporal and regional changes in DNA methylation in the embryonic, extraembryonic and germ cell lineages during mouse embryo development, *Development*, 99:371-382.

Pesce, M., Farrace, M.G., Dolci, S., Piacentini, M. and De Felici, M., 1993, Stem cell factor and leukemia inhibitory factor promote primordial germ cell survival by suppressing programmed cell death (apoptosis). *Development*, 118: 1089-1094.

Resnick, J.L., Bixler, L.S., Cheng, L., and Donovan, P.J., 1992, Long-term proliferation of mouse primordial germ cells in culture. *Nature*: 359:550-551.

Stewart, C.L., 1994, Stem cells from primordial germ cells can re-enter the germ line. *Dev. Biol*, 161:626-628.

Vigier, B., Watrim, F., Magre, S. et al., 1987, Purified AMH induces a characteristic freemartin effect in fetal rat prospective ovaries exposed to it *in vitro*. *Development*, 100:43-55.

Wabik-Sliz, B., and McLaren, A., 1984, Culture of mouse germ cells isolated from fetal gonads. *Exp. Cell Res.*, 154:530-536.

Witte, O.N., 1994, Steel locus defines new multipotent growth factor. *Cell*, 63:5-6.

Zamboni, L. and Upadhyay, S., 1983, Germ cell differentiation in mouse adrenal glands, *J. Exp. Zool.*, 228:178-193.

EPIGENETIC PROGRAMMING OF GENE EXPRESSION AND IMPRINTING IN MOUSE AND HUMAN DEVELOPMENT

Marilyn Monk

Molecular Embryology Unit
Institute of Child Health
30 Guilford Street
London WC1N 1EH

SUMMARY

Gene expression has been extensively studied in mouse embryonic development. The studies have been directed towards the determination of the timing and tissue-specificity of activation of specific embryonic genes, the timing of the transition from the maternal gene products (inherited from the egg cytoplasm) to embryonic gene products, and the molecular mechanisms regulating these events. Originally gene expression was monitored by activity of protein products. Recently, sensitive PCR techniques, which discriminate the expression of individual alleles of specific genes, have allowed the study of hemizygous expression associated with X-chromosome inactivation and imprinting in development.

Global changes in methylation of the DNA occurring in early development are involved in programming different patterns of gene expression. Methods which detect methylation of specific CpG sites in X-linked and imprinted genes are helping to elucidate the role of methylation in the initiation and maintenance of the differential gene activity. These sensitive molecular analyses, established in the mouse, are now being applied to the study of early human development.

INTRODUCTION

Very little is known about the patterns of differential gene expression at different times and in different tissues in early development and the molecular mechanisms which initiate and maintain gene activity or inactivity. This is because the very few embryonic cells available make molecular studies difficult. Much of our work has been directed towards the development of highly sensitive procedures to detect enzyme activity, mutation or modification of specific genes, and transcription of particular alleles of a gene, in small numbers of cells. We have used X-chromosome inactivation during mouse development as a model system to apply these sensitive procedures to study gene activation and inactivation and the role of methylation in these processes. In this paper, I will review these studies in the mouse and indicate how they are being extended to the molecular analysis of human embryonic development.

X-INACTIVATION IN DEVELOPMENT

In the somatic cells of female mammals one of the two X chromosomes is inactivated (Lyon, 1961). This mechanism of dosage compensation ensures that females are equivalent to males with respect to the level of X-linked gene products. The X chromosomes of all eutherian mammals carry homologous genes although their arrangement may be different. Although most genes on the inactivated X chromosome are silent, there are several genes which escape inactivation (reviewed in Lyon, 1988; Monk, 1992).

Initially, biochemical assays for X-linked gene activity were used to establish the timing and tissue-specificity of X-inactivation in development (Monk and Harper, 1978; Kratzer and Gartler, 1978; Epstein et al. 1978). In the mouse female embryo, two X chromosomes are active during oocyte growth and maturation, whilst in sperm, the single X chromosome is inactive. Both X chromosomes are active in the preimplantation embryo. Inactivation of an X chromosome occurs first in the trophectoderm, then in the primary endoderm and finally in the fetal precursor cells of the egg cylinder (Monk and Harper, 1979). In the fetal precursor cells, X-inactivation is random. In the extra-embryonic tissues, it is always the paternally-inherited X chromosome from the sperm that is inactivated (one of the first examples of imprinting in mammals; reviewed in Monk and Grant, 1990). The memory distinguishing the gametic origin of the X chromosomes is thus retained for at least four or five cleavage divisions. In the female germ line, re-activation of the inactive X chromosome occurs around the time of meiosis (reviewed in Monk, 1981).

We have used the well-characterised cycle of changes in X-chromosome activity in female mouse embryo development to study DNA methylation as a mechanism of regulation of gene expression. We know that, in somatic cells, the silencing of genes on the inactive X chromosome is correlated with methylation of CpG islands at their 5' ends. Other sites in the body of an X-linked gene may be more methylated on the active X chromosome (reviewed in Monk, 1986). In some instances, methylation of CpG sites within the island of the inactive X chromosome may be variable. Nevertheless, there are certain key sites whose methylation is always associated with inactivity·

We have seen that methylation plays a role in the maintenance of gene activity and inactivity in somatic cells. Does methylation also play a role in the *initiation* of activation and inactivation of X-linked and autosomal genes in development?

METHYLATION IN DEVELOPMENT

Since it was known that tissue-specific genes were fully methylated in sperm, it was originally assumed that this was also the case in oocytes and that the "ground state" in development would be fully methylated. We examined this suggestion in two ways. At a time when techniques were not sufficiently sensitive to analyse specific genes in development, we looked at methylation of multiple copy sequences and overall genomic methylation (Monk et al. 1987). In the latter procedure, total DNA isolated from embryos and different embryonic lineages was cut with either MspI or HpaII and the resulting fragments end-labelled with P^{32} and electrophoresed on agarose gels. More methylated genomic DNA is more resistant to HpaII digestion. Hence, the degree of overall methylation could be determined by the proportion of larger to smaller fragments in densitometry tracings of the gels. These early experiments clearly established the following pattern of methylation in development:

1) Oocyte DNA is markedly undermethylated overall.

2) Sperm DNA is also relatively undermethylated compared with somatic tissue but is more methylated than the oocyte.

3) There is a striking loss of methylation between the 8-cell embryo and the blastocyst stage.

4) *De novo* methylation occurs at the time of implantation, detectable first in the ICM; it occurs over the period of gastrulation and to a greater extent in somatic lineages than in the extra-embryonic lineages.

5) The primordial germ cells initially escape detectable *de novo* methylation.

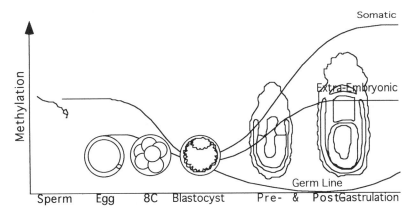

Figure 1. Diagram showing the global changes in DNA methylation during early development of the mouse.

These early observations, illustrated in Figure 1, were important in that they highlighted a number of possibilities with respect to the role of methylation in gene expression in development. These are:

1) The differential methylation of sperm and egg is a candidate mechanism for imprinting.

2) The marked loss of methylation in preimplantation development (and continuing into the germ line?) presents itself as a mechanism of erasure of gametic genetic programming in the somatic lineages and erasure of imprinting in the germ line.

3) The *de novo* methylation occurs independently in the extra-embryonic lineages and in the three definitive germ layers and thus provides a basis for differential programming of these different lineages,

4) The germ line escapes *de novo* methylation and this may be a requirement for retention of developmental potency

One reasonable hypothesis to explain the marked loss of methylation between the 8-cell and blastocyst stages is that the maintenance methylase may be absent in preimplantation embryos. Measurement of methylase activity showed that activity inherited in the egg cytoplasm was indeed rapidly diluted out during preimplantation development and, in addition, enzyme was actively degraded (other maternally inherited gene products are also degraded from the 8-cell stage, reviewed in Harper and Monk, 1983). However, although this profile of methylase activity follows the loss of methylation, this hypothesis is not entirely satisfactory. This is because the activity of methylase in the egg is extraordinarily high, over 400 times the normal level in somatic cells and, despite the loss of methylase activity, the residual activity is not markedly deficient compared to somatic cells (Monk et al. 1991).

METHYLATION OF THE X CHROMOSOME IN DEVELOPMENT

Subsequently, we developed sensitive techniques to study methylation of specific genes as X-inactivation occurs in the preimplantation embryo. We wanted to know whether the paternal (sperm) X chromosome was differently methylated from the maternal (egg) X chromosome (as a basis of imprinting and paternal X-inactivation), and whether methylation preceded or followed inactivation of X-linked genes. There was some early evidence (Lock et al. 1987) that the Hprt gene on the inactive X chromosome was not methylated fully until about day 10 of development, although we, and others, had shown earlier that this gene was already inactivated by day 4. We also wanted to know whether methylation was correlated with inactivation in both the embryonic and extra-embryonic tissues and whether it occurred at all on the inactive X chromosome in the primordial germ cells.

To answer these questions, the methylation status of specific HpaII (CCGG) sites of X-linked genes (Pgk-1, G6pd) was examined throughout the time of X-inactivation. This required highly sensitive PCR amplification of gene sequences containing the informative CpG site, before and after HpaII digestion (Singer-Sam et al. 1990; Grant et al. 1992). When the site is methylated on the inactive X chromsome in the female, PCR amplification is resistant to HpaII digestion. Thus, amplification cannot occur after HpaII digestion of male DNA (single active X chromosome with unmethylated CpG sites) or female DNA at a time when both X chromosomes are active (with unmethylated CpG sites). However, when a CpG site becomes methylated on the inactive X chromosome a PCR product is obtained. This work showed that these X-linked genes are not methylated in sperm, eggs and preimplantation embryos up to the blastocyst stage, and that methylation of their CpG sites occurs on one X chromosome in female implanting embryos at the time of X-inactivation (we do not know whether before or after). In addition, there is some evidence of spreading of methylation along the X chromosome in that the *Pgk-1* gene, which is closer to the X-inactivation centre, is methylated prior to the *G6pd* gene (Grant et al. 1992).

METHYLATION OF THE XIST LOCUS IN DEVELOPMENT

The above experiments provide the first evidence of methylation associated with gene silencing in preimplantation development and represent the earliest time *de novo* methylation has been detected. These experiments again raise the question as to whether methylation may be one of the causes of transcriptional silencing. Methylation could be causal to X-inactivation in somatic cells although it is clear that, for the sites studied, methylation of the X chromosome is not necessary for its inactivation in the primordial germ cells. An important question is whether differential methylation of the gametes is the cause of imprinting of the paternal X chromosome. More precise information on this question has been obtained by examining the methylation of a locus involved in the initiation of X-inactivation in mouse and human, the XIST locus (X-Inactivation Specific Transcript, Brockdorff et al 1991; Brown et al 1991).

Norris et al (1994) have shown that all CpG sites analysed in the Xist locus in somatic tissues are methylated on an active X chromosome and unmethylated on an inactive X chromosome and thus methylation at this locus is correlated with its inactivity. Recently, we have investigated whether differential methylation of Xist in sperm and eggs is the primary imprint distinguishing the paternal and maternal X chromosomes leading to preferential paternal X-inactivation in the extra-embryonic lineages. It was shown that three sites just downstream from the putative TATA box in the promoter region of the Xist gene are methylated in eggs but not in sperm (Zuccotti and Monk, 1994). These results are consistent with the expectation that the Xist locus on the paternal X will be exclusively expressed in preimplantation embryos (Kay et al. 1994) and that the paternal X chromosome is destined to be preferentially inactivated in the extra-embryonic tissues as they are delineated (Harper et al. 1982). In this case, the methylation difference in sperm and eggs is a primary gametic imprint *causal* to the differential expression of the maternal and paternal alleles.

Further methylation differences are now being sought for different alleles of imprinted genes in the adult, in gametes, and in development, in both the mouse and human. Like Xist, the Igf2r gene is differentially methylated in gametes, although not in a region normally associated with control of expression (Stoger et al. 1993). Unlike Xist, the maternal Igf2r allele, which is methylated in the egg, is the allele expressed.

EPIGENETIC PROGRAMMING OF DEVELOPMENT

What about the pattern of methylation of autosomal genes? Kafri et al. (1992), using methylation-sensitive restriction enzyme digest followed by PCR, have shown that the methylation profiles of several genes tested in early development follow the profiles established for overall methylation (Monk et al. 1987), i.e., oocyte sequences are less methylated than sperm, demethylation occurs between the 8-cell and the blastocyst stage, *de novo* methylation occurs around the time of implantation, and the germ line initially escapes *de novo* methylation.

Although there is extensive demethylation between the 8-cell and the blastocyst stages, it is possible that not all CpG sites are demethylated by the blastocyst stage. In this way, epigenetic imprints would be perpetuated into soma. Chaillet et al (1991) showed that an imprinted transgene is methylated in mature oocytes and unmethylated in sperm. The oocyte methylation imprint appears to be stable in preimplantation development since it is still present in the blastocyst. However, we do not know whether demethylation and remethylation of the transgene have both already occurred by this time, or whether the transgene escaped demethylation. It is notable that the transgene is unmethylated in primordial germ cells which implies, if the imprint is stable throughout preimplantation development, a demethylation event in the germ cell lineage. It is possible that the general loss of methylation (erasure), commenced in preimplantation embryos, continues in the germ cell lineage. According to this hypothesis, we would assume that the transgene studied by Hadchouel et al. (1987), which becomes permanently methylated after transmission through the oocyte, remains methylated in subsequent generations due to the fact that the methylation imprint escapes erasure, not only in preimplantation development, but also in the germ line.

The observations we have made on methylation, X-linked gene activity and imprinting have led to the following hypothesis (see also Figure 1) concerning epigenetic programming in development:

1)   Epigenetic programmes (DNA modification) of gametes govern early events in the zygote; there is potential imprinting where modifications differ between egg and sperm, or

where "homologous" alleles diverge in chromatin conformation or base sequence (Monk, 1990).

2)   Erasure of gametic programmes (de-programming) occurs during preimplantation development and continues in the embryonic germ cells.

3)   The "ground" state is approached at the ICM stage where DNA is unmethylated, two X chromosomes are active, and the cells are developmentally totipotent (as ES cells or in chimaeras).

4)   Programming of the zygote occurs from the ICM stage. De-programming of gametic modification and re-programming of the zygote may overlap in time.

5)   Imprints occur in soma when differing epigenetic modifications in egg and sperm fail to be erased by the time of "setting" (see below) and are perpetuated into the somatic lineage.

6)   Permanent imprints, or epimutations, occur when epigenetic modifications fail to be erased in the germ line. Epimutations show no further switching in activity imposed by passage through sperm and egg in future generations.

Setting. I have hypothesised an event in development which I call "setting". This event decides which genes are potentially "on" and which genes are permanently "off" in a particular cell. Setting is an irreversible event linked to cell differentiation at gastrulation. Setting may vary cell-to-cell in the chromosome region, gene, or allele, inactivated, thus giving rise to cellular and allelic mosaicism. Such mosaicism may be observed as a mottled phenotype where there is heterozygosity in the region concerned, e.g., in random X-chromosome inactivation, chromosome position-effect variegation and, possibly, random hemizygous allele expression. The mosaicism may be generated by a variation in timing of setting from one chromosome region to another in different cells.

MOLECULAR ANALYSES IN THE HUMAN

The similarity of mouse and human embryos in the first week of development means that studies in the mouse pave the way for similar studies with human embryos.   However, there are important differences that should be noted. First, cleavage is less synchronous in the human than in the mouse and the products of cleavage are less regular (some human blastomeres may even lack a nucleus). Second, although mouse embryos can be cultured *in vitro* throughout the preimplantation period to the blastocyst stage (albeit more slowly in the chemically defined medium than in the reproductive tract), human embryo culture is far less efficient. Only about 30 per cent of human 4- to 8-cell embryos reach the morula and blastocyst stages (Edwards and Holland, 1988).   This is partly due to the very high incidence of chromosome abnormalities in human oocytes and embryos (up to 30 per cent of human oocytes are chromosomally abnormal; Plachot et al, 1988).   In addition to these morphological differences, there are likely to be differences at the molecular level. Just how useful the mouse model will be for molecular studies of human development is not yet known.

The elegant molecular analyses applied to studies on mouse development are now being extended to investigations of molecular aspects of human development. Little is known about the timing and tissue specificity of X-inactivation in humans and whether paternal X-inactivation occurs in the extra-embryonic tissues of the human conceptus. Harrison (1989) found that the paternal X chromosome is inactive in most chorionic villus cells, whereas Migeon et al (1985) found random X-inactivation (expression of both G6PD alleles in tissue from an heterozygous female conceptus) if these cells are cultured in vitro. It is important to know if paternal X-inactivation does occur in the human, as this may influence prenatal diagnosis in mothers heterozygous for X-linked disease genes. We are currently involved in establishing profiles of expression of a number of X-linked and autosomal genes, determining the pattern of X-inactivation, investigating possible imprinting of the paternal X chromosome and looking at the molecular mechanisms of silencing of X-linked genes (including XIST), and imprinted autosomal genes. This work requires the development of highly sensitive single cell procedures for the human genes of interest. Although we still know very little about gene expression and its regulation in human preimplantation embryos, one of the major developments arising from single cell analyses developed for the human so far is preimplantation diagnosis of genetic disease.

Preimplantation Diagnosis of Genetic Disease

The access to human preimplantation embryos afforded by IVF and embryo transfer procedures has led to the development of preimplantation diagnosis as an option for couples who have a high probability of transmitting a severe genetic disease to their children (reviewed in Monk, 1993). This involves testing for the particular genetic defect in a single cell removed from an 8-cell embryo and then placing only those embryos without the defect into the uterus of the woman to initiate pregnancy. Another approach to the prevention of inherited genetic disease is the diagnosis of individual oocytes before fertilisation. In this case, the cell sampled for analysis is the first polar body produced by the first meiotic division of the oocyte. The mouse has been used for the development of the procedures of embryo biopsy and genetic analysis of the single cell and to establish the feasibility of preimplantation diagnosis in the human.

Monk et al. (1987) initially used a mouse model for Lesch-Nyhan syndrome. The embryos of female mice heterozygous for hypoxanthine phosphoribosyl transferase (HPRT) deficiency, were tested by assay of HPRT activity in a single cell biopsied from each embryo. Progeny derived from the diagnosed embryos were either normal or mutant, exactly as predicted by testing the biopsied cell. However, when these procedures were performed on spare human embryos, the danger of extrapolating too closely from mouse models to the human became apparent. Although the HPRT activity could be readily assayed in a single cell from a human embryo, it turned out that this activity was derived from the oocyte, i.e., it was not synthesised under the direction of the embryo's own Hprt gene.

Currently, the diagnostic test that is most applicable to preimplantation diagnosis is the direct detection of the altered gene sequence using the polymerase chain reaction (PCR) to obtain sufficient numbers of copies of the gene sequence under test from a single cell so that it can be directly studied for the presence or absence of a mutation. The procedures for rapid PCR detection of a specific gene sequence in a single cell were also first developed in a mouse model (for beta-thalassaemia; Holding and Monk, 1989). The specificity, and therefore the sensitivity, of the PCR reaction was increased by the introduction of nested primers so that PCR amplification of a beta-globin gene sequence could be detected in a single blastomere of an 8-cell embryo. In this way, mutant thalassaemic mouse embryos (carrying a deletion of the beta-globin gene) were readily distinguished from the normal embryos. Soon after this, it was shown that the sickle cell mutation site could be analysed in the first polar body of the human oocyte (Monk and Holding, 1990). This work, and the work of Strom et al.(1990) established the feasibility of "preconception diagnosis". If this approach proved to be sufficiently reliable, it has the advantage that the diagnosis of individual oocytes before fertilisation avoids the need to interfere with the embryos themselves.

Gene Expression Studies in the Human

Although most of the molecular analyses in the human so far have had clinical preimplantation diagnosis as their aim, the sensitive PCR techniques developed may now be applied to studies on gene expression, X-chromosome inactivation, methylation and imprinting in the human. Recently, we have developed single cell procedures for the study of a number of genes which show instability of triplet repeat sequences within their transcribed regions, namely FMR-1 associated with fragile X mental retardation, MPK (myotonin protein kinase) associated with myotonic dystrophy, and AR (androgen receptor) associated with Kennedy's disease. We are interested in the timing and mechanisms of instability (mutation) of the repeat sequences in the germ line and in early development. However, these genes are also of value to human embryo research because of their high degree of polymorphism for the number of repeats in the normal alleles in the population. Hence, allele-specific PCR procedures can be used to monitor the modification and expression of the maternal (egg) and the paternal (sperm) alleles separately in the zygote.

Using allele-specific RT-PCR (reverse-transcriptase PCR), we have recently demonstrated, by the technique of rapid amplification of cDNA ends (RACE), the expression of the myotonic dystrophy gene in the human zygote at the 1-cell and early cleavage stages (Daniels et al. 1994). The separate detection of the parental alleles allows identification of the expression of the paternally-inherited allele which thus confirms the onset of embryonic gene transcription. We do not know the significance of this extremely early expression of the myotonic dystrophy gene. It may be indicative of a generalised derepression of gene activity in the early embryo, at least in the paternal gene complement.

CONCLUDING REMARKS

Sensitive techniques are now available to establish the profiles of expression of specific genes and to study the molecular mechanisms regulating their transcription in both mouse and human preimplantation embryos. In particular, we have investigated DNA methylation as an epigenetic modification associated with gene silencing. We have demonstrated genome-wide changes in methylation in early development. In addition, we have shown that methylation occurs very close in time to gene inactivation in preimplantation development and that differential methylation in the gametes may be the primary cause of imprinted gene expression. In the future, we can expect that the elucidation of gene regulation and genomic imprinting in mice will bring new insight and understanding on the mechanisms of inheritance of certain genetic diseases and childhood tumours whose patterns of inheritance are unusual and difficult to explain.

REFERENCES

Brockdorff, N., Ashworth, A., Kay, G.F., Cooper, P., Smith, S., McCabe, V.M., Norris, D.P., Penny, G.D., Patel, D. and Rastan, S. (1991) Conservation of position and exclusive expression of mouse Xist from the inactive X chromosome. Nature 351: 329-31.

Brown, C.J., Ballabio, A., Rupert, J.L., Lafreniere, R.G., Grompe, M., Tonlorenzi, R. and Willard, H.F. (1991) A gene from the region of the human X inactivation centre is expressed exclusively from the inactive X chromosome. Nature 349: 38-44.

Daniels, R., Kinis, T., Serhal, P. and Monk, M. (1994) Expression of the myotonin protein kinase gene in preimplantation human embryos. Submitted to Hum. Molec. Genet..

Chaillet, J. R., Vogt, T. F., Beier, D. R. and Leder, P. (1991) Parental-specific methylation of an imprinted transgene is established during gametogenesis and progressively changes during embryogenesis. Cell 66: 77-83.

Edwards, R.G. and Holland, P. (1988) New advances in human embryology: implications of the preimplantation diagnosis of genetic disease. Hum Reprod 3: 549-56

Epstein, C. J., Smith, S., Travis, B. and Tucker, G. (1978) Both X chromosomes function before visible X-chromosome inactivation in female mouse embryos. Nature 274, 500-3.

Grant, M., Zuccotti, M. and Monk, M. (1992) Methylation of CpG sites of two X-linked genes coincides with X-inactivation in the female mouse embryo but not in the germ line. Nature Genet. 2, 161-6.

Hadchouel, M., Farza, H., Simon, D., Tiollais, D. and Purcel, C. (1987) Maternal inhibition of paternal hepatitis B surface antigen gene expression in transgenic mice correlates with *de novo* methylation. Nature 329: 454-6.

Harper, M. I., Fosten, M. and Monk, M. (1982) Preferential paternal X-inactivation in extra-embryonic tissues of early mouse embryos. J. Embryol. Exp. Morphol. 67: 127-38.

Harper, M. and Monk, M. (1983) Evidence for translation of HPRT enzyme on maternal mRNA in early mouse embryos. J. Embryol. Exp. Morphol. 74: 15-28.

Harrison, K.B. (1989) X-chromosomal inactivation in the human cytotrophoblast. Cytogenet. Cell Genet. 52, 37-41.

Holding, C. and Monk, M. (1989) Diagnosis of beta-thalassaemia by DNA amplification in single blastomeres from mouse preimplantation embryos. Lancet ii, 532-5.

Kafri, T., Ariel, M., Brandeis, M., Shemer, R., Urven, L., Mc Carrey, J., Cedar, H. and Razin, A. (1992) Developmental pattern of gene-specific DNA methylation in the mouse embryo and germ line. Genes Develop. 6: 705-14.

Kay, G. F., Penny, G. D., Patel, D., Ashworth, A., Brockdorff, N. and Rastan, S. (1993) Expression of Xist during mouse development suggests a role in the initiation of X chromosome inactivation. Cell 72: 171-82.

Kratzer, P. G. and Gartler, S. M. (1978) HGPRT activity changes in preimplantation mouse embryos. Nature 274: 503-4.

Lock, L. F., Takagi, N. and Martin, G. R. (1987) Methylation of the mouse Hprt gene on the inactive X occurs after chromosome inactivation. Cell 48: 39-46.

Lyon, M. (1961) Gene action on the X chromosome of the mouse (*Mus musculus L*). Nature 190: 392-3.

Lyon, M. (1988) X-chromosome inactivation and the location and expression of X-linked genes. Am. J. Hum. Genet. 42: 8-16.

Migeon, B.R., Wolf, S.F., Axelman J., Kaslow, D.C. and Schmidt, M. (1985) Incomplete X chromosome dosage compensation in chorionic villi of human placenta. Proc. Nat. Acad. Sci. 82: 3390-4.

Monk, M. (1981) A stem line model for cellular and chromosomal differentiation in early mouse development. Differentiation 19: 71-6.

Monk, M. (1986) Methylation and the X chromosome. BioEssays 4: 204-8.

Monk, M. (1990) Variation in epigenetic inheritance. Trends in Genet. 6: 110-4.

Monk, M. (1992) The X chromosome in mouse and Man. J. Inher. Metab. Dis. 15: 499-513.

Monk, M. (1993) In "Prenatal Diagnosis and Screening", edited by D.J.H. Brock, C.H. Rodeck and M.A. Ferguson-Smith, pp 627-638, Churchill-Livingstone.

Monk, M., Adams, R. L. P. and Rinaldi, A. (1991) Decrease in DNA methylase activity during preimplantation development in the mouse. Develt. 112: 189-92.

Monk, M., Boubelik, M. and Lehnert, S. (1987) Temporal and regional changes in DNA methylation in the embryonic, extra-embryonic and germ cell lineages during mouse embryo development. Develt. 99:371-82.

Monk, M. and Grant, M. (1990) Preferential X-chromosome inactivation, DNA methylation and imprinting. Develt. Suppl. (1990).

Monk, M., Handyside, A., Hardy, K. and Whittingham, D. (1987) Preimplantation diagnosis of deficiency of hypoxanthine phosphoribosyl transferase in a mouse model for Lesch-Nyhan syndrome. Lancet ii, 423-6.

Monk, M. and Harper, M. (1978) X-chromosome activity in preimplantation mouse embryos from XX and XO mothers. J. Embryol. Exp. Morphol. 46: 53-64.

Monk, M. and Harper, M. (1979) Sequential X-chromosome inactivation coupled with cellular differentiation in early mouse embryos. Nature 281: 311-3.

Monk, M. and Holding, C. (1990) Amplification of a B-haemoglobin sequence in individual human oocytes and polar bodies. Lancet 335: 985-8.

Norris, D. P., Patel, D., Kay, G. F., Penny, G, D., Brockdorff, N., Sheardown, S. A. and Rastan, S. (1994) Evidence that random and imprinted Xist expression is controlled by preemptive methylation. Cell 77: 41-51.

Plachot, M., Veiga, A., Montagut, J., et al. (1988) Are clinical and biological IVF parameters correlated with chromosomal disorders in early life: a multicentric study. Hum. Reprod. 3: 627-35.

Singer-Sam, J., Grant, M., Le Bon, J.M., Okuyama, K., Chapman, V.M., Monk, M. and Riggs, A. (1990) Use of a HpaII-polymerase chain reaction assay to study DNA methylation in the Pgk-1 CpG island of mouse embryos at the time of X-chromosome inactivation. Molec. Cell Biol. 10: 4987-9.

Stoger, R., Kubicka, P., Lui, C.G., Kafri, T., Razin, A., Cedar, H. and Barlow, D.P. (1993) Maternal -specific methylation of the imprinted mouse Igf2r locus identifies the expressed locus as carrying the imprinting signal. Cell 738: 61-71.

Strom, C. M., Verlinsky, Y., Milayeva, S., et al. (1990) Preconception genetic diagnosis of cystic fibrosis. Lancet 336: 306-7.

Zuccotti, M. and Monk, M. (1994) Differential methylation of the Xist gene in sperm and eggs - a primary imprint for paternal X-inactivation in extra-embryonic lineages of the mouse embryo. Submitted to Nature Genetics.

COMPARISON OF POLY(A)-DEGRADING ACTIVITY
IN THE AVIAN AND MAMMALIAN OOCYTES

Bożenna Olszańska and Urszula Stępińska

Institute of Genetics and Animal Breeding, Polish Academy of Sciences
Jastrzębiec n. Warsaw, Poland 05 551

INTRODUCTION

It is generally known that maternal RNA contained in oocytes is much more stable than RNA of somatic cells, and that it becomes destabilized at maturation and fertilization. The high stability of the maternal RNA in oocytes may be due to several factors such as: 1) the protective effect of proteins complexed with the RNA within RNP (Blobel, 1973; Bernstein and Ross, 1989; Darnbrough and Ford, 1981; Jackson and Standart, 1990), 2) specific structures of mRNA conferred by poly(A) tracts (Bernstein et al., 1989; Spirin, 1994; Xing and Worcel, 1989) or specific sequences present at the 3' and/or 5' ends (Bouvet et al., 1991; Brewer and Ross, 1988; Jackson and Standart, 1990; Rosenthal and Ruderman, 1987; Shaw and Kamen, 1986; Treisman, 1985; Wickens, 1993), 3) presence/absence of RNA-degrading enzymes.

The first two aspects have been investigated by many authors. The lack of data concerning the last aspect induced us to verify the presence and significance of ribonucleases in animal oocytes. The experiments were mainly performed with quail oocytes and early embryos containing large amounts of maternal RNA (1.1 µg/germinal disc, Olszańska and Borgul, 1993) which is predestinated to be eliminated and replaced by the new embryonic transcripts in the course of embryogenesis. Some of the experiments were also done with the mouse eggs and 2-cell embryos, containing a low store of maternal RNA (Bachvarova et al., 1985, Piko and Clegg, 1982; Olszańska and Borgul, 1993).

We report here the absence of a detectable RNase A activity in the quail oocytes and early embryos, and the presence of a high poly(A)-specific RNase activity contained mainly in the oocyte nucleus. This poly(A)-specific enzyme is similar or identical to endoribonuclease IV

Organization of the Early Vertebrate Embryo
Edited by N. Zagris *et al.*, Plenum Press, New York, 1995

(EC 3.1.26.6) found in chick and quail oviducts (Müller, 1976; Müller et al., 1976). Detectable poly(A)-degrading activity was not observed in the mouse egg or 2-cell embryo.

MATERIALS

The quail oocytes and embryos at different stages were obtained from birds kept at the experimental farm of the Institute. The mouse eggs and 2-cell embryos were obtained after standard PMSG and hCG- induced superovulation from the randomly bred mouse colony.

Roman numerals are used for designation of the developmental stages of quail embryos from the uterine and laid eggs (Eyal-Giladi and Kochav, 1976) and arabic numerals - for the embryos from incubated eggs (Hamburger and Hamilton, 1951).

Previtellogenic and small vitellogenic quail oocytes were isolated manually under a dissecting microscope, after digestion with collagenase (0.5 mg/ml, Sigma). From the largest vitellogenic oocytes only the germinal discs were cut out and used after washing off the adhering yolk. The oocyte nuclei were manually isolated from the largest vitellogenic oocytes in 0.25M sucrose-0.01% sodium citrate-3mM $CaCl_2$ under a dissecting microscope. All material was frozen in 10mM Tris(HCl, pH 8.2)-140mM NaCl and kept at -70°C until used.

METHODS

The presence of RNase A and poly(A)-degrading activity was tested in vitro using as a substrate, respectively, [^{3}H]poly(U) or [^{3}H]poly(A) incubated with homogenates or $(NH_4)_2SO_4$ protein precipitates obtained from the quail or mouse material according to Müller (1976). The incubation was usually for 30 min at 37°C in 0.1 SSC for [^{3}H]poly(U) and 100mM Tris(HCl, pH 8.7)-0.5mM $MnCl_2$-0.2mM dithiothreitol for [^{3}H]poly(A). In the control reactions, the substrate was incubated with BSA (Sigma, fraction V) instead of biological material. Protein concentration in the incubation mixture, determined by the Lowry method, was 200 µg/ml, if not stated otherwise. However, the protein concentrations in the quail germinal vesicle, the mouse egg and 2-cell embryo homogenates were too low to be measured by this method. Therefore, as a measure of protein amount in the reaction mixture, the number of nuclei per milliliter was used (60 nuclei/ml). Since the detection limit of the Lowry method was about 20 µg/ml, we estimate that the protein concentration in this case was less than 20 µg/ml. For the mouse egg and 2-cell embryo homogenates we assumed the amount of 23 ng of protein per 1 egg (or 2-cell embryo) (Hogan et al., 1986). Thus, to obtain a protein concentration of 100 µg/ml, about 440 mouse eggs (or 2-cell embryos) were used in 100µl of the reaction mixture.

Poly(U) and poly(A) degradation assays were performed, after incubating one of the substrates with homogenate or protein precipitate, by the following methods:

1. Adsorption of the undegraded substrate on Whatman DEAE-cellulse (DE-81) discs and the radioactivity measurements (Müller, 1976). The difference between the radioactivity

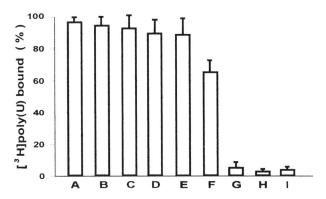

Figure 1. Degradation of poly(U) by homogenates of quail oocytes and embryos. [³H]poly(U) bound to DE-81 discs (undegraded substrate) after incubation with homogenates. Control = [³H]poly(U) bound to the discs after incubation with BSA = 100%. (**A**)- previtellogenic oocytes ø ≤ 1mm; (**B**)- germinal discs of vitellogenic oocytes ø ≈ 2cm; (**C**)- embryos from the laid eggs (stage X/XI); (**D**)- embryos at gastrulation (stage 4); (**E**)- 10-somite embryos; (**F**)- magnum; (**G**)- pancreas; (**H**)- RNase A (0.1 µg/ml); (**I**)- RNase A (0.1 µg/ml) + homogenate of germinal discs of vitellogenic oocytes (ø ≈ 2 cm).

fixed to the filters after reaction with BSA and the radioactivity left on the filters after reaction with the quail/mouse homogenates gave a measurement of the substrate degradation (%).

2. Sephadex G-50 column chromatography of the degradation products, without previous purification (Müller, 1976).

3. Agarose gel electrophoresis of the purified degradation products followed by transfer to a Zeta-Probe membrane (Biorad) and fluorography.

An RNA degradation assay was used to reveal the activity of other ribonucleolytic enzymes, if present. For this purpose, the total RNA extracted from gastrulating quail embryos was incubated either with homogenates of the quail material or with BSA (control reaction). Since, in this case, the unlabelled substrate was used, the RNA degradation process was monitored solely by electrophoresis.

RESULTS

Estimation of RNase A Activity in the Quail Oocytes and Early Embryos

The percentage of the undegraded poly(U) radioactivity left after incubation with the quail material, measured by adsorption of poly(U) to DE-81 filters, is shown in Fig. 1. There is practically no poly(U) degradation by the homogenates of the oocytes and X-XI stage embryos from the laid eggs (Fig. 1A,B,C) and a very slight, statistically insignificant, tendency for degradation in the case of gastrulating 4-stage and 10-somite stage embryos (Fig. 1D,E). A significant poly(U) degradation is seen in the case of oviduct and pancreas homogenates, used here as positive controls (Fig. 1F,G).

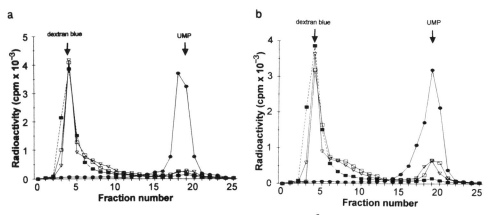

Figure 2. Sephadex distribution of the reaction products released from [^{3}H]poly(U) after incuabtion with quail homogenate. Control - BSA (■-■).

a) germinal discs of vitellogenic oocytes ø ≈ 2cm (❑-❑); embryos from the laid eggs, stage X/XI, (∇–∇); RNase A, 1 µg/ml, (•-•).

b) embryos at gastrulation, stage 4 (∇-∇); 10-somite embryos, stage 10, (❑-❑); pancreas (•–•).

Much more sensitive seems to be the Sephadex chromatography method which revealed an assymetric shoulder on the substrate peak and some radioactivity displaced to the mononucleotide peak in the case of gastrulating and 10-somite embryos (Fig. 2b). However, even this method revealed no poly(U) degradation by the oocyte and X-stage embryos – all the radioactivity was found in the substrate peak, just as in the control reaction with BSA alone (Fig. 2a). The control reactions with commercial RNase A and the quail pancreas homogenate resulted, as expected, in a complete degradation of the substrate and accumulation of the radioactivity in the mononucleotide peak.

The electrophoretic patterns of the poly(U) degradation products failed to reveal any diffusive change in the substrate bands after incubation with the quail oocyte and embryo homogenates (results not shown). However, the eventual traces of mononucleotides, visible after Sephadex separation, would not have been precipitated with ethanol during the purification procedure.

To verify the effect of some inhibitory factors which could be present in the quail homogenates or the formation of RNase-resistant poly(A)-poly(U) hybrids between [^{3}H]poly(U) substrate and poly(A+)RNA contained in the homogenate, we incubated the labelled poly(U) substrate with commercial RNase A in the presence of the homogenate from germinal discs of vitellogenic oocytes. Degradation of the substrate was determined by adsorption on DE-81 filters. As shown in Fig. 1(H,I) the addition of the germinal disc homogenate failed to prevent poly(U) degradation by RNase A (0.1 µg/ml). This would suggest the absence of such inhibitory factors in the investigated material. At the same time, the incubation conditions (0.1 SSC) would prevent the hybrid formation.

RNA-degrading Activity in the Quail Oocytes and Early Embryos

Since the above results pointed to the absence of a detectable RNase A activity in the quail oocytes, it seemed interesting to see whether any other RNase activity is present. For this purpose, total quail embryonic RNA (as a substrate) was incubated with homogenates of quail oocytes and early embryos. After 30 min of incubation the RNA was purified by phenol-chloroform and electrophoresed in 1% agarose gel. The electrophoretic patterns (Fig. 3) show disappearance of 28 and 18 S ribosomal bands and the presence of a diffuse RNA smear after incubation with any quail material. At the same time, the rRNA bands in the control reactions (lanes a and b) were intact.

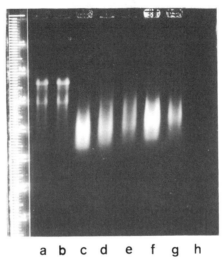

a   b   c   d   e   f   g   h

Figure 3. Gel electrophoresis of total RNA after incubation with the homogenates of quail oocytes and embryos. (**a**)- control: unincubated total RNA; (**b**)- control: total RNA incubated with BSA; (**c**)- previtellogenic oocytes ≤1mm; (**d**)- germinal discs of vitellogenic oocytes ø ≈ 2cm; (**e**)- embryos from the laid eggs (stage X/XI); (**f**)- embryos at gastrulation (stage 4); (**g**)- 10-somite embryos; (**h**)- RNase A (1 µg/ml).

Presence of Poly(A)-degrading Activity

Determination of poly(A)-degrading activity by adsorption to DE-81 filters, by Sephadex column chromatography and by agarose gel electrophoresis gave similar results.

DE-81 filter assay data (shown for the oldest vitellogenic oocyte only, Fig 4D,E) show that the highest activity was found in the previtellogenic and vitellogenic oocytes as well as in the germinal vesicles (isolated from the oldest vitellogenic oocytes) where degradation of 75-80% of the substrate was observed. In the case of the cytoplasm the degradation was 60%, and for the oocyte vitellus, embryos at cleavage stage I-III, embryos from the laid eggs, gastrulating (stage 4) embryos and oviduct tissue (magnum) it was 40%. These data show the high poly(A)-degrading activity in the oocytes (especially in the oocyte nucleus) and much lower activity – in the early embryos.

Poly(A)-degrading activity seems to decrease after fertilization and in the embryos from cleavage to gastrulation it remains at same level, lower than in the oocytes (40% of the substrate degradation).

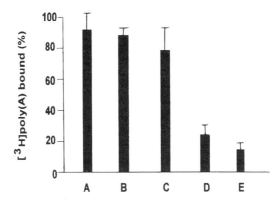

Figure 4. Undegraded [^{3}H]poly(A) bound to DE-81 discs after incubation with homogenates of the mouse and quail material. Control (poly(A) bound to the discs after incubation with BSA) = 100%. (**A,B**)- mouse eggs (100 and 200 μg protein/ml, respectively); (**C**)- 2-cell mouse embryo (100 μg protein/ml); (**D,E**)- germinal discs of quail vitellogenic oocytes ø ≈ 2cm (100 and 200 μg protein/ml, respectively).

For the homogenates of mouse eggs, 8 and 13% degradation was observed at the protein concentrations of 100 and 200 μg/ml, respectively, and 20% degradation – in the case of 2-cell embryos (100 μg protein/ml). For comparison, the degradation by the quail oocyte homogenate was ≈ 80% (Fig. 4D,E). Even if there was a slight tendency for higher poly(A) degradation at the increased protein concentration (200 μg/ml) in the case of the mouse eggs and 2-cell embryos, the differences were not statistically significant.

There was no difference in the degrading activity between quail homogenates and (NH$_4$)$_2$SO$_4$ protein precipitates at equivalent protein concentrations (data not shown).

The patterns of Sephadex separation of the poly(A) degradation products (Fig. 5a) and electrophoresis on 6% Nu Sieve agarose (Fig. 6) show that the high poly(A)-degrading activity present in the quail oocytes resulted in complete disappearance of the poly(A) substrate and accumulation of the degradation product in a peak of about (A)$_{10}$, at the same time no distinct AMP peak appeared. In the case of early quail embryos, in addition to some oligo(A)$_{10}$, there was still a lot of undegraded and partially degraded poly(A) substrate left (Fig. 5b). In the control reaction incubated with BSA practically all of the radioactivity was contained in the substrate peak (Fig. 5a,b).

The chromatographic distribution patterns of the poly(A) degradation products by mouse eggs and 2-cell embryos is shown in Fig. 7 in comparison with quail oocyte, at the corresponding protein concentration.

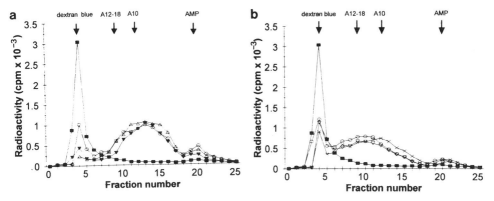

Figure 5. Sephadex chromatography of $[^3H]$poly(A) degradation products after incubation with the quail material. Control - BSA (■-■): **a)** oocytes: previtellogenic ≤1 mm (Δ–Δ); small vitellogenic ø = 3-4 mm (O-O); germinal discs of vitellogenic oocytes ø ≈ 2cm (▼–▼); **b)** embryos from: cleavage stage I-III (O-O); the laid eggs, stage X-XI (∇-∇); gastrulation stage 4 (x-x).

For eggs and 2-cell stage mouse embryos, practically all radioactivity was found in the substrate peak and there was no accumulation of the degradation product in the A_{10} fractions. However, an assymetric shoulder (the right one) on the substrate peak, especially for the 2-cell mouse embryo homogenate might suggest the presence of a very low poly(A)-degrading activity, negligible in comparison to that of the quail oocyte (Fig. 7).

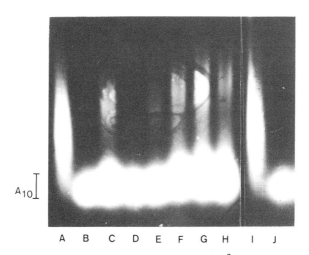

Figure 6. Gel electrophoresis of the degradation products of $[^3H]$poly(A) incubated with quail material. (**A**)- control, $[^3H]$poly(A) incubated with BSA; (**B**)- nuclei of vitellogenic oocytes (60 nuclei/ml); (**C**)- cytoplasm of vitellogenic oocytes; (**D**)- germinal discs of vitellogenic oocytes; (**E**)- vitellus of vitellogenic oocytes; (**F**)- embryos at cleavage, stage I-III; (**G**)- embryos from laid eggs, stage X/XI; (**H**)- embryos from laid eggs, 400 µg/ml; (**I**)- control, unincubated $[^3H]$poly(A); (**J**)- previtellogenic oocytes.

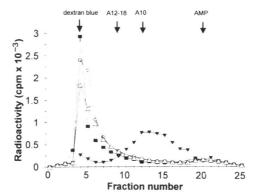

Figure 7. Sephadex chromatographic distribution of [³H]poly(A) degradation products obtained after incubation with the mouse material. Control, BSA (■-■); mouse eggs (○-○); 2-cell mouse embryos (Δ-Δ); germinal discs of the quail vitellogenic oocytes $\emptyset \approx 2$ cm (▼-▼) , at concentration 100 µg protein/ml.

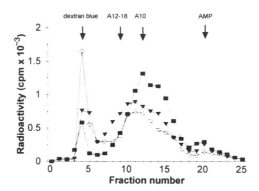

Figure 8. Sephadex chromtography of [³H]poly(A)-degradation products released after incubation with homogenates prepared from vitellogenic quail oocytes ($\emptyset \approx 2$cm): cytoplasm of the germinal discs (▼-▼); the nucleus (60 nuclei/ml) (■-■); $(NH_4)_2SO_4$ protein precipitate of the vitellus (○-○).

Localization of the Poly(A)-degrading Activity within the Quail Oocyte

The distribution of the poly(A)-degrading activity in quail oocyte is shown in Fig. 8. The highest activity was found in the germinal vesicle, then in the cytoplasm and in the vitellus of the oocyte.

One should take into account that in the germinal vesicle homogenate the protein concentration was estimated to be less than 20 µg/ml (60 nuclei/ml). On this basis we estimate the enzyme activity in the nucleus to be at least 10 times more than in the cytoplasm.

Effect pH and Mn^{2+} Concentration on the Poly(A)-degrading activity of Quail Oocytes

Dependance of the enzyme activity on pH and $MnCl_2$ concentration was tested with DE-81 filter assay, using the homogenate of germinal discs of the vitellogenic oocytes. The enzyme activity was strongly affected by pH, with the optimum at pH 8.7 (85% of poly(A) degradation) and it was almost completely inactivated in the presence of 5 mM EDTA (not shown).

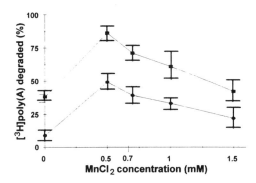

Figure 9. Effect of Mn^{2+} concentration on poly(A) degradation. Assay by DE-81 disc adsorption after incubation with: homogenate of germinal discs of the quail vitellogenic oocytes ø ≈ 2cm (■-■); $(NH4)2SO4$ protein precipitate from the oviduct (•-•).

Poly(A) degradation depended also on $MnCl_2$ concentration with the optimum at 0.5 mM, and showed a similar relation to Mn^{2+} concentration as the poly(A)-degrading enzyme of the quail oviduct homogenate, found by Müller (1976) and used in our experiments as a positive control (Fig. 9). Note that the two curves in Fig. 9 are parallel, but the oocyte enzyme activity is about twice as high as that of the oviduct.

DISCUSSION

The main results reported here are the lack of a detectable RNase A activity and the presence of a very high poly(A)-degrading activity in the quail oocytes, localized mainly in the germinal vesicle and being similar or identical to endoribonuclease IV (EC 3.1.26.6.), described earlier by Müller et al. (Müller, 1976; Müller et al., 1976) - see table 1.

The presence of RNase A in the mouse eggs or early embryos has not been checked as yet, and the poly(A)-degrading activity is extremely weak, if any.

Table 1. Similarity of the poly(A)-degrading activity from the quail oocyte and endoribonuclease IV from avian oviduct

	Poly(A)-degrading activity from quail oocytes	Endoribonuclease IV (Müller, 1976; Müller et al., 1976)
Source	Quail oocytes and early embryos	Chick and quail oviduct
Substrate	Poly(A)	Poly(A)
Localization	Mainly nucleus, less in cytoplasm and vitellus	Mainly nucleus, less in cytoplasm
Product	$Oligo(A)_{\overline{10}}$	$Oligo(A)_{\overline{10}}$
pH optimum	8.7	8.7
Mn^{2+} optimum	0.5 mM	0.5 mM
Inhibitor	EDTA	EDTA

RNase A Activity in the Quail Oocytes and Embryos

The stability of maternal RNA is somehow broken at maturation and fertilization of oocytes, and for many mRNAs it seems to be connected with shortening of their poly(A) tails (see review by Bachvarova, 1992; Jackson and Standart, 1990; Wickens, 1990).

The maternal stock of rRNA in Xenopus oocytes is sufficient to supply the protein synthesis until midblastula stage ($\approx$ 4000 cells) when the new rRNA synthesis starts (review in Davidson, 1986). We have seen that rRNA synthesis in quail embryos starts in uterine stages VI-VII (Olszańska et al., 1984) at the period of area pellucida and nucleoli formation in about 5000 ± 1000 cell embryo (Stępińska and Olszańska, 1983; Raveh et al., 1976, note the coincidence with the Xenopus embryo as to the cell number). Since that time, the large amount of the contained rRNA ($\sim$ 1.1 µg/germinal disc, Olszańska and Borgul, 1993) are progressively replaced by the new transcripts. Maternal rRNA, as not protected by poly(A) tails, could be more susceptible to ribonucleolytic enzymes, e.g. RNase A, if present. Thus the lack of detectable RNase A activity in the quail oocytes might be a prerequisite condition preventing rRNA damage during the long-time storage in the course of oogenesis. The appearance of the enzyme activity at gastrulation correlates with the appearance of the new rRNA transcripts (Olszańska et al., 1984). Other authors (Bouvet et al., 1991; Duval et al., 1990) pointed also to the necessity of the novo protein synthesis (but not transcription) for degradation of several deadenylated mRNAs in Xenopus embryo and to the absence of some degradation-specific factors or their activators during early Xenopus development. Assuming that one of the factors is a ribonuclease, this would be consistent with our data. In that context it would be interesting to check the eventual accumulation of the RNase A mRNA in the stock of maternal RNA of Xenopus or quail oocytes.

Poly(A)-degrading Activity

Numerous reports concerning an intensive poly(A) degradation at maturation and ferti-lization of animal oocytes (Bachvarova, 1992; Dolecki et al., 1977; Paynton et al., 1988; Piko and Clegg, 1982; Rosenthal and Ruderman, 1987; Varnum et al., 1992; Wilt, 1977) could suggest the presence of a highly active poly(A)-degrading enzyme at these developmental stages.

Our results presented here give evidence that such an enzyme exists in quail oocyte and is localised mainly in the germinal vesicle. The characteristics of the poly(A)-degrading enzyme found in the quail oocyte correspond well to those of the enzyme (endoribonuclease IV) found by Müller et al. (1976 a,b) in the chick and quail oviduct (see table 1). The parallel course of the Mn^{2+} dependance for the both enzymes points also to their similarity (Fig. 9).

It would be interesting to verify whether such an enzyme operates also in oocytes of other animals, where intensive deadenylation processes were observed, like in Xenopus (Cabada et al., 1977; Darnbrough and Ford, 1979; Varnum et al., 1992), sea urchin (Dolecki et al., 1977; Wilt, 1977) or Spisula (Rosenthal and Ruderman, 1987). In the Xenopus oocyte there is strong evidences pointing to the presence of an unidentified factor(s), localized in the nucleus and required for deadenylation of some mRNAs at maturation. The deadenylation taking place in the cytoplasm after GVBD required de novo protein synthesis and did not occured in enucleated oocytes (Varnum et al., 1992).

We would argue that endoribonuclease IV-like activity found in quail oocyte could be the ubiquitous poly(A)-degrading enzyme present in the oocytes containing important amo-unts of maternal poly(A+)RNA and is responsible for the intensive degradation of the poly(A) tails. Such an enzyme accumulated in the germinal vesicle during oogenesis would be released into the cytoplasm during nuclear membrane dissolution at the oocyte maturation, and its increased concentration in the cytoplasm would enable the digestion of poly(A) tracts of selected poly(A+)RNA. The selectivity of the digestion might be conferred by the presen-ce/absence of specific mRNA sequences (see review by Jackson and Standart, 1990; Wickens, 1990). Taking into account that the stored maternal mRNAs have usually short poly(A) tracts (Bachvarova, 1992; Jackson and Standart, 1990), it is possible that the poly(A)-degrading enzyme contained in the cytoplasm might destroy the part of a longer poly(A) tract which is not protected by poly(A)-binding protein (Jackson and Standart, 1990; Spirin, 1994). If, after nuclear envelope reconstitution, the enzyme is again sequestered into the nucleus, its amount per nucleus would decrease by half. At that time the process of polyadenylation could proceed in the cytoplasm between the consecutive nuclear membrane breakdowns. Repetitive cycles of cell divisions at cleavage with successive dissolution and reconstruction of the nuclear membrane might be responsible for establishing a sort of equilibrium between the cytoplasmic polyadenylating and nuclear deadenylating activities. This kind of mechanism could be observed as the apparent "turnover" of poly(A) tracts noted in cleaving sea urchin embryos (Dolecki et al.,1977, Wilt, 1977). Nuclear localization of the high poly(A)-degrading activity would enable accumulation of poly(A+) RNA in the growing oocyte cytoplasm without damaging its poly(A) segments.

In the mouse oocytes, containing much lower amounts of maternal RNA, where cleavage divisions last much longer, there is probably no need for such high accumulation of the deadenylating enzyme in oogenesis. Though the present data do not show this convincingly, the picture obtained for mouse eggs and 2-cell embryos is rather compatible with the idea of de novo synthesis of the enzyme, and not with its accumulation and storage in the oocyte.

The degradation of rRNA bands in the total embryonic RNA incubated with the oocyte homogenates (Fig. 3) could not be a result of RNase A activity as it was absent in the oocyte material. This means that it was caused either by some other RNase present in the oocytes or by the endoribonuclease IV-like activity found in the oocytes — if in the rRNA there are oligo(A) tracts long enough to serve as a substrate for the enzyme. Both possibilities remain to be tested.

REFERENCES

Bachvarova, R., De Leon, V., Johnson, A., Kaplan, G., Paynton, B.V., 1985, Changes in total RNA, polyadenylated RNA and actin mRNA during meiotic maturation of mouse oocytes, Dev. Biol. 108:325.

Bachvarova, R., 1992, A maternal tail of poly(A): The long and short of it, Cell. 69:895.

Bernstein, P., Peltz, S.T., Ross, J., 1989, The poly(A)-poly(A) binding protein complex is a major determinant of mRNA stability in vitro, Mol. Cell. Biol. 9:659.

Bernstein, P., Ross, J., 1989, Poly(A), poly(A)binding protein and the regulation of mRNA stability, Trends Bioch. Sci. 14:373.

Blobel, G., 1973, A protein of molecular weight 78 000 bound to the polyadenylated region of eucaryotic messenger RNAs, Proc. Natl. Acad. Sci. USA 70:927.

Bouvet, P., Paris, J., Philippe, M., Osborne, H.B., 1991, Degradation of a developmentally regulated mRNA in Xenopus embryos is controlled by the 3' region and requires the translation of another maternal mRNA, Mol. Cell. Biol. 11:3115.

Brewer, G., Ross, J., 1988, Poly(A) shortening and degradation of the 3' A+U-rich sequences of human c-myc mRNA in a cell- free system, Mol. Cell. Biol. 8:1697.

Cabada, M.O., Darnbrough, C., Ford, P.J., Turner, P.C., 1977, Differential accumulation of two size classes of poly(A) associated with messenger RNA during oogenesis in Xenopus laevis, Dev. Biol. 56:427.

Darnbrough, C., Ford, P.J., 1979, Turnover and processing of poly(A) in full-grown oocytes and during progesterone- induced oocyte maturation in Xenopus laevis, Dev. Biol. 71:323.

Darnbrough, C.H., Ford, P.J., 1981, Identification in Xenopus laevis of a class of oocyte-specific proteins bound to messenger RNA, Eur. J. Biochem. 113:415.

Davidson, E.H., 1986, "Gene Activity in Early Development," Academic Press, Orlando, FL.

Dolecki, G.J., Duncan, R.F., Humphreys, T., 1977, Complete turnover of poly(A) on maternal mRNA of sea urchin embryos, Cell 8:51.

Duval, C., Bouvet, P., Omilli, F., Roghi, C., Dorel, C., LeGuellec, R., Paris, J., Osborne, H.B., 1990, Stability of maternal mRNA in Xenopus embryos: Role of transcription and translation, Mol. Cell. Biol. 10:4123.

Eyal-Giladi, H., Kochav, S., 1976, From cleavage to primitive streak formation: A complementary normal table and new look at the first stages of the development of the chick. I. General morphology, Dev. Biol. 49:321.

Hamburger, V., Hamilton, H.L., 1951, A series of normal stages in the development of the chick embryo, J. Morph. 88:49.

Hogan, B., Constantini, F., Lacy, E., 1986, "Manipulating the Mouse embryo," Cold Spring Harbor Laboratory, Cold Spring Harbor NY.

Jackson, R.J., Standart, N., 1990, Do the poly(A) tail and 3' untranslated region control mRNA translation, Cell 62:15.

Müller, W.E.G., 1976, Endoribonuclease IV. A poly(A)-specific ribonuclease from chick oviduct. 1. Purification of the enzyme, Eur. J. Biochem. 70:241.

Müller, W.E.G., Seibert, G., Steffen, R., Zahn, R.K., 1976, Endoribonuclease IV . 2. Further investigation on the specificity, Eur. J. Biochem. 70:249.

Olszańska, B., Kłudkiewicz, B., Lassota, Z., 1984, Transcription and polyadenylation processes during early development of quail embryo, J. Embryol. Exp. Morph. 79:11.

Olszańska, B., Borgul, A., 1993, Maternal RNA content in oocytes of several mammalian and avian species, J. Exp. Zool. 265:317.

Paynton, B.V., Rempel, R., Bachvarova, R., 1988, Changes in state of adenylation and time course of degradation of maternal mRNAs during oocyte maturation and early embryonic development in the mouse, Dev. Biol. 129:304.

Piko, L., Clegg, K.B., 1982, Quantitative changes in total RNA, total poly(A), and ribosomes in early mouse embryos, Dev. Biol. 89:362.

Raveh, D., Friedlander, M., Eyal-Giladi, H., 1976, Nucleolar ontogenesis in the uterine chick germ correlated with morphological events, Exp. Cell. Res. 100:195.

Rosenthal, E.T., Ruderman, J.V., 1987, Widespread changes in the translation and adenylation of maternal messenger RNAs following fertilization of Spisula oocytes, Dev. Biol. 121:237.

Shaw, G., Kamen, R., 1986, A conserved AU sequence from the 3' untranslated region of GM-CSF mRNA mediates selective mRNA degradation, Cell 46:659.

Spirin, A.S., 1994, Storage of messenger RNA in Eukaryotes: Envelopment with protein, translational barrier at 5' side or conformational masking by 3' side?, Mol. Reprod. Dev. 38:107.

Stępińska, U., Olszańska, B., 1983, Cell multiplication and blastoderm development in relation to egg envelope formation during uterine development of quail (Coturnix coturnix japonica), J. Exp. Zool. 228:505.

Treisman, R., 1985, Transient accumulation of c-fos RNA following serum stimulation requires a conserved 5' element and c-fos 3' sequences, Cell 42:889.

Varnum, S.M., Hurney, C.A., Wormington, W.M., 1992, Maturation-specific deadenylation in Xenopus oocytes requires nuclear and cytoplasmic factors, Dev. Biol. 153:283.

Wickens, M., 1990, In the beginning is the end: regulation of poly(A) addition and removal during early development, Trends Biochem. Sci. 15:320.

Xing, Y.Y., Worcel, A., 1989, A 3' exonuclease degrades the pseudogene 5S RNA transcript in Xenopus oocytes, Genes Dev. 3:1008.

MICROTUBULE PATTERN DURING CYTOPLASMIC REORGANIZATION OF THE FERTILIZED *XENOPUS* EGG : EFFECT OF UV IRRADIATION

J.Paleček

Department of Physiology and Developmental Biology
Faculty of Science, Charles University
Viničná 7, 128 00 Prague 2
Czech Republic

INTRODUCTION

The origin and mechanisms of establishment of the various embryonic polarities is one of the attractive problems in developmental biology.

After fertilization the *Xenopus* egg becomes bilaterally symmetrical before first cleavage. Interaction of egg and sperm triggers a number of processes including the cortical reaction, pigment movement, rearrangement of yolk platelets and cytoplasm, dynamic changes of cytoskeletal structures, migration and fusion of pronuclei, dorsoventral and anteroposterior polarity determination and first cleavage (Paleček et al., 1978; Ubbels et al., 1983; Dent and Klymkowsky, 1989; Danilchik and Denegre, 1991; Houliston and Elinson, 1992; Schroeder and Gard, 1992; Elinson and Paleček, 1993; Sardet et al., 1994).

The egg acquires bilateral symmetry due to global rotation of the eggs cortex relative to subcortical cytoplasm (Vincent et al., 1986; Vincent and Gerhart, 1987). The cortical cytoplasmic rotation is thought to be the earliest determining event in the specification of the embryo's dorsal-ventral axis (Danilchik and Denegre ,1991). The movement of the cortical cytoplasm with respect to the inner cytoplasm causes profound movement within the inner cytoplasm. Details of the deep cytoplasmic movements associated with cortical rotation have been studied in eggs vitally stained during oogenesis with trypan blue (TB), which labels the outer layer of yolk platelets. During the middle part of the first cell cycle, a mass of central egg cytoplasm flows from the ventral to the future dorsal side of the embryo. By the end of first division, the cytoplasm of the prospective dorsal side of the embryo is distictly different than that of the prospective ventral side (Danilchik and Denegre, 1991).

A role for microtubules (MTs) in the embryonic developmental of *Xenopus* has been suspected for some time after the experimental demonstration that both gray crescent formation (Manes at al., 1978) and cytoplasmic firmness (Elinson 1983) were sensitive to colchicine. More recent studies have investigated an impressive array of aligned microtubules which were found beneath the entire vegetal surface of the egg. The timing of their appearance and their position suggested their role as a tracks for the rotation (Elinson and Rowning, 1988; Houliston and Elinson, 1991 a,b, 1992). Cortical rotation and the following cytoplasmic rearrangement can be blocked by microtubule (MT) depolymerizing agents (Manes at al., 1978; Scharf and Gerhart, 1983; Vincent et al., 1987) or by UV irradiation (Vincent et al., 1987; Rowning and Elinson, 1988).

In this study, we have examined in detail the simultaneous pattern of cortical and cytoplasmic MTs related to spatial changes of TB labelled yolk platellets of control and ultraviolet irradiated fertilized *Xenopus* eggs, mainly in the second part of the first cell cycle.

MATERIAL AND METHODS

Animals, Eggs and Embryos

Procedures for maintaining *Xenopus laevis*, induction of ovulation, insemination and dejellying were made as described by Zisckind and Elinson (1990). After insemination, fertilized eggs were dejellied with 2.5 % cysteine-HCl, pH 7.8 and washed and cultivated in 20 % Steinberg solution. To compare separate experiments, all results are expressed on a normalized time (NT) scale (Scharf and Gerhart, 1983). Insemination is time 0 NT and first cleavage 1.0 NT.

Labelling of Egg Yolk Platelets by Trypan Blue (TB)

Eggs with labelled yolk platelets were obtained by the method described by Danilchik and Gerhart (1987) and Danilchik and Denegre (1991). Ten days after gonadotropin - induced spawning, female frogs were injected in the dorsal lymph sac with 300 ml / frog of 0.1 % (TB) in 20 % Steinberg solution. Frogs were induced to spawn 10-14 days after injection of the label.

Fixation for Immunofluorescence

Several different fixation protocols were tested for their effectiveness in preserving fluorescence in TB labelled yolk platelets and permiting immunofluorescent detection of MTs. Significant differences in the preservation of both structures after different protocols will be discussed under Results. To visualize both structures simultaneously, three procedures were tested: 1. Eggs and embryos were fixed overnight at -20 °C or +20 °C in absolute methanol containing 1 % formaldehyde. 2. Eggs and embryos were fixed 24 hrs in Bouin Hollande or Bouin solution and washed with 5mM NH$_4$OH in 50 % ethanol (Danilchik and Denegre 1991). 3. Eggs and embryos were fixed 4 hrs according to Gard (1991) with minor modifications in 60 mM Pipes, 25 mM Hepes, 10 mM EGTA, 1 mM MgCl$_2$, 0.25 % glutaraldehyde, 0.37 % formaldehyde and 0.15 % TRITON - X-100, pH 6.8, followed by postfixation in absolute methanol at -20°C overnight. Methanol and glutaraldehyde fixed eggs were then rehydrated through methanol/PBS to phosphate - buffered saline (PBS: 128

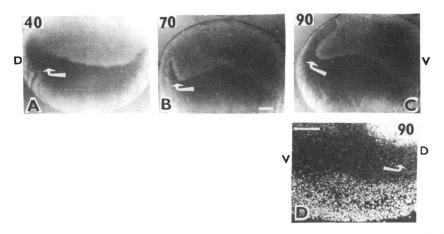

Fig.1. **Cytoplasmic redistribution of trypan blue.** Fluorescent pattern during the first cell cycle. Eggs were sectioned parallel to the meridian defined by the SEP and the animal-vegetal axis. Eggs show here are oriented with animal pole up, and the SEP on the right (A,B,C) and left (D) side. Eggs were fixed at 0.40 (A), 0.70 (B) and 0.90 NT (C,D) in methanol with 1 % formaldehyde and embedded in polyester wax. One cell cycle unit (c.c.u.) is defined as 1/100 of the time to first cleavage. Numbers in pictures thus indicate time of cell cycle 0.40 NT = 40 c.c.u. etc. Nonfluorescent central cytoplasm (white arrows) begins shiftig towards the prospective dorsal side (D). Scale bar = 50 μm

mM NaCl, 2 mM KCl, 8 mM Na_2HPO_4, 2 mM KH_2PO_4) and extracted in PBS containing 0.25 % TRITON -X-100 (Houliston and Elinson, 1991a). Glutaraldehyde / formaldehyde fixed eggs were washed in PBS and incubated for 15 hrs in PBS containing 100 mM $NaBH_4$.

Eggs were then rinsed extensively in PBS and transfered to polyester wax (Steedman, 1957, Houliston and Elinson, 1991a). Bouin fixed eggs were embeded into JB - 4 or polyester wax. Eggs were embedded with known orientation in the plane defined by the sperm entry point (SEP) and the animal pole. 7 mm section were cut either sagitally (parallel to the plane defined by the SEP and the animal pole) or horizontally (perpendicular to the SEP and the animal pole).

Trypan Blue and Microtubule Visualization

Microtubules were detected immunochemically using a mouse monoclonal anti-beta -tubulin antibody (N.357 Amersham) and a DTAF -labelled goat antimouse antibody (Bio Can).Specimen preparation generally followed Houliston and Elinson (1991 a).Sections were mounted in 50 % v/v glycerol/PBS pH 8.3 containing 1.5 % N-propyl gallate and viewed on a Leitz Ortoplan microscope using A and N2 filters - fluorescein excitation (microtubules - indirect immunofluorescence) and rhodamine excitation (trypan blue - direct fluorescence).

Ultraviolet Irradiation

Dejellied, fertilized eggs were irradiated with an ultraviolet light (UV) using a Mineralight Lamp (Scharf and Gerhart, 1980). The vegetal sides of the fertilized eggs were exposed through a quartz slide to UV at different NT and for different periods. UV irradiated

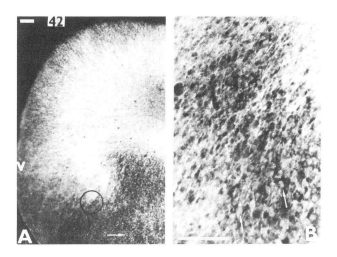

Fig.2 **Simultaneous visualization of trypan blue labelled yolk platelets** (small white arrows) **and microtubule fluorescence.** Egg shown here is oriented with animal pole up and ventral side (V) left. Egg was fixed in methanol and formaldehyde. Fig.B is higher magnification of sperm aster area (black circle in Fig.A). Scale bar = 50 μm

eggs were then fixed at different NT and / or kept in 20 % Steinberg solution for observation of dorsoanterior structures (Kao and Elinson, 1988).

RESULTS

Labelling of Egg Cytoplasm with Trypan Blue

A large number of fertilizable, developmentally normal eggs were obtained by injecting of TB into frogs. The resulting eggs were slightly smaller (800 - 1000 mm in diameter) than normal,with visible blue color on their surface. The dye has no teratogenic activity when confined to yolk platelets as evidenced by normal development up to stage 37/38 (Nieuwkoop and Faber, 1967) in our experiments. TB is fluorescent (Fig.1A, B,C,D 3D) with a broad emission spectrum similar to that of rhodamine in endosomes and yolk platelets (Danilchik and Denegre, 1991) and is visible partly with the fluorescein exitation filter as well.

Simultaneous Visualization of TB and MT Fluorescence

The simultaneous preservation of TB labelled yolk platelet fluorescence and immunocytochemical -MT fluorescence required the use of either formaldehyde - or glutaraldehyde - containing fixatives and polyester wax embedded specimens. Bouin fixative used by Danilchik and Denegre (1991) exhibited much weaker fluorescence of MTs than methanol or glutaraldehyde but perfectly preserved TB fluorescence. No MT fluorescence was observed, on the other hand, after embedding the eggs in JB4 (Polysciences). The crucial

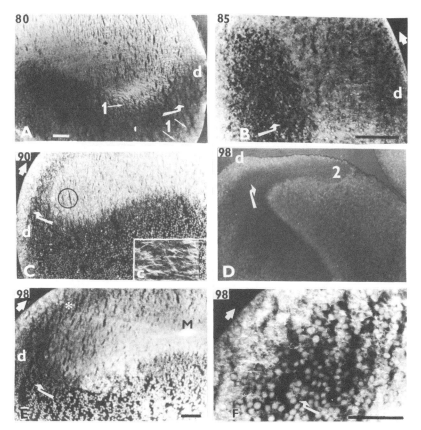

Fig.3 **Microtubules and trypan blue fluorescent pattern.** Eggs were fixed in glutaraldehyde (C,D,E,F) and methanol-formaldehyde fixation (A,B). Animal pole is up (in Fig.D right), prospective dorsal side = d. TB unlabelled central cytoplasm is shifted towards the prospective dorsal side and to the animal pole direction (bent arrows). The rest of the cytoplasmic MTs (1-small white arrows) is interrupted by moving zone of TB unlabelled yolk mass near the egg's equator (A). Cytoplasmic MTs disappear later from the cytoplasm (B). The newly formed interphase MTs grow within the animal half (C,E) but they are also interrupted by continuation of yolk movement (F = higher magnification from Fig.E area = asterisk). In Fig.D, where only fluorescence of TB is visible, apparent the upper limit of the central unlabelled yolk mass cytoplasm. Fig.Cc higher magnification of interphase MTs (black circle in Fig.C). Full arrows-direction of cortex movement. M - lobed cleavage nucleus. Scale bars = 50 μm

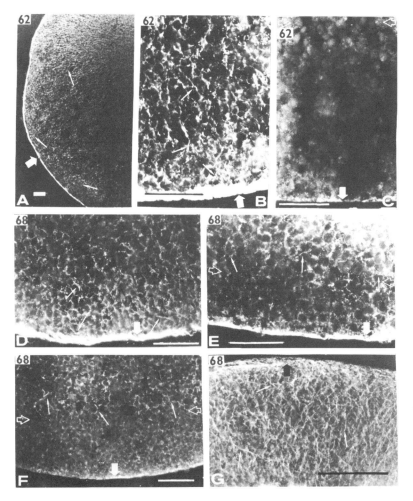

Fig.4 **Microtubule pattern of ultraviolet irradiated eggs.** Cortical (large) and cytoplasmic (small arrows) MTs within control (A,B,D) and UV irradiated (C,E,F,G) fertilized eggs fixed in methanol/formaldehyde. Central saggital sections. Interval of UV irradiation was 165 sec.(C,G) 120 sec.(F) and 75 sec.(E). Depth of UV penetration is approximately 300 μm (C), 100 μm (F) and 50 μm (E) after UV irradiation at O.43 NT (C,E) and 0.38 NT (F).The boundaries between the affected and unaffected zone are indicated by unfilled white arrows. Compare with the Fig.6.A,B. No change in MT pattern was observed within the animal half after UV irradiation (G).Eggs were sectioned with animal pole up (black arrow) and vegetal pole down (white arrow). Scale bar = 50 mm

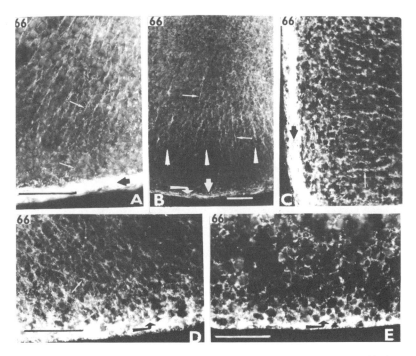

Fig.5 **Microtubule pattern of control and UV irradiated eggs.** Vegetal cortical MTs (black arrow) and cytoplasmic MTs (small arrows) within the control (A,C) and UV irradiated (B, D, E) fertilized eggs fixed at O.66 NT after UV irradiation at O.38 NT. Eggs were sectioned sagitally (B) or horizontally (A, C, D, E). Horizontal sections were prepared approximately 200 (C), 100 (D), and 50 μm (A, E) from the vegetal pole up. Interval of UV irradiation was 120 sec. No microtubules ended approximately 100 μm from the vegetal pole (white triangles - B) and only remnants from vegetal cortical array of parallel microtubules were present black or white (black bent arrows).Glutaraldehyde fixation. Scale bar = 50 μm

step was rehydration to PBS, when 30-100 % of TB fluorescence disappeared, depending on the fixative. Among different fixatives compared with regard to TB and MT fluorescence, optimal fixatives were methanol/formaldehyde (Houliston and Elinson, 1991a) or glutaraldehyde/formaldehyde (Gard, 1991) with small modifications (Fig. 2A,B).

Cytoplasmic TB Redistribution and MT Pattern During the First Cell Cycle

To study rearrangement of cytoplasmic components and a possible role of MTs in this phenomenon during the first cell cycle, labelled eggs were fixed at various times folowing fertilization. They were sectioned in known orientations relative to the sperm entry point-SEP (Paleček et al.,1978). During the first cell cycle the central cytoplasmic mass shifts progressively toward the prospective dorsal side (1A,B,C,D 3D). The formation and growth of two independent MT systems takes place during this period within the egg cortex and cytoplasm (Elinson and Paleček, 1993). These two systems are the astral microtubules (Fig.2A,B, 4A, 6A) and a vegetal cortical array of parallel microtubules (Fig. 4A,B,D).

Shortly before and during the cortical rotation (0.45-0.75 NT) short and longer nonastral microtubules begin to appear throughout the cytoplasm. These form together with

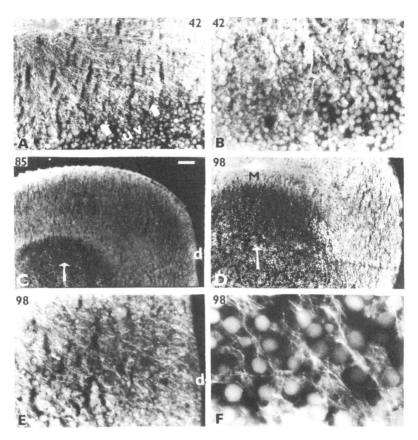

Fig.6 **Microtubules within TB labelled and UV irradiated eggs.** Double fluorescence of MTs and trypan blue treated yolk platelets after UV irradiation in sagital sections of fertilized eggs (A-F).Eggs were UV irradiated (165 sec.)at 0.40 NT and fixed in glutaraldehyde. Cytoplasmic MTs disappeared from the vegetal zone of eggs which were UV irradiated (A,B) No cortical or cytoplasmic MTs were present near the prospective dorsal (D) side (C) after UV irradiation. New interphase MTs formed within the animal cytoplasm when mitosis (M - nuclear vesicles) occurred (D,E, F) but their pattern differed from non-UV'ed ones (F). UV ed eggs produced only a central conical upwelling of cytoplasm (white arrows) at the end of the first cell cycle (C,D) which was surrounded by interphase MTs (D, E, F). Unfilled whitw arrow - male pronucleus. Glutaraldehyde fixation. Scale Bar = 50 μm

astral MTs, a dense network of radially organized cytoplasmic MTs. Most of them probably originate in the aster. They merge gradually to the band of parallel MTs (Fig.4B,D,5A,C) in the vegetal cortex. At 0.78-0.95 NT, cytoplasmic MTs disappear gradually first on the prospective ventral side and then on the dorsal side (Fig. 3A,B).The vegetal cortical array of parallel MTs persists until the time of first cleavage in the vegetal cortex. The egg cytoplasm becomes almost free of MTs with only the radial organization of mitotic MTs present. Interphase MTs, which are newly formed at about 0.90 NT, gradually fill the prospective ventral half, but are present only partly on the dorsal side (Fig.3E). The TB unlabelled cytoplasmic area shift to the future dorsal side in the animal pole direction (1A-D, 3D) as a result of the cortical rotation. Cytoplasmic MTs are interrupted by moving yolk platelets (Fig.3A) at 0.80 NT. These cytoplasmic MTs disappearing after 0.85 NT (Fig.3B), reflecting the continued yolk movement. Clearly the movement is not dependent on cytoplasmic MTs around this time. The newly formed interphase MTs, which grow radially from the vicinity of the lobed cleavage nucleus, are interrupted by a moving zone of central yolk (Fig.3C,E,F).The changing pattern of pigment TB, and MTs in the fertilized egg is summarized in Fig.7. Note that this diagram indicates only the general distribution and orientation of the Mts as visualized in sagittal sections.

Microtubule Pattern of Control and UV Irradiated Eggs

Fertilized *Xenopus* eggs were selected according to the visibility of SEP. They were UV irradiated on the vegetal side at 0.25-0.30 NT, 0.38-0.43 NT, and 0.48-0.50 NT for 30, 75,120 or 165 sec. Control and UV irradiated eggs were either allowed to develop until stage 37/38 (Nieuwkoop and Faber, 1967) or were fixed at 0.62-0.66 NT or 0.75-1.0 NT. Control and UV irradiated embryos were scored according to the dorsoanterior index (DAI) where a normal embryo is DAI grade 5 and a symmetric, ventralized embryo is scored as a grade 0 (Kao and Elinson, 1988). DAI's of the UV irradiated embryos were between 0.13 (0.38 NT, 120 sec.) and 1.4 (0.26 NT, 75 sec.). These results suggest that doses of UV irradiation prevent the formation of most dorsoanterior structures.

Control and UV irradiated *Xenopus* eggs were examined shortly before (0.29 , 0.42 NT) formation of the vegetal cortical parallel MTs and at the time of cortical rotation (0.62-0.68 NT). The vegetal cortical MTs and cytoplasmic MTs (Fig. 4A,B, D 5A) were present in control eggs (0.62-0.68 NT), but only remnants remain after UV irradiation of the vegetal side of the egg (Fig. 4C,E,F). Moreover different times of UV exposure (30, 75,120,165 sec.) appears to correlate with different depths of UV penetration within the vegetal half of the egg (Fig. 4C,E,F). Both cytoplasmic and cortical MTs were present (Fig. 5A) in horizontal sections of control eggs sectioned 50 mm up from the vegetal pole. In comparison UV irradiated eggs (0.38 NT, 120 sec.UV), sectioned at the same distance (Fig.5E)) retained only remnants of vegetal parallel MTs. Short, atypical cytoplasmic MTs were visualized 100 mm up from the vegetal pole (Fig.5D). No difference was visible between the MT pattern of control and UV irradiated eggs at 200 mm distance from the vegetal pole (Fig.5C). It thus appears that the effective depth of UV penetration was approximately 100 mm after 120 sec. of irradiation (Fig.5B) in the 1200 mm egg.

Microtubules in TB Labelled and UV Irradiated Eggs

Fertilized, TB labelled eggs (diameter 800-1000 mm) were UV irradiated (120 sec.) at 0.39-0.42 NT and fixed either immediately or at 0.80-1.00 NT. The change of astral MT length due to the UV effect is visible (Fig.6A,B) compared to control eggs. (Fig. 2A,B) at 0.42 NT. Cytoplasmic MTs disappeared shortly after 0.85 NT (Fig.6C). Newly formed interphase MTs radiate from the vicinity of the nuclear vesicles (Fig.6D,E,F) to the future ventral and dorsal side shortly before first cleavage time. These MTs never grow to the expected zone of UV penetration on the vegetal side. No apparent movement of cytoplasm or formation of a "swirl" are observed in UV irradiated eggs between 0.80-1.00 NT towards

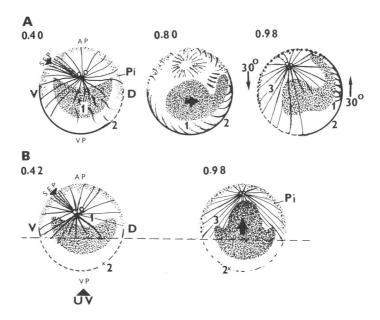

Fig.7 Diagram summarizing the general patterns of pigment (Pi), cytoplasmic (1), vegetal (2), interphase (3) MTs and cytoplasmic rearrangements through the first cell cycle in control (A) and ultraviolet (UV) irradiated (B) eggs. Note the changes position of pigment granules and central TB unlabelled cytoplasm after cortical rotation, about 30° to the prospective dorsal (D) side, and the lack of MTs in the area a of swirl. In UV'ed eggs, only the cortical upwelling of unlabelled central cytoplasm, surrounded by MTs is visible. Interphase MTs did not enter the UV zone within the vegetal half, and only a few parallel vegetal MTs were present there. V- ventral side, SEP - sperm entry point, AP - animal pole, VP - vegetal pole.

the prospective dorsal or ventral side direction (Fig.6C,D) as compared with control eggs (Fig.3B,C,E). The compact central unlabelled yolk mass is shifted in the animal pole direction and interphase MTs never enter this area. They form only "a cup formation " (Fig.6D) in prospective "dorsal or ventral" sides of this area. We were not able to see the interruption of the newly formed interphase MTs by the moving mass of central yolk cytoplasm (Fig.6D,E) in the same area as in control eggs (Fig.3E,F). MTs present here show only a radial network like pattern (Fig.6F). The altered pattern of pigment, TB and MTs in the fertilized eggs control and UV irradiated eggs is sumarized in Fig,7.

DISCUSSION

Two related movements were described during the first cell cycle. Vegetal yolk cytoplasm rotates relative to the plasma membrane and cortex by about 30° towards the SEP, that is away from the prospective dorsal side of the embryo (Vincent et al., 1986; Vincent and Gerhart, 1987). Danilchik and Denegre (1991) described the development of a complicated swirl of cytoplasm which correlates in position with the future dorsal side of the embryo. One part of this swirl originates from a mass of unlabelled deep cytoplasm which flows progressively toward the prospective dorsal side of the fertilized egg, during the second part of the first cell cycle . Experiments by Wakahara (1986) and Yuge et al. (1990)

suggest that dorsal activity is near of the equator and probably in deep cytoplasm. Moreover a maternal protein that reacted with the monoclonal antibody MoAb Xa 5B6 was distributed asymetrically along the dorso-ventral axis in the upper region of the equatorial zone of the fertilized egg. This protein was concentrated in the prospective dorsal side by the movement of inner cytoplasm (Suzuki et al., 1991). More recent studys of Holowacz and Elinson (1993) indicates that dorsalizing cytoplasm is close to the vegetal surface and is not deep or animal. That means that Yuge´s cytoplasm is probably located near the surface. Furthermore, three vegetally localized mRNAs, Vg1,X cat-2, and X cat-3, were all found to be specifically retained within the isolated vegetal cortices (Elinson et al., 1993). Because the process of converting an egg to a multicellular organism often begins with the polarization of information within the egg, the cytoskeleton plays the crucial role in maintaining such assymetries. Among vertebrates,the vegetal cortex of the *Xenopus* oocyte is the best candidate for a cytoplasmic domain important for developmental patterning (Elinson et al.,1993). Our observations of the MT pattern during rearrangement of the cytoplasm to the future dorsal side suggests that the vegetal cortical MTs are probably responsible for this movement. To show this, we used different times of ultraviolet irradiation of the egg´s vegetal side, followed by a detailed examination of MTs, displacement of deep cytoplasmic mass, and scoring of dorsoanterior index (DAI).

Embryos lacked all dorsoanterior structures after different doses of UV irradiation before cortical rotation. DAI was scored as 1.0 ± 0.5 in all experiments. Similar results were published by Elinson and Rowning (1988).

A crucial question has been the real depth of UV penetration inside of eggs (Youn and Malacinski,1980) and the influence of UV on formed MTs. No difference was visualized within the animal half. Astral MTs had the same pattern as in control eggs. Cytoplasmic MTs of UV irradiated vegetal halves never reached the vegetal cortex as in controls . A zone free of MTs varied approximately from 100 μm to 300 μm up from vegetal pole, depending on the UV dose. Few MTs were seen within the vegetal cortex. Unability of MT grow can correspond with the linkage of cytoplasmic tubulin to the others present structures or with denaturation of tubulin dimers after UV irradiation. Because the cytoplasmic MTs were formed within the vegetal half after the time of UV irradiation, the probable depth of UV penetration is reflected by the absence of MTs. These results suggest, that UV has either slowed down the arrival and growth of cytoplasmic MTs (inactivation of tubulin) or has partly destroyed MTs already present within the vegetal half. No apparent time delay of development was observed in TB labelled or unlabelled eggs after UV irradiation during the first cell cycle. The formation of nuclear vesicles or cleavage furrow was seen at the same time as in control eggs, but the tubulin pattern and TB fluorescent was different. There was no pigment displacement, no cortical MTs array, and no apparent movement before 1.0 NT. The central TB unlabelled yolk cytoplasmic mass was shifted as a whole in the animal pole direction and MTs surrounded this zone . UV´ed eggs often produced a central conical upwelling of cytoplasm at the end of the first cell cycle as was observed by Danilchik using confocal microscopy (unpublished observation). Vegetal cortical array of parallel microtubules was not formed. No apparent cytoplasmic swirl and only changed microtubule pattern were visible before the end of the first cell cycle. Because two microtubule-containing structures are implicated in dorsoventral polarization of the frog egg (sperm aster and vegetal cortical array) we can conclude that cytoplasmic swirls are a consequence of vegetal subcortical movement since UV did not affect microtubule pattern more animally.

ACKNOWLEDGMENTS

This work was partly made in Rick Elinson`s Laboratory, Toronto University, and was supported by a grant of NSERC, Canada. I thank Dr.R.P.Elinson for his scientific help and critical reading and extensive revision of the manuscript. May thanks also Mrs.M.Nohýnková for helpful technical assistence.

REFERENCES

Danilchik, M.V., and Denegre. J.M., 1991, Deep cytoplasmic rearrangements during early development in *Xenopus laevis*. Development.111:845-856.

Denegre, J.M., and Danilchik, M.V., 1993, Deep cytoplasmic rearrangements in axis-respecified *Xenopus* embryos. Dev.Biol. 160: 157-164.

Danilchik, M.V., and Gerhart, J.C. 1987, Differentiation of the animal-vegetal axis in *Xenopus laevis* oocytes. I.Polarized intracellular translocation of platelets establishes the yolk gradient. Dev.Biol. 122:101-112.

Dent, J.A., and Klymowsky, M.W. 1989, Whole mount analyses of cytoskeletal reorganisation and function during oogenesis and embryogenesis in *Xenopus*. In "The Cell Biology of Fertilization" (H. Schatten and G. Schatten eds.) pp. 63-102. Academic Press, San Diego, California.

Elinson, R.P., 1983, Cytoplasmic phases in the first cell cycle of activated frog egg. Dev.Biol. 100: 440-451.

Elinson, R.P., King, M.L., and Forristall, C., 1993, Isolated vegetal cortex from *Xenopus* oocytes selectively retains localized m RNAs. Dev.Biol. 160: 554-562.

Elinson, R.P., and Paleček, J., 1993, Independence of two microtubule systems in fertilized frog eggs: The sperm aster and the vegetal parallel array. Roux`s Arch.Dev.Biol. 202: 224-232.

Elinson, R.P., and Rowning, B., 1988, A transient array of parallel microtubules in frog eggs: Potential tracks for a cytoplasmic rotation that specifies the dorso-ventral axis. Dev.Biol. 128: 185-197.

Gard, D.L., 1991, Organization, nucleation and acetylation of microtubules in: *Xenopus laevis* oocytes: A study by confocal immunofluorescence microscopy. Dev.Biol. 143: 346-362.

Gerhart, J.C., Danilchik, M., Doniach, T., Roberts, S., Rowning, B., and Steward, R. 1989, Cortical rotation of the *Xenopus* egg: Consequences for the anteriposterior pattern of embryonic dorsal development. Development 107: Supplement 37-51.

Holowacz, T., and Elinson, R.P., 1993, Cortical cytoplasm, whith induces dorsal axis formation in *Xenopus* is inactivated by UV irradiation of the oocyte. Development 119: 277-285.

Houliston, E., 1994, Microtubule translocation and polymerization during cortical rotation in *Xenopus* eggs. Development 120: 1213-1220.

Houliston, E., and Elinson, R.P., 1991a, Patterns of microtubule polymerization relating to cortical rotation in *Xenopus laevis* eggs. Development 112: 107-117.

Houliston, E., and Elinson, R.P., 1991b, Evidence for the involvement of microtubules, ER, and kinesin in the cortical rotation of fertilized frog eggs. J.Cell Biol. 114: 1017-1028.

Houliston, E., and Elinson, R.P., 1992, Microtubules and cytoplasmic reorganisation in the frog egg. Curr.Top.Dev.Biol. 26: 53-70.

Kao, K.R., and Elinson, R.P., 1988, The entire mesodermal mantle behaves as Spemann`s organizer in dorsoanterior enhanced *Xenopus laevis* embryos. Dev.Biol. 127: 64-77.

Manes, M.E., and Barbieri, F.D., 1977, On the possibility of sperm aster involvement in dorsoventral polarisation and pronuclear migration in the amphibian egg. J.Embryol.Exp.Morphol. 42: 187-197.

Manes, M.E., Elinson, R.P., and Barbieri F.D., 1978, Formation of the amphibian egg grey crescent: Effects of colchicine and cytochalasin B. Roux`s Arch.Dev.Biol. 185: 99-104.

Nieuwkoop, P.D., and Faber, J., 1967, Normal Table of *Xenopus laevis* (Daudin). North-Holland, Amsterdam.

Paleček, J., Ubbels, G.A., and Rzehak, K., 1978, Changes of the external and internal pigment pattern upon fertilization in the egg of *Xenopus laevis*. J.Embryol.Exp.Morph. 45: 203-218.

Sardet, Ch., McDougall, A., and Houliston, E., 1994, Cytoplasmic domains in eggs. Trends in Cell Biol. 4:

of colchicine and cytochalasin B. Roux`s Arch.Dev.Biol. 185: 99-104.

Nieuwkoop, P.D., and Faber, J., 1967, Normal Table of *Xenopus laevis* (Daudin). North-Holland, Amsterdam.

Paleček, J., Ubbels, G.A., and Rzehak, K., 1978, Changes of the external and internal pigment pattern upon fertilization in the egg of *Xenopus laevis*. J.Embryol.Exp.Morph. 45: 203-218.

Sardet, Ch., McDougall, A., and Houliston, E., 1994, Cytoplasmic domains in eggs. Trends in Cell Biol. 4: 166-171.

Scharf, S.R., and Gerhart, J.C., 1980, Axis determination in eggs of *Xenopus laevis*: A critical period before first cleavage identified by the common effects of cold, pressure and UV irradiation. Dev.Biol. 99: 75-87.

Schroeder, M.M., and Gard, D.L., 1992, Organization and regulation of cortical microtubules during the first cell cycle of *Xenopus* eggs. Development 114: 699-709.

Steedman, H.P., 1957, Polyester wax. A new ribboning embedding medium for histology. Nature 179: 1345-1346.

Stewart-Savage, J., and Grey, R.D., 1982, The temporal and spatial relationship between cortical contraction sperm trail formation and pronuclear migration in fertilized *Xenopus* eggs. Roux`s Arch.Dev.Biol. 191: 241-245.

Suzuki, A.S., Manabe, J., and Hirakawa, A., 1991, Dynamic distribution of region-specific maternal protein during oogenesis and early embryogenesis of *Xenopus laevis*. Roux`s Arch.Dev.Biol. 200: 231-222.

Ubbels, G.A., Harra, K., Koster, C.H., and Kirschner, M.W., 1983, Evidence for a functional role for cytoskeleton in determination of the dorsoventral axis in *Xenopus laevis*. J.Embryol.Exp.Morphol. 77: 15-37.

Vincent, J.P., and Gerhart, J.C., 1987, Subcortical rotation in *Xenopus* eggs: An early step in embryonic axis specification. Dev.Biol. 123: 526-539.

Vincent, J.P., Scharf, S.R., and Gerhart, J.C., 1987, Subcortical rotation in *Xenopus* eggs: A prelimnary study of its mechanochemical basis. Cell Motil.Cytoskelet. 8: 143-154.

Vincent, J.P., Oster, G.F., and Gerhart, J.C., 1986, Kinematic of gray crescent formation in *Xenopus* eggs: The displacement of subcortical cytoplasm relative to the egg surface. Dev.Biol. 113: 484-500.

Wakahara, M., 1986, Modification of dorso-ventral polarity in *Xenopus laevis* embryos following withdrawal of egg content before first cleavage. Develop.Growth Differ. 31: 197-207.

Yuge, M., Kabayakawa, Y., Fujisue, M., and Yamana, K., 1990, A cytoplasmic determinant for dorsal axis formation in an early embryo of *Xenopus laevis*. Development 119: 1051-1056.

Zisckind, N., and Elinson, R.P., 1990, Gravity and microtubules in dorsoventral polarization of the *Xenopus* egg. Dev.Growth Differ. 32: 575-581.

THE FORMATION OF MESODERM AND MUSCLE IN *XENOPUS*

J. B. Gurdon

Wellcome CRC Institute
Tennis Court Road
Cambridge CB2 1QR
England

INTRODUCTION

The mesoderm is important in vertebrate development because it contributes cells to nearly all organs of the complete animal and because it emits signals which influence the differentiation of many other cells and tissues during development. The formation, function, and activities of the mesoderm have been more fully investigated in *Xenopus* than in any other organism. The conclusions which come from its study are therefore likely to be a guide to the understanding of equivalent tissues and activities in the other vertebrates.

Nearly all the work described below has been published, and in many cases reviewed. The references supplied at the end of this article are only those needed to enable a student to find other more primary sources of information. In addition, I have chosen to include a reference to many of the publications from this laboratory, since this may be convenient for those interested in our particular contributions to this subject area.

DEFINITION OF MESODERM

Traditionally, the mesoderm is the middle group of cells which are inside a gastrula and which give rise directly to tissues such as notochord, somites, and various "ventral mesodermal" cell types. In my view, it would be more helpful now to redefine the mesoderm as those cells which express the gene *Xbrachyury* (*Xbra*) (Smith *et al.*, 1991) during the early gastrula stage.

Organization of the Early Vertebrate Embryo
Edited by N. Zagris *et al.*, Plenum Press, New York, 1995

PREFORMATION OF THE MESODERM

Lineage experiments (Dale and Slack, 1987; Moody, 1987) trace the gastrula mesoderm back mainly to cells of the B and C tiers of the 32-cell embryo. This information provides the lineage history of the mesoderm but does not tell us about the mechanism by which cells become committed to the mesoderm. One idea is that the mesoderm arises because a signal from the vegetal pole of the embryo moves upwards towards the animal pole and commits cells in the middle equatorial region to become mesodermal. The other idea is that cells of the B and C tiers of the 32-cell embryo have inherited cytoplasm from the equatorial region of the egg and this cytoplasm commits cells of those tiers to become mesoderm.

To distinguish these ideas, the fertilized but undivided Xenopus egg can be ligated by a hair loop and the developmental fate of individual parts of that egg tested. These results (Gurdon et al., 1985a) showed that muscle, and therefore part of the mesoderm, can be formed from a sub-equatorial region of the undivided egg cytoplasm. Therefore some degree of preformation seems to exist in this case. This result is complemented by experiments on the 32-cell embryo in which it is shown that tiers A and B or D are not essential for muscle and mesoderm formation (Gurdon et al., 1985b).

Recent results (Lemaire and Gurdon, 1994) have related the concept of preformation to the activation of individual genes. It was previously shown that dissociated *Xenopus* cells can be cultured up to the early gastrula stage in the absence of calcium, and that these cells retain their developmental potentiality if reaggregated before stage 10 (Gurdon et al., 1984); furthermore, the timing with which these dissociated cells divide remains normal up to stage 10 (Grainger and Gurdon, 1989). If *Xenopus* embryos are cultured in the absence of calcium, and if these cells are widely dispersed every half hour, as was done by Lemaire (above), it was found that the genes *Xgoosecoid (Xgscd)* and *Xwnt 8* are expressed to about one third of their full extent in such dissociated cells when analyzed at stage 10. This activation involves a 6-7 fold increase over maternal level for *Xgscd* and a 25-36 fold increase for *Xwnt 8*. Therefore, there is a very large transcriptional activation of these genes in fully dissociated cells which have not interacted with each other. In contrast, *Xbra* completely fails to be activated at all in the dissociated cells. It was furthermore found that the activation of *Xgscd* and *Xwnt 8* take place in dissociated cells in the right part of the embryo, both along the dorso-ventral and anterior-posterior axes. These results show that at least two genes can be activated to much of their full extent in the complete absence of cell interactions and therefore must presumably depend upon the activity of materials prelocalized in the equatorial region of the fertilized egg.

IMPORTANCE OF INDUCTION FOR MESODERM FORMATION

A simple experiment indicating the importance of induction for mesoderm formation comes from the lineage experiments referred to above of Slack and Moody, combined with blastomere separation. In a normal embryo, tiers B and C contribute directly to mesoderm. If however tiers A and B are separated from tiers C and D at the 32-cell stage, mesoderm is not formed by tier B (or, of course, by tier A) (Gurdon *et al.*, 1985a). The conclusion is that tier B will form mesoderm in normal development, but only if it is in proximity to tier C from which it is presumed to receive a mesoderm-forming signal.

MESODERM FORMING MOLECULES

In recent years, an enormous amount of work and great progress has been made in this field by identifying candidate-inducing and receptor molecules for many of these factors. This area has been reviewed (Kimmelman *et al.*, 1992; Smith, 1993; Sive, 1993; Slack, 1994). Furthermore, the contribution to this Course by D. Kessler (this volume) deals directly with this topic and it will not therefore be covered here.

DETERMINATION OF THE AMOUNT OF MESODERM FORMED

The total amount of mesoderm formed, that is the number of embryonic cells which give rise to it, needs to be carefully regulated. This is because all of the cells in the animal cap of an embryo are capable of responding to mesoderm-forming signals. If these signals progress from the vegetal pole too far towards the animal pole, all of the animal cells would be converted into mesoderm and this would leave the embryo deficient for all ectodermal and neural derivatives.

Experiments conducted some time ago involve combining animal and vegetal tissues of different ages (Gurdon *et al.*, 1985b). As a result, it was possible to conclude that the total amount of mesoderm formed depends on how long the signal from the vegetal part of the egg is allowed to influence animal hemisphere cells. The amount of this time is regulated by the time at which the competence of animal cells to respond to the signal is lost.

To summarize so far, we envisage a process in which tier C cells are committed to mesoderm formation as a result of inheriting components from the egg. In addition, mesoderm forming factors are released by daughters of cells in tiers C and D and these factors spread upwards towards the daughters of tiers B and A. However, cells derived from tiers A and B eventually lose competence to respond to the mesoderm forming signal coming from vegetal cells of tiers C and D. Therefore mesoderm formation by induction is limited, approximately, to cells derived from tier B but not from tier A. Of course, it is also necessary for the strength of the inducing signal to be regulated in normal development, and we do not know how this is done; it may be by the provision of a finite amount of maternal mesoderm inducer in the egg.

DEMARCATION WITHIN THE MESODERM

In normal development, the sheet of mesoderm divides itself during gastrulation into a thin rod of central notochord, flanked on each side by future somite material and, more ventral to that, by the so-called ventral mesoderm. The mechanism by which this demarcation within the mesodermal sheet occurs is not well understood. However, a strong candidate explanation assumes that there is a morphogen gradient throughout the mesoderm at the early gastrula stage, according to which a high amount of the morphogen is present in the most dorsal region of the mesoderm destined to form notochord. A middle amount would be present in the part of the mesoderm which will eventually form somites and muscle, and a low amount of morphogen would be present in a part of the mesoderm which will eventually form ventral structures. The presumed morphogen would be formed both by preformation and by the Nieuwkoop induction (see contribution by D. Kessler). It is supposed that the cells within the mesodermal sheet sense the concentration of morphogen to which they are exposed and that they then respond to this by forming high (notochord), middle (somite), or low grade (ventral) mesoderm.

The evidence in favour of this idea has come very largely from the work of Green and Smith (1990) and Green *et al.* (1992). Their work involves incubating animal cap cells in different concentrations of the candidate mesoderm inducer activin and showing that different kinds of mesoderm genes are activated according to the concentration of activin used. Most recently, two papers have been published (Green *et al.*, 1994; Wilson and Melton, 1994) in which it is pointed out that the early responses of these animal cap cells to activin protein does not result in differential responses, but rather to all genes showing a similar dose response. Only during the subsequent time, that is from the early gastrula to the neurula (stage 17), do cells reveal responses of the previously described kind.

Very recently, work of another kind has been published (Gurdon *et al.*, 1994) which approaches this problem in another way. They have made Nieuwkoop conjugates with vegetal tissue containing different amounts of activin messenger RNA and attached these to sheets of animal cap tissue. As a result, they find that the location of cells which express the mesodermal gene *Xbra* can be regulated according to the amount of activin stimulation which they receive. This work appears to add substantial support to the concept of a morphogen gradient and the response of cells to it. Therefore, the demarcation of the mesoderm into its component structures may well depend upon their response to a morphogen gradient.

POST MESODERMAL CELL INTERACTIONS ARE REQUIRED FOR MUSCLE DIFFERENTIATION

Is the mesoderm-forming induction sufficient to commit cells to muscle or notochord differentiation, or is some further cell interaction required?

The best test of when a cell no longer needs further interaction with its normal neighbours is to transplant a cell to another site in embryos, so that it is surrounded by cells of a different developmental fate. When this was done for *Xenopus* mesoderm cells (Kato and Gurdon, 1992), it was found that dorso-lateral cells of an early gastrula (DL cells) must be in association with each other until the end of gastrulation (6 hours) for them to be able to differentiate as muscle cells after transplantation, singly, to the future gut region. However the mesoderm-forming induction, which is sufficient for *Xbra* expression is complete by the early gastrula stage; after this stage, animal cap cells lose their competence to respond to mesoderm-inducing signals (Jones and Woodland, 1987). Therefore some of the cells which have already become mesoderm must interact with their neighbours, that is with other DL cells of the early gastrula. As normally dissected, about 40% of the DL cells will become muscle; most of the others are endoderm or ectoderm cells which cannot easily be distinguished or separated from the muscle progenitor cells.

A COMMUNITY EFFECT FOR MUSCLE DIFFERENTIATION

Early experiments involved placing blastula animal cap cells, singly or in groups, into sandwiches of blastula vegetal cells (Gurdon, 1988). These experiments established the need for competent animal cap cells to communicate with each other, in order for them to achieve muscle differentiation, but did not determine whether this requirement was during or after mesoderm formation. Later experiments (Gurdon *et al.*, 1993c) implanted DL cells of an early gastrula (after mesoderm induction) into blastula ectoderm sandwiches, and used XMyoD as a marker of muscle differentiation. With DL cells from an early gastrula, 100-200 cells needed to be in a group, or community, for muscle differentiation to be achieved (by 40% of them). This led, together with other experiments, to the concept that the community effect is mediated by cells secreting a factor which must exceed a threshold concentration for muscle gene activation to be achieved.

The mechanism of the community effect was shown not to depend on gap junction communication among DL cells, since this took place in the presence of cytochalasin without preventing XMyoD expression (reference above). However the use of a dominant negative cadherin construct showed that cadherin contacts between cells are required for the community effect (Holt *et al.*, 1994). The identity of molecules able to mediate the *Xenopus* gastrula community effect is not known. Combinations of blastula vegetal or animal cap cells, with early gastrula DL cells, indicate that the myogenic community factor is not the same as factors normally involved in the preceding mesoderm induction (Gurdon *et al.*, 1993c). It is possible that *noggin* (Smith and Harland, 1992; Smith *et al.*, 1993) may be able to provide the community effect, though this has yet to be tested with normal concentrations of *noggin*.

NOTOCHORD COMMUNITY EFFECT AND VENTRAL INHIBITION

Using exactly the same experimental design as above, and a notochord marker MZ15 (Smith and Watt, 1985), we have found that notochord differentiation requires a post-mesodermal community effect, which precedes in time the muscle-forming community effect (Weston *et al.*, 1994). We have also identified a strong inhibitory influence of the ventral ectoderm on the muscle community effect (Kato and Gurdon, 1994).

We propose that post-mesodermal differentiation includes three major steps. First the most dorsal mesodermal cells would respond to a high concentration of morphogen by releasing a notochord community factor. This would affect only those cells in the most dorsal (future notochord) region. An hour or more later, we suppose that dorso-lateral mesodermal cells would respond to a middle concentration of the morphogen by secreting a muscle-forming community factor. Soon after that, we believe that ventral ectodermal cells would release an inhibitory factor working against the muscle community factor. As a result, the mesodermal morphogen gradient could be converted to a stepped distribution of notochord, muscle and ventral mesoderm. Thus we see the function of community effects as converting a gradient to a stepped distribution of cell response, and as a means of promoting uniformity within populations of cells which need to differentiate coordinately (Gurdon *et al.*, 1993a).

DORSALIZATION AND THE MYOGENIC COMMUNITY EFFECT

For some years, there has been evidence of a "dorsalization" process in *Xenopus* development. This is regarded as a third induction signal emanating from the early gastrula dorsal lip and spreading horizontally towards the ventral side. Slack, Smith and colleagues (Smith and Slack, 1983; Dale and Slack, 1987; Lettice and Slack, 1993) found that dorsal lip mesoderm could induce ventral mesoderm to form muscle, that is to raise its developmental fate to a more dorsal level.

It could be suggested that the muscle community effect is a consequence of the dorsalization process. One idea is that our DL cell population is contaminated by a few prospective notochord cells normally present only in the most dorsal region. This is unlikely because we do not find notochord cells in our DL cultures. Another idea is that, by the early gastrula stage, dorsalizing factors have already spread from the most dorsal region into the DL region.

I believe that the best way of trying to relate the myogenic community effect to dorsalization is to consider the development of lateral equatorial tissue from a mid-blastula. The normal developmental fate of cells in this region is to become muscle, but not notochord. When this tissue is excised at the mid-blastula stage and cultured, it forms muscle (Dale and Slack, 1987; Smith *et al.*, 1985). One interpretation of this is that the dorsalization signal has already spread to the lateral region by this stage, a significant modification of the original dorsalization hypothesis. In my view, it is not important to know whether or not dorsalization has to some extent taken place by the mid-blastula stage. What is clear is that the most dorsal cells of an early gastrula are not, in normal embryos, in contact with the most ventral mesoderm cells. The normal neighbours of early gastrula DL cells are cells nearest to them, that is the cells which provide the community effect. It therefore seems to me that the principle of the community effect is likely to be valid, even if this is regarded as part of a dorsalization process.

It is interesting that a similarity seems to exist when mesoderm induction is compared to dorsalization or the muscle community effect. In each case, cells at one extreme location (vegetal pole of a blastula or the dorsal lip of a gastrula) can induce cells in the other extreme position (animal pole of a blastula and the ventral region of a gastrula) to form cell-types characteristic of cells in the intervening areas. In the case of a blastula, mesoderm is well formed in the absence of tier D cells of a 32-cell stage, and, similarly, muscle is efficiently formed in the absence of the gastrula dorsal lip. Thus cells in the extreme positions (vegetal and dorsal) possess signalling capacities which are not essential, and which are normally provided by cells in more intermediate positions.

OTHER EXAMPLES OF A COMMUNITY EFFECT

Several examples are given by Gurdon *et al.*, (1993). In several species and tissues, the numbers of cells required to form a community is much smaller than for early gastrula *Xenopus* mesoderm. This might be expected if one function of the community effect is to provide uniformity of differentiation among that number of cells which normally differentiate in the same way, and this number will vary from species to species and from tissue to tissue.

Acknowledgement

Since 1983, my research has been supported by the Cancer Research Campaign.

REFERENCES

Dale, L. and Slack, J.M.W. (1987). Fate map for the 32-cell stage of *Xenopus laevis*. *Develop*. 99, 527-551.

Dale, L. and Slack, J.M.W. (1987). Regional specification within the mesoderm of early embryos of *Xenopus laevis*. *Development* 100, 279-295.

Grainger, R. and Gurdon, J.B. (1989). Loss of competence in amphibian induction can take place in single non-dividing cells. *Proc. Nat. Acad. Sci. US* 86, 1900-1904.

Green, J.B.A. and Smith, J.C. (1990). Graded changes in dose of a *Xenopus* activin A homologue elicit stepwise transitions in embryonic cell fate. *Nature* 347, 391-394.

Green, J.B.A., New, H.V. and Smith, J.C. (1992). Responses of embryonic Xenopus cells to activin and FGF are separated by multiple dose thresholds and correspond to distinct axes of the mesoderm. *Cell* 71, 731-739.

Green, J.B.A., Smith, J.C. and Gerhart, J.C. (1994). Slow emergence of a multithreshold response to activin requires cell-contact-dependent sharpening but not prepattern. *Development* 120, 2271-2278.

Gurdon, J.B. (1988). A community effect in animal development. *Nature* 336, 772-774.

Gurdon, J.B., Kato, K. and Lemaire, P. (1993a). The community effect, dorsalization, and mesoderm induction. *Curr. Op. Gen. Devel.* 3, 662-667.

Gurdon, J.B., Lemaire, P. and Kato, K. (1993b). Community effects and related phenomena. *Cell* 75, 831-834.

Gurdon, J.B., Brennan, S., Fairman, S. and Mohun, T.J. (1984). Transcription of muscle-specific actin genes in early *Xenopus* development: nuclear transplantation and cell dissociation. *Cell* 38, 691-700.

Gurdon, J.B., Mohun, T.J., Fairman, S. and Brennan, S. (1985a). All components required for the eventual activation of muscle-specific actin genes are localized in the subequatorial region of an uncleaved Amphibian egg. *Proc. Nat. Acad. Sci. US* 82, 139-142.

Gurdon, J.B., Fairman, S., Mohun, T.J. and Brennan, S. (1985b). The activation of muscle-specific actin genes in *Xenopus* development by an induction between animal and vegetal cells of a blastula. *Cell* 41, 913-922.

Gurdon, J.B., Tiller, E., Roberts, J. and Kato, K. (1993c). A community effect in muscle development. *Curr. Biol.* 3, 1-11.

Gurdon, J.B., Harger, P., Mitchell, A. and Lemaire, P. (1994). Activin signalling and response to a morphogen gradient. *Nature*, in press.

Holt, C.E., Lemaire, P. and Gurdon, J.B. (1994). Cadherin-mediated cell interactions are necessary for the activation of MyoD in *Xenopus* mesoderm. *Proc. Nat. Acad. Sci. US*, in press.

Jones, E.A. and Woodland, H.R. (1987). The development of animal cap cells in *Xenopus*: a measure of the start of animal cap competence to form mesoderm. *Development* 101, 557-563.

Kato, K. and Gurdon, J.B. (1992). Single-cell transplantation determines the time when *Xenopus* muscle precursor cells acquire a capacity for autonomous differentiation. *Proc. Nat. Acad. Sci. US* 90, 1310-1314.

Kato, K. and Gurdon, J.B. (1994). An inhibitory effect of *Xenopus* gastrula ectoderm on muscle cell differentiation and its role for dorsoventral patterning of mesoderm. *Devel. Biol.* 163, 222-229.

Kimmelman, D., Christian, J.L. and Moon, R.T. (1992). Synergistic principles of development: overlapping patterning systems in *Xenopus* mesoderm induction. *Development* 116, 1-9.

Lemaire, P. and Gurdon, J.B. (1994). A role for cytoplasmic determinants in mesoderm patterning: cell-autonomous activation of the *Goosecoid* and *Xwnt8* genes along the dorso-ventral axis of early *Xenopus* embryos. *Development* 120, 1191-1199.

Lettice, L.A. and Slack, J.M.W. (1993). Properties of the dorsalizing signal in gastrulae of *Xenopus laevis*. *Development* 117, 263-271.

Moody, S.A. (1987). Fates of the blastomere of the 32-cell-stage Xenopus embryo. *Devel. Biol.* 122, 300-319.

Sive, H.L. (1993). The frog prince-ss: a molecular formula for dorsoventral patterning in Xenopus. *Genes Dev.* 7,1-12.

Slack, J.M.W. (1994). Inducing factors in *Xenopus* early embryos. *Curr. Biol.* 4, 116-126.

Smith, J.C. (1993). Mesoderm inducing factors in early vertebrate development. *EMBO J.* 12, 4463-4470.

Smith, J.C. and Slack, J.M.W. (1983). Dorsalization and neural induction: properties of the organizer in *Xenopus laevis*. *J. Embryol. exp. Morph.* 78, 299-317.

Smith, J.C. and Watt, F.M. (1985). Biochemical specificity of Xenopus notochord. *Differentiation* 29, 109-115.

Smith, W.C. and Harland, R.M. (1992). Expression cloning of *noggin*, a new dorsalizing factor localized to the Spemann organizer in Xenopus embryos. *Cell* 70, 829-840.

Smith, J.C., Dale, L. and Slack, J.M.W. (1985). Cell lineage labels and region-specific markers in the analysis of inductive interactions. *J. Embryol. exp. Morph. Suppl.* 89, 317-331.

Smith, W.C., Knecht, A.K., Wu, M. and Harland, R.M. (1993). Secreted *noggin* protein mimics the Spemann organizer in dorsalizing *Xenopus* mesoderm. *Nature* 361, 547-549.

Smith, J.C., Price, B.M.J., Green. J.B.A., Weigel, D., and Herrmann, B.G. (1991). Expression of a Xenopus homolog of *Brachyury (T)* is an immediate-early response to mesoderm induction. *Cell* 67, 79-87.

Weston, M.J.D., Kato, K. and Gurdon, J.B. (1994). A community effect is required for amphibian notochord differentiation. *Roux's Arch. Dev. Biol.* 203, 250-253.

Wilson, P.A. and Melton, D.A. (1994). Mesodermal patterning by an inducer gradient depends on secondary cell-cell communication. *Curr. Biol.* 4, 676-686.

REGULATION OF CELL FATE BY PROCESSED VG1 PROTEIN

Daniel S. Kessler

Department of Molecular and Cellular Biology
Howard Hughes Medical Institute
Harvard University
Cambridge, MA 02138

SUMMARY

Mesoderm induction during *Xenopus* development has been extensively studied, and two members of the transforming growth factor-β family, activin βB and Vg1, have emerged as strong candidates for the natural inducer of dorsal mesoderm. Analysis of Vg1 activity has relied on injection of hybrid Vg1 molecules, which have not been shown to direct efficient secretion of active ligand and therefore, the mechanism of mesoderm induction by processed Vg1 is unclear. Injection of *Xenopus* oocytes with a chimeric activin-Vg1 mRNA, encoding the pro-region of activin βB fused to the mature region of Vg1, directed the processing and secretion of mature Vg1, resulting in soluble preparations with a concentration of 100-500 ng/ml. Treatment of animal pole explants with mature Vg1 resulted in formation of dorsal mesodermal tissues and dose-dependent activation of both dorsal and ventrolateral mesodermal markers. At high doses mature Vg1 induced formation of "embryoids" with a rudimentary axial pattern, head structures including eyes, and a functional neuromuscular system. Furthermore, truncated forms of the activin and FGF receptors, which block mesoderm induction in the intact embryo, fully inhibited mature Vg1 activity. Follistatin, a specific inhibitor of activin βB which does not block endogenous mesoderm induction, failed to inhibit Vg1. The results support a role for endogenous Vg1 in dorsal mesoderm induction during *Xenopus* development.

INTRODUCTION

At the blastula stage of *Xenopus* development, vegetal blastomeres produce signals that induce mesoderm formation by overlying equatorial cells (Nieuwkoop, 1969a, b; Sudarwati and Nieuwkoop, 1971). Much effort has been directed at the molecular analysis of mesoderm induction and, consequently, a number of molecules with mesoderm-inducing or patterning activities have been identified (Klein and Melton, 1994; Sive, 1993; Slack, 1994). Among the described mesoderm-inducing factors, three endogenous secreted proteins, activin and bone morphogenetic protein (BMP) (members of the TGF-β family), and fibroblast growth factor (FGF), are implicated in the formation of dorsal, ventral and ventrolateral mesoderm, respectively (Asashima et al., 1990a, b; Dale et al., 1992; Jones, 1992; Kimelman and Kirschner, 1987; Koster et al., 1991; Slack et al., 1987, 1988; Smith et al., 1990; Sokol et al., 1990; Thomsen et al., 1990; van den Eijnden-Van Raaij et al., 1990). Expression of dominant inhibitory activin, FGF or BMP receptors, lacking the intracellular kinase domain, supports the involvement of these factors in endogenous mesoderm induction and patterning. In embryos, the truncated activin receptor fully blocks mesoderm formation (Hemmati-Brivanlou and Melton, 1992), the truncated FGF receptor blocks formation of trunk and posterior mesoderm (Amaya et al., 1991), and the truncated BMP receptor converts ventral mesoderm to dorsal mesoderm (Graff et al., 1994; Suzuki et al., 1994). Additional molecules, noggin and members of the wnt family, can alter the dorsoventral pattern of mesoderm while lacking intrinsic mesoderm-inducing activity (Christian et al., 1991, 1992; Christian and Moon, 1993; Ku and Melton, 1993; Moon and Christian, 1992; Smith and Harland, 1991, 1992; Smith et al., 1993; Sokol et al., 1991; Sokol, 1993; Sokol and Melton, 1992). Despite the potent activities observed for these molecules, each fails to fulfill a strong expectation for an endogenous mesoderm inducer: localization to the vegetal pole blastomeres responsible for mesoderm induction and patterning during normal development. Restricted localization of maternal activin, BMP or FGF protein has yet to be observed (Asashima et al., 1991; Dohrmann et al., 1993; Rebagliati and Dawid, 1993; Fukui et al., 1993; Nishimatsu et al., 1993; Ueno et al., 1992).

Vg1 is a maternal mRNA that is localized to the vegetal pole of *Xenopus* oocytes and early embryos (Rebagliati et al., 1985; Weeks and Melton, 1987). Vg1 mRNA is synthesized early in oogenesis and becomes tightly localized to the vegetal cortex by the end of oogenesis (Melton, 1987; Mowry and Melton, 1992; Yisraeli and Melton, 1988; Yisraeli et al., 1990), and therefore, Vg1 mRNA and protein become partitioned within vegetal pole blastomeres (inducing tissue) in the early embryo (Dale et al., 1989; Tannahill and Melton, 1989). A member of the TGF-β family, Vg1 is a strong candidate for an endogenous mesoderm inducer. TGF-β-like proteins are synthesized as inactive precursors, which form dimers and are proteolytically cleaved, releasing a mature C-terminal dimer (Kingsley, 1994; Massague et al., 1994). However, endogenous Vg1 protein accumulates as an unprocessed precursor (46kD) and little or no mature Vg1 ligand (18kD) is detectable (Dale et al., 1989; Tannahill and Melton, 1989; Thomsen and Melton, 1993). Consistent with this observation, injection of Vg1 mRNA produced high levels of Vg1 precursor, but no processed protein is formed and, consequently, neither mesoderm induction nor any other developmental effect is observed (Dale et al., 1989, 1993; Tannahill and Melton, 1989; Thomsen and Melton, 1993). In contrast, injection of activin mRNA directs efficient production of mature protein and animal pole explants are induced to form mesoderm (Thomsen et al., 1990). Therefore, despite a suggestive localization to the *in vivo* source of mesoderm-inducing signals, an inability to produce processed protein has hindered the analysis of mature Vg1 activity. These observations indicate that processing of endogenous Vg1 is tightly regulated, and perhaps a key step in mesoderm induction is a localized and/or transient production of mature Vg1.

Recent studies using chimeric BMP-Vg1 molecules have succeeded in demonstrating mesoderm-inducing activity for mature Vg1 (Dale et al., 1993; Thomsen and Melton, 1993). A fusion protein, consisting of the N-terminal pro-region of BMP and the C-terminal mature region of Vg1, was shown to direct precursor processing, and permitted the analysis of mature Vg1 function. Injection of BMP-Vg1 mRNA resulted in dorsal mesoderm induction in animal pole explants and dorsal axis formation in both ultraviolet-ventralized and normal embryos, demonstrating the inductive activity of the Vg1 hybrids. While processing is greatly enhanced in the fusion constructs, the secretion of soluble mature Vg1 by BMP4-Vg1 injected oocytes is very inefficient in the absence of polyanions (Dale et al., 1993), bringing into question the mechanism of action in injected embryos. Specifically, it is possible that the activity of hybrid Vg1 molecules is due to an inhibitory activity upon other endogenous TGF-βs, such as BMPs. Furthermore, using injected hybrids it is not possible to precisely control the dosage or period of treatment, and therefore issues of competence, mesoderm induction versus dorsalization, specific activity and receptor binding specificity cannot be examined. In an attempt to elucidate Vg1 ligand activity and its role during normal development we have undertaken the production of soluble mature Vg1.

Utilizing a hybrid activin βB-Vg1 molecule, expressed in *Xenopus* oocytes, we have obtained efficient secretion of soluble, biologically active processed Vg1. Mature Vg1 concentration as low as 10pM was sufficient to induce mesoderm in animal pole explants. Induced explants often formed "embryoids" displaying a rudimentary axial pattern, complete with head structures and a functional neuromuscular system. In addition, both the truncated activin type II receptor and the truncated FGF receptor fully inhibited the inducing activity of mature Vg1. Moreover, follistatin, a specific inhibitor of activin which does not block endogenous mesoderm formation, fails to inhibit mature Vg1. These results establish the mesoderm-inducing activity of secreted Vg1 ligand. Furthermore, our observations suggest that inhibition of endogenous mesoderm induction by truncated receptors may, in fact, be due to an inhibition of signaling by endogenous Vg1.

Vg1 is the only known mesoderm-inducing factor which is localized to vegetal pole blastomeres, the source of endogenous inducing signals. Here I describe a system for the production of secreted, biologically active mature Vg1, and demonstrate the potent mesoderm inducing activity of this preparation. Furthermore, inhibitors of endogenous mesoderm induction are shown to block the function of mature Vg1. These results establish Vg1 as a strong candidate for the natural dorsal mesoderm inducer during *Xenopus* development.

MATERIALS AND METHODS

Embryological Procedures and Histology

Xenopus eggs were obtained, fertilized and cultured using standard procedures (Thomsen and Melton, 1993). For animal cap experiments, embryos were injected at the 2-cell stage, animal cap explants prepared at stage 8 and cultured in 0.5X MMR or 0.5X MMR supplemented with purified FGF or oocyte-conditioned media. Oocytes were surgically removed from anesthetized females, follicle cells removed by digestion with collagenase (2 mg/ml) in OR2 (Ca^{2+}, Mg^{2+} free), injected with 50 ng of capped, *in vitro* transcribed RNA and cultured at 19 °C in OR2 (with Ca^{2+} and Mg^{2+}) supplemented with BSA (0.5 mg/ml). For histology, samples were fixed in Bouin's fixative and processed for Paraplast sectioning and Giemsa staining. For solutions and general methods see (Peng, 1991).

Construction of Activin βB-Vg1 Hybrid and mRNA Synthesis

The activin βB-Vg1 hybrid was constructed by PCR amplification of a fragment of the activin βB gene (Thomsen et al., 1990) encoding the N-terminal pro-region, the tetrabasic cleavage site and the first four amino acids of the C-terminal mature region. This fragment was ligated, in frame, to a PCR amplified fragment of the Vg1 gene (Weeks and Melton, 1987) encoding the C-terminal mature region lacking the first four amino acids. The resulting ligation product was reamplified, digested at terminal restriction sites and cloned into a derivative of pSP64T (Krieg and Melton, 1987). This hybrid is analogous to the BMP2-Vg1 construct previously described (Thomsen and Melton, 1993; Figure 1).

For *in vitro* synthesis of capped mRNA, linearized templates for activin βB-Vg1, BMP2-Vg1 (Thomsen and Melton, 1993), activin βB (Sokol et al., 1991), Vg1 (unpublished construct), or follistatin (Hemmati-Brivanlou et al., 1994) were used to program the Megascript, SP6 transcription kit (Ambion).

Western Blotting

Supernatants of injected oocytes were resolved on 15% SDS-PAGE gels, electroblotted to nylon membranes and probed using a mature region-specific Vg1 monoclonal antiserum (Tannahill and Melton, 1989) or an affinity purified activin βB polyclonal antiserum specific for the mature region (provided by P. Klein, unpublished). Cross-reacting peptides were detected using alkaline phosphatase-conjugated secondary antibodies, visualized by chemiluminescence (Bio-Rad).

Reverse Transcription-Polymerase Chain Reaction

RNA isolation, reverse transcription, PCR conditions and primer sequences have been previously described (Wilson and Melton, 1994). PCR products were resolved on a 5% non-denaturing acrylamide gel and visualized by autoradiography. The ubiquitous EF1α mRNA (Krieg et al., 1989) was used as an RNA extraction and reverse transcription control. Xbra is a pan-mesodermal marker expressed throughout the gastrula marginal zone (Smith et al., 1991) and Xwnt8 is a marker of ventrolateral mesoderm at the gastrula stage (Christian et al., 1991). Cardiac actin is a marker of somitic muscle (Stutz and Sphor, 1986) and goosecoid and noggin are markers of dorsal mesoderm (Blumberg et al., 1991; Smith and Harland, 1992). NCAM is a general marker of neural tissue (Kintner and Melton, 1987).

RESULTS

Secretion of Processed Vg1 by *Xenopus* Oocytes

Expression of hybrid BMP-Vg1 molecules in embryos directs precursor processing, but little evidence for efficient secretion of processed Vg1 is available. BMP2-Vg1 expression in COS cells or *Xenopus* oocytes resulted in low or undetectable levels of secreted mature Vg1 (Figure 2; D. Kessler, G. Thomsen and D. Melton, unpublished), although intracellular processing of precursor was apparent. BMP4-Vg1 expression in oocytes directed secretion of little mature protein in the absence of polyanions, and no activity assessment was attempted (Dale et al., 1993). Consistent with these observations, secretion of BMP2 is not detected in these systems (D. Kessler and D. Melton, unpublished), suggesting that the pro-region may be a target for the regulation of secretion (Gray and Mason, 1990). These observations suggest that the effects of BMP-Vg1 in embryos may not be due to secretion of mature protein, but rather the response is due to an intracellular activity of the hybrid. On the other hand, functional activin βB ligand is efficiently secreted in both COS cells and oocytes (Thomsen et al., 1990; Figure 2). Therefore, in an attempt to obtain active mature Vg1, an additional hybrid molecule was prepared containing the activin βB signal sequence, pro-region, tetrabasic cleavage site and four amino acids of the activin βB mature region fused to the mature region of Vg1 (Figure 1).

To test the function of the activin-Vg1 hybrid, animal pole explants of injected embryos were prepared at the blastula stage, cultured in isolation and scored for mesoderm formation. Injection of 100pg of activin-Vg1 mRNA resulted in tissue elongation, indicative of mesoderm induction, and expression of muscle-specific cardiac actin, while injection of 5pg of BMP2-Vg1 mRNA resulted in an equal degree of mesoderm induction. The dosage difference reflects a difference in ligand processing efficiency in the early embryo (data not shown).

Defolliculated oocytes were injected with 50ng of capped, *in vitro* transcribed RNA and cultured for 3-4 days. Conditioned supernatants were analyzed by western blotting using antisera specific for the mature region of activin βB or Vg1 (Figure 2). Injection of activin βB mRNA resulted in efficient secretion of processed protein, detected as a reduced monomer of ~14kD. No secreted mature Vg1 was detected following injection of Vg1 mRNA, and a small amount of mature Vg1 secretion was detected with BMP2-Vg1 injection. Injection of activin-Vg1 mRNA resulted in secretion of substantial amounts of processed Vg1 ligand. The processed Vg1 migrated as a series of bands of ~18kD. The different mature species presumably reflect glycosylated forms, as described for the endogenous precursor (Tannahill Melton, 1989; Dale et al., 1989, 1993). Secretion of mature protein directed by activin-Vg1 was comparable in efficiency to activin βB, and in both cases a lesser amount of secreted precursor (~46kD) is also detected. Metabolic labeling of injected oocytes indicated that the highly abundant, newly synthesized proteins present in the supernatant corresponded to the injected mRNA (data not shown).

The concentration of mature protein was determined by western blotting and quantitation of supernatants using known amounts of *in vitro* translated Vg1 or activin βB (data not shown). Oocyte-conditioned supernatants contained soluble, processed Vg1 and activin βB at concentrations of 100-500 ng/ml. It

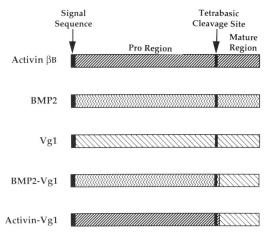

Figure 1. Chimeric Vg1 constructs. Schematic representation of activin βB, BMP2, Vg1, and chimeric BMP2-Vg1 (Thomsen and Melton, 1993) and activin-Vg1 (see material and methods). Members of the TGF-β superfamily, these genes encode a signal sequence, pro-region, tetrabasic cleavage site and mature region. Activin βB and BMP2, but not Vg1, form disulfide-linked dimers that are subsequently cleaved, releasing the mature C-terminal peptide as a secreted bioactive dimer. The chimeric constructs, fused downstream of the cleavage site, are designed to facilitate processing and secretion of mature Vg1.

should be noted that, if active, this quantity of protein is sufficient to perform a variety of functional tests, including receptor binding assays.

Mature Vg1 Induces Dorsal Mesoderm in Animal Pole Explants

Oocyte supernatants were tested for mesoderm inducing activity on animal pole explants. Blastula-stage animal poles (presumptive ectoderm) were

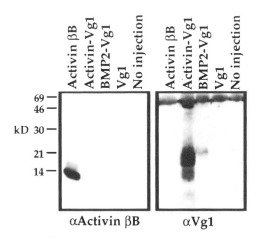

Figure 2. Western blot analysis of oocyte supernatants. Oocytes were injected with the indicated mRNA and conditioned supernatants analyzed using activin (left) or Vg1 (right) antisera. Activin βB mRNA directs secretion of mature protein. While Vg1 mRNA directs no secretion and BMP2-Vg1 directs little secretion, activin-Vg1 mRNA results in secretion of abundant mature Vg1 protein.

explanted, incubated with supernatant, cultured to the neurula stage and scored for the formation of mesodermal tissues and expression of mesodermal markers. Supernatants of uninjected, Vg1 or BMP2-Vg1 injected oocytes all failed to induce morphogenetic movements and expression of mesodermal markers (Figure 3; data not shown). Supernatants of activin-Vg1 or activin βB injected oocytes strongly induced morphogenetic movements indicative of mesoderm induction (Figure 3), expression of mesodermal markers and differentiation of the mesodermal tissues, muscle and notochord.

Control

Mature Vg1

Activin βB

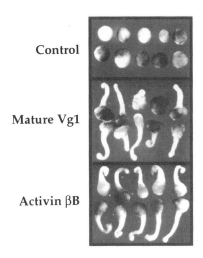

Figure 3. Mature Vg1 induces morphogenetic movements in ectodermal explants. Blastula stage animal pole explants were treated with 10% supernatant (1/10 dilution) and cultured to the neurula stage (stage 15). Mature Vg1 results in a dramatic elongation indicative of mesoderm induction (Symes and Smith, 1987). Activin βB results in similar elongation, while supernatant of uninjected oocytes (control) have no effect.

Low doses of mature Vg1 induced expression of the general mesodermal marker Xbra (Smith et al., 1991). Intermediate doses induced the ventrolateral marker Xwnt8 (Christian et al., 1991) and the dorsal marker cardiac actin (Stutz Sphor, 1986). High doses of mature Vg1 induced expression of the dorsal mesodermal markers goosecoid (Blumberg et al., 1991) and noggin (Smith and Harland, 1992), and, by secondary induction, the neural marker NCAM (Kintner and Melton, 1987). These markers were also induced by activin βB, which was a positive control, but not by supernatant of uninjected oocytes (Figure 4). Expression of the ventral marker globin was not induced by either mature Vg1 or activin (data not shown). Induction of morphogenetic movements in Vg1-treated explants was consistent with the response profile of mesodermal markers. At low doses (Xbra positive) no elongation was apparent, and at higher doses (cardiac actin positive) marked elongation is observed (data not shown). The dosage limit for mesoderm induction by mature Vg1 supernatant is approximately 1% (1/100 dilution), and based on the quantitation of mature Vg1 in the supernatant discussed above, this corresponds to a concentration of 2 ng/ml or approximately 10pM. This specific activity is consistent with that found for activin βB and other growth factors.

Examination of mature Vg1-treated explants cultured to the late neurula stage by immunohistochemistry and histology revealed the differentiation of mesodermal tissues, notochord and somitic muscle, as well as neural tissue (data not shown). Explants cultured to the late tadpole stage, following treatment with mature Vg1, often developed a high degree of axial organization. These "embryoids" displayed a rudimentary anterior-posterior pattern with organized head structures, including eyes, and a functional neuromuscular system. Histological sections showed the presence of muscle, notochord and neural tissue organized in a rudimentary axial pattern, and a differentiated eye (Figure 5). These results demonstrate the potent mesoderm-inducing activity of this preparation of mature Vg1.

Inhibiton of Mature Vg1 Activity by the Truncated Activin and FGF Receptors

To address the relationship between the activity of mature Vg1 and endogenous mesoderm-inducing signals we tested the ability of two dominant inhibitory receptors to block the activity of mature Vg1. A truncated form of the activin type II receptor (tAR), lacking the intracellular kinase domain, blocked formation of detectable mesoderm and expression of mesodermal markers in intact embryos (Hemmati-Brivanlou and Melton, 1992). Animal pole explants of embryos injected with 4ng of tAR mRNA were treated with mature Vg1 or

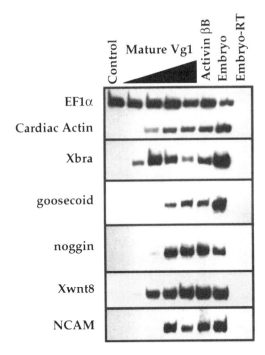

Figure 4. Dose-dependent induction of mesodermal markers by mature Vg1. Blastula stage animal pole explants were treated with increasing doses of mature Vg1 supernatant (1%, 3%, 10%, 30%), or a single dose of control (30%) or activin βB supernatant (10%). Treated explants were cultured to the neurula stage and analyzed by RT-PCR (see materials and methods). At low doses (1%) the general marker Xbra is induced; at intermediate doses the ventrolateral marker Xwnt8 and the dorsal marker cardiac actin are induced; and at high doses (10-30%) the dorsoanterior markers goosecoid and noggin, and the neural marker NCAM are induced. Activin βB treatment results in a similar response and control supernatants have no effect. EF1α is a quantitation control and the embryo and embryo-RT are additional positive and negative controls.

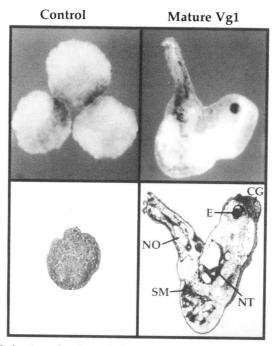

Control   **Mature Vg1**

Figure 5. Induction of embryoids by mature Vg1. Blastula animal pole explants treated with a high dose of mature Vg1 and cultured to the late tadpole stage (stage 40) differentiation into embryoids displaying a rudimentary axial organization, with anterior-posterior pattern and head structures. The embryoid shown (top right) has a clear head to tail pattern and a pigmented eye. A histological section (bottom right) reveals differentiated notochord (NO), somitic muscle (SM), neural tube (NT), eye (E), and cement gland (CG). Treatment with supernatant of uninjected oocytes (control) has no effect and explants form atypical epidermis (left).

activin βB, and mesoderm induction was assessed. tAR expression fully inhibited induction of morphogenetic movements and expression of cardiac actin by mature Vg1 (also described in Schulte-Merker et al., 1994). As expected, tAR also blocked activin βB activity and resulted in NCAM expression regardless of treatment with supernatants (Figures 6, 7). A dominant inhibitory FGF receptor (tFGFR), lacking the kinase domain, blocked formation of trunk and posterior mesoderm, resulting in loss of axial mesoderm and tail structures (Amaya et al., 1991). In addition to blocking FGF activity, tFGFR also blocks mesoderm induction by activin βB in animal pole explants (LaBonne and Whitman, 1994; Cornell and Kimelman, 1994). Injection of 4ng of tFGFR mRNA resulted in a complete block of mesoderm induction by both mature Vg1 and activin βB (Figure 6, 7). These observations suggest that the ability of the truncated activin and FGF receptors to perturb mesoderm induction in the embryo may, in fact, be due to inhibition of signaling by endogenous Vg1. Moreover, the described characteristics of mature Vg1 activity fulfill those expected for a natural mesoderm inducer.

Mature Vg1 is Not Inhibited by Follistatin

A natural inhibitor of activin function is the activin-binding protein follistatin (Kogawa et al., 1991; Nakamura et al., 1990). *Xenopus* follistatin is maternally expressed and blocks activin mediated induction of animal pole explants (Fukui et al., 1993, 1994; Hemmati-Brivanlou et al., 1994; Tashiro et al., 1991). However, unlike the truncated activin receptor, overexpression of follistatin in early embryos does not block mesoderm induction (Schulte-Merker et al., 1994; A. Hemmati-Brivanlou, O. Kelly, and D. Melton, unpublished). Furthermore, in recombinants of animal and vegetal pole explants, an assay recapitulating endogenous mesoderm induction, addition of follistatin in a transfilter experiment did not block induction (Slack, 1991b). These results raise questions about the role of activin in endogenous mesoderm induction.

To further examine the relationship of mature Vg1 activity to endogenous mesoderm induction, the ability of follistatin to inhibit mature Vg1 was examined. Mature Vg1 or activin supernatants were combined with oocyte-expressed follistatin prior to addition to blastula animal pole explants. While follistatin fully blocked activin-induced morphogenetic movements and expression of cardiac actin, mature Vg1 activity was unaffected by follistatin (Figures 8, 9). Even at a 10-fold excess of follistatin to mature Vg1 no inhibition was observed (data not shown). These observations are consistent with a role for processed Vg1 in endogenous mesoderm induction.

DISCUSSION

Vertebrate development relies on inductive interactions between cells which regulate the progressive spatial and functional organization of information contained within the egg, resulting in tissue differentiation and embryonic patterning (Slack, 1991a). The earliest induction in *Xenopus leavis* results in formation of the mesoderm, which arises from equatorial cells as a consequence of signals produced by vegetal pole blastomeres (Nieuwkoop, 1969a, 1969b; Sudarwati and Nieuwkoop, 1971). This initial mesoderm induction results in further refinements of patterning through interactions within the mesoderm and induction of ectoderm to form neural tissue (Hamburger, 1988). Advances in the identification of molecules capable of inducing mesoderm have implicated members of the transforming growth factor-β (Vg1, activin and BMP) and FGF families as endogenous mesoderm-inducing factors. Additional factors, wnts and noggin, can modify the character

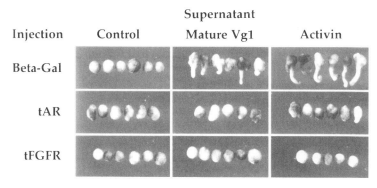

Figure 6. Inhibition of mature Vg1 activity by truncated activin and FGF receptors. At the 2-cell stage embryos were injected with 4ng of the truncated activin type II receptor (tAR), the truncated FGF receptor (tFGFR) or β-galactosidase (Beta-Gal) mRNA. Blastula stage animal pole explants were prepared and treated with mature Vg1, activin βB or control supernatants (10%). Both truncated receptors blocked morphogenetic movements induced by mature Vg1 and activin.

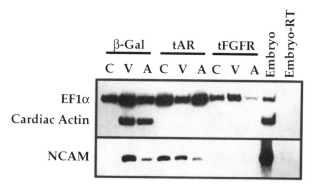

Figure 7. Truncated activin and FGF receptors block cardiac actin induction by mature Vg1. At the 2-cell stage embryos were injected with 4 ng of mRNA encoding the truncated activin type II receptor (tAR), the truncated FGF receptor (tFGFR) or β-galactosidase (Beta-Gal). Blastula stage animal pole explants were prepared and treated with mature Vg1, activin βB or control supernatants (10%). Following incubation to the neurula stage samples were examined by RT-PCR. The truncated receptors fully block cardiac actin induction by both mature Vg1 and activin, and tAR stimulates NCAM expression regardless of treatment. β-galactosidase is an injection control.

of mesoderm while failing to induce mesoderm directly (Sive, 1993; Klein and Melton, 1994; Slack, 1994). The studies presented here address the activity of processed Vg1, the only described molecule which is maternal, induces dorsal mesoderm, organizes a complete body axis, and is localized to vegetal blastomeres, the source of endogenous mesoderm-inducing signals (Thomsen and Melton, 1993; Dale et al., 1993). These observations establish Vg1 as a strong candidate for the natural inducer of dorsal mesoderm.

Here I report the successful preparation of soluble, biologically active mature Vg1 by expression of an activin βB-Vg1 chimera in *Xenopus* oocytes. The results demonstrate that the mesoderm-inducing activity of Vg1 does reside in the secreted mature region, establishing the mechanism of action of injected BMP-Vg1 hybrids (Dale et al., 1993; Thomsen and Melton, 1993). Mature Vg1 is a potent inducer of mesoderm in animal pole explants, resulting in expression of dorsal markers (goosecoid, noggin and cardiac actin), a ventrolateral marker (Xwnt8) and a general mesodermal marker (Xbra). Formation of mesodermal tissues (muscle and notochord) and secondary induction of neural tissue is observed, and in many cases these tissues form a well-organized axial pattern, including head structures. Based on dosage and extent of response it can be concluded that mature Vg1 and activin βB have markedly similar activities on animal pole explants. Furthermore, we find that truncated forms of the activin and FGF receptors, which interfere withendogenous mesoderm induction, fully inhibit mesoderm induction by mature Vg1, similar to inhibition of activin function. In addition, follistatin, an activin inhibitor that does not block endogenous mesoderm induction, fails to block mature Vg1 activity. The observed correspondence of the inhibition profile of mature Vg1 activity and endogenous mesoderm induction strongly support a role for processed Vg1 in natural mesoderm induction, and indicate that the inhibitory effects of truncated receptors may actually reflect an inhibition of Vg1 signaling in the developing embryo.

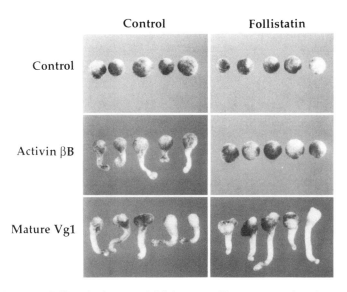

Figure 8. Follistatin does not inhibit mature Vg1 activity. Blastula stage animal pole explants were treated with mature Vg1 or activin βB (left) or with these supernatants following mixture and pre-incubation with follistatin supernatant. While follistatin fully blocks activin induced morphogenetic movements, induction by mature Vg1 was unaffected.

The behavior of the hybrid Vg1 molecules in this and other studies reveals the complexity of processing and secretion in the TGF-β family. The described differences in the effective dose and extent of induction with the two BMP-Vg1 hybrids points to an informative difference in the design of these molecules. The BMP2-Vg1 fusion, which incorporates the BMP2 tetrabasic cleavage site, induces both notochord and muscle in animal pole explants and rescues a complete axis at a dose of 50pg (Thomsen and Melton, 1993). The BMP4-Vg1 fusion, which retains the Vg1 cleavage site, induces only muscle in animal pole explants and a partial secondary axis at doses of 500pg or more (Dale et al., 1993). The stimulation of Vg1 processing by replacement of the pro-region indicates a critical role for the pro-region in the regulation of Vg1 processing (Gray and Mason, 1990). In addition, the difference in effective dose for the two BMP-Vg1

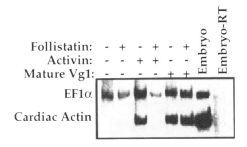

Figure 9. Follistatin fails to inhibit cardiac actin induction by mature Vg1. Blastula animal pole explants were treated with mature Vg1 or activin only, or with these supernatants pre-incubated with follistatin. Upon reaching the neurula stage samples were analyzed by RT-PCR. While follistatin fully inhibits cardiac actin induction by activin, no effect on mature Vg1 activity is detected.

fusions implicates the cleavage site in the control of processing. While both hybrids are efficiently processed in the early embryo and the oocyte (Dale et al., 1993; G. Thomsen, D. Kessler, C. Dohrmann and D. Melton, unpublished), neither is secreted from oocytes in substantial quantities (Dale et al., 1993; this study). The activin-Vg1 hybrid, on the other hand, is abundantly secreted by oocytes. Clearly, the specific pro-region and cleavage site can influence processing, but, in addition, regulation at the level of secretion is apparent, perhaps in a stage-specific manner. It has been suggested that localized Vg1 processing is a likely point of regulation in mesoderm induction and axial patterning (Thomsen and Melton, 1993), and it is important to note that no direct evidence for endogenous processed Vg1 is available. The described behavior of the Vg1 hybrids suggests the potential complexity of Vg1 regulation, which may be controlled at the level of dimerization, processing and/or secretion.

The mechanism of mature Vg1 inhibition by the truncated activin type II receptor is not clear. The simplest proposal is that the activin type II receptor binds mature Vg1 as well as activin with high affinity. However, the TGF-β-related receptor family is a diverse group of related molecules which form heterodimeric complexes (Kingsley, 1994; Massague et al., 1994). Given the heterodimeric nature of these receptors, inhibition of signaling by overexpression of a truncated receptor may result from direct ligand binding or inactivation of an associated subunit within an existing or novel receptor complex. Therefore, in the absence of information regarding the structure of receptor complexes, it is difficult to draw any conclusions about the specificity of a truncated receptor for an individual TGF-β-related ligand. The activin type II receptor could be a high affinity Vg1 receptor, a subunit in a Vg1 receptor complex, or have no involvement in an endogenous Vg1 receptor complex. Preliminary results using metabolically-labeled mature Vg1 in receptor binding assays suggest that the activin type II receptor does not directly bind Vg1 (D. Kessler and D. Melton, unpublished). Alternatively, mature Vg1 could direct increased synthesis or secretion of endogenous activin, resulting in activin-dependent mesoderm induction which, of course, would be inhibited by the truncated receptors. However, no increase in activin βB mRNA was observed following treatment of animal pole explants with mature Vg1 (data not shown).

The requirement for a functional FGF signaling pathway for induction of animal pole explants by both mature Vg1 and activin suggests that this receptor-tyrosine kinase activity is necessary for productive signaling by members of the TGF-β family (LaBonne and Whitman, 1994; Cornell and Kimelman, 1994). FGF signaling could establish a permissive state in which a common intracellular signaling molecule is in an active state. Alternatively, FGF signaling may alter the fate of cells, rendering them competent to respond to mesoderm induction by Vg1 or activin. It appears that FGF and TGF-β signaling do not share a common pathway since individual intracellular targets are not shared (LaBonne and Whitman, 1994), and a subset of mesodermal markers are still responsive to activin in the presence of the truncated FGF receptor (LaBonne and Whitman, 1994; Cornell and Kimelman, 1994). In addition, expression of the truncated activin receptor potentiates the response to FGF in animal pole explants (Hemmati-Brivanlou and Melton, 1992). Clearly a complex relationship exists between these signaling pathways, and both appear to be required for natural mesoderm induction and mature Vg1 signaling.

Follistatin, a specific activin inhibitor, acts by directly binding activin protein and forming an inactive complex (Kogawa et al., 1991; Nakamura et al., 1990). Overexpression in embryos can stimulate neural markers, but has no substantial effect on endogenous mesoderm induction (Hemmati-Brivanlou et al., 1994; Schulte-Merker et al., 1994; A. Hemmati-Brivanlou, O. Kelly and D. Melton, unpublished). This observation is inconsistent with a role for activin in natural mesoderm induction. In this study, and another (Schulte-Merker et al., 1994), it has been demonstrated that follistatin does not inhibit the mesoderm inducing activity of processed Vg1. This result is consistent with the function of endogenous Vg1 in mesoderm induction.

An oriented cortical rotation, during the first cell cycle, establishes the initial dorsoventral pattern in *Xenopus* (Elinson and Rowning, 1988; Vincent and Gerhart, 1987). The dorsoventral pattern is reflected in the capacity of dorsal vegetal cells to induce notochord and muscle, while ventral vegetal cells induce mesenchyme, blood and small amounts muscle (Nieuwkoop, 1969b; Boterenbrood and Nieuwkoop, 1973; Dale and Slack, 1987). In addition, as early as the 32-cell stage dorsal vegetal blastomeres can organize a complete body axis following ventral vegetal transplantation (Gimlich and Gerhart, 1984; Kageura, 1990). The dorsal vegetal cells which comprise this organizing center (Nieuwkoop center) have an endodermal fate and induce dorsal mesoderm, resulting in formation of Spemann's organizer and axial organization (Elinson and Kao, 1989; Gerhart et al., 1989).

It is proposed that cortical rotation directs the localized processing of Vg1 precursor, resulting in formation of the Nieuwkoop center and, subsequently, the Spemann organizer. The proteolytic activation of Vg1 could be dependent on a locally-activated protease or restricted release of a cleavage site inhibitor. It has been shown that ventral vegetal injection of ultraviolet-ventralized embryos with BMP2-Vg1 results in complete axial rescue and injected cells exclusively populate the endoderm, consistent with the formation of a Nieuwkoop center (Thomsen and Melton, 1993). While both noggin and Xwnt8 injection result in axial rescue and Nieuwkoop center formation (Smith and Harland, 1991, 1992; Sokol el al., 1991), these factors, unlike Vg1, are not localized to vegetal pole blastomeres and do not induce dorsal mesoderm. Vg1 is the strongest candidate for an endogenous factor required for Nieuwkoop center formation or function. The induction of dorsal mesoderm in explants by mature Vg1 may dependent on formation of an endodermal organizing center. Consistent with this suggestion would be the induction of endodermal markers by mature Vg1 and the detection of endodermal tissues in apposition to mesodermal tissues in induced explants and these predictions are currently being tested. It should be noted that the induction of an endoderm-specific antigen by mesoderm-inducing factors has been reported (Jones et al., 1993), consistent with the proposed expectations.

Acknowledgments

I am grateful to Doug Melton for providing support and encouragement and to the entire Melton Laboratory for invaluable interactions. The follistatin experiments were performed with the help of Olivia Kelly. This work was supported by a postdoctoral fellowship from the Jane Coffin Childs Memorial Fund for Medical Research. Support was also received from the Howard Hughes Medical Institute and the NIH.

REFERENCES

Amaya, E., Musci, T. J. and Kirschner, M. W. (1991) Expression of a dominant negative mutant of the FGF receptor disrupts mesoderm formation in *Xenopus* embryos. Cell 66: 257-270.

Asashima, M., Nakano, H., Shimada, K., Kinoshita, K., Ishii, K., Shibai, H. and Ueno, N. (1990a) Mesodermal induction in early amphibian embryos by activin A (erythroid differentiation factor). Roux's Arch. Dev. Biol. 198: 330-335.

Asashima, M., Nakano, H., Uchiyama, H., Davids, M., Plessow, S., Loppnow-Blinde, B., Hoppe, P., Dau, H. and Tiedemann, H. (1990b) The vegetalizing factor belongs to a family of mesoderm inducing proteins related to erythroid differentiation factor. Naturwissenschaften 77: 389-391.

Asashima, M., Nakano, H., Uchiyama, H., Sugino, H., Nakamura, T., Eto, Y., Ejima, D., Nishimatsu, S. I., Ueno, N. and Kinoshita, K. (1991) Presence of activin (erythroid differentiation factor) in unfertilized eggs and blastulae of *Xenopus laevis*. Proc. Natl. Acad. Sci. USA 88: 6511-6514.

Blumberg, B., Wright, C. V., De Robertis, E. M. and Cho, K. W. (1991) Organizer-specific homeobox genes in *Xenopus laevis* embryos. Science 253: 194-196.

Boterenbrood, E. C. and Nieuwkoop, P. D. (1973) The formation of the mesoderm in urodelean amphibians. V. Its regional induction by the endoderm. Roux Arch. EntwMech. Org. 173: 319-332.

Christian, J. L., McMahon, J. A., McMahon, A. P. and Moon, R. T. (1991) Xwnt-8, a Xenopus Wnt-1/int-1-related gene responsive to mesoderm inducing factors may play a role in ventral mesodermal patterning during embryogenesis. Development 111: 1045-1056.

Christian, J. L. and Moon, R. T. (1993) Interactions between Xwnt-8 and Spemann organizer signaling pathways generate dorsoventral pattern in the embryonic mesoderm of *Xenopus*. Genes Dev. 7: 13-28.

Christian, J. L., Olson, D. J. and Moon, R. T. (1992) Xwnt-8 modifies the character of mesoderm induced by bFGF in isolated *Xenopus* ectoderm. EMBO J. 11: 33-41.

Cornell, R. and Kimelman, D. (1994) Activin-mediated mesoderm induction requires FGF. Development 120: 453-462.

Dale, L., Howes, G., Price, B. M. J. and Smith, J. C. (1992) Bone morphogenetic protein 4: a ventralizing factor in *Xenopus* development. Development 115: 573-585.

Dale, L., Matthews, G. and Colman, A. (1993) Secretion and mesoderm-inducing activity of the TGF-β-related domain of *Xenopus* Vg1. EMBO J. 12: 4471-4480.

Dale, L., Matthews, G., Tabe, L. and Colman, A. (1989) Developmental expression of the protein product of Vg1, a localized maternal mRNA in the frog *Xenopus laevis*. EMBO J. 8: 1057-1065.

Dale, L. and Slack, J. M. W. (1987) Regional specification within the mesoderm of early embryos of *Xenopus laevis*. Development 100: 279-295.

Dohrmann, C. E., Hemmati, B. A., Thomsen, G. H., Fields, A., Woolf, T. M. and Melton, D. A. (1993) Expression of activin mRNA during early development in *Xenopus laevis*. Dev. Biol. 157: 474-83.

Elinson, R. P. and Kao, K. R. (1989) The location of dorsal information in frog early development. Dev. Growth Diff. 31: 423-430.

Elinson, R. P. and Rowning, B. (1988) A transient array of parallel microtubules in frog eggs: potential tracks for a cytoplasmic rotation that specifies the dorso-ventral axis. Dev. Biol. 128: 185-97.

Fukui, A., Nakamura, T., Sugino, K., Takio, K., Uchiyama, H., Asashima, M. and Sugino, H. (1993) Isolation and characterization of *Xenopus* follistatin and activin. Dev. Biol. 159: 131-139.

Fukui, A., Nakamura, T., Uchiyama, H., Sugino, K., Sugino, H. and Asashima, M. (1994) Identification of activins A, AB, and B and follistatin proteins in *Xenopus* embryos. Dev. Biol. 163: 279-281.

Gerhart, J., Danilchik, M., Doniach, T., Roberts, S., Rowning, B. and Stewart, R. (1989) Cortical rotation of the *Xenopus* egg: consequences for the anteroposterior pattern of embryonic dorsal development. Development 107 (Suppl.): 37-51.

Gimlich, R. L. and Gerhart, J. C. (1984) Early cellular interactions promote embryonic axis formation in *Xenopus laevis*. Dev. Biol. 104: 117-130.

Graff, J. M., Thies, R. S., Song, J. J., Celeste, A. J. and Melton, D. A. (1994) Studies with a *Xenopus* BMP receptor suggest that ventral mesoderm-inducing signals override dorsal signals in vivo. Cell 79: 169-179.

Gray, A. M. and Mason, A. J. (1990) Requirement for activin A and transforming growth factor-β1 pro-regions in homodimer assembly. Science 247: 1328-1330.

Hamburger, V. (1988) "The Heritage of Experimental Embryology: Hans Spemann and the Organizer," Oxford University Press, New York.

Hemmati-Brivanlou, A., Kelley, O. G. and Melton, D. A. (1994) Follistatin, an antagonist of activin, is present in the Spemann organizer and displays direct neuralizing activity. Cell 77: 283-295.

Hemmati-Brivanlou, A. and Melton, D. A. (1992) A truncated activin receptor dominantly inhibits mesoderm induction and formation of axial structures in *Xenopus* embryos. Nature 359: 609-614.

Jones, C. M., Lyons, K. M., Lapan, P. M., Wright, C. V. E., Hogan, B. J. M (1992) DVR-4 (bone morphogenetic protein-4) as a postero-ventralizing factor in *Xenopus* mesoderm induction. Development 115: 639-647.

Jones, E. A., Abel, M. H. and Woodland, H. R. (1993) The possible role of mesodermal growth factors in the formation of endoderm in *Xenopus laevis*. Roux's Arch. Dev. Biol. 202: 233-239.

Kageura, H. (1990) Spatial distribution of the capacity to initiate a secondary embryo in the 32-cell embryo of *Xenopus laevis*. Dev. Biol. 142: 432-8.

Kimelman, D. and Kirschner, M. (1987) Synergistic induction of mesoderm by FGF and TGFβ and the identification of FGF in the early *Xenopus* embryo. Cell 51: 869-877.

Kingsley, D. M. (1994) The TGF-β superfamily: new members, new receptors, and new genetic tests of function in different organisms. Genes Dev. 8: 133-146.

Kintner, C. R. and Melton, D. A. (1987) Expression of *Xenopus* N-CAM RNA in ectoderm is an early response to neural induction. Development 99: 311-325.

Klein, P. S. and Melton, D. A. (1994) Hormonal regulation of embryogenesis: the formation of mesoderm in *Xenopus laevis*. Endo. Rev. 15: 326-340.

Kogawa, K., Nakamura, T., Sugino, K., Takio, K., Titani, K. and Sugino, H. (1991) Activin-binding protein is present in pituitary. Endocrinology 128: 1434-1440.

Koster, M., Plessow, S., Clement, J. H., Lorenz, A., Tiedemann, H. and Knochel, W. (1991) Bone morphogenetic protein 4 (BMP4), a member of the TGF-β family, in early embryos of *Xenopus laevis*: analysis of mesoderm inducing activity. Mech. Dev. 33: 191-200.

Krieg, P., Varnum, S., Wormington, M. and Melton, D. A. (1989) The mRNA encoding elongation factor 1α (EF1α) is a major transcript at the mid-blastula transition in *Xenopus*. Dev. Biol. 133: 93-100.

Krieg, P. A. and Melton, D. A. (1987) In vitro RNA synthesis with SP6 RNA polymerase. Meth. Enzymol. 155: 397-415.

Ku, M. and Melton, D. A. (1993) Xwnt-11: a maternally expressed *Xenopus* wnt gene. Development 119: 1161-1173.

LaBonne, C. and Whitman, M. (1994) Mesoderm induction by activin requires FGF-mediated intracellular signals. Development 120: 463-472.

Massague, J., Attisano, L. and Wrana, J. L. (1994) The TGF-β family and its composite receptors. TICB 4: 172-178.

Melton, D. A. (1987) Translocation of a localized maternal mRNA to the vegetal pole of *Xenopus* oocytes. Nature 328: 80-82.

Moon, R. T. and Christian, J. L. (1992) Competence modifiers synergize with growth factors during mesoderm induction and patterning in *Xenopus*. Cell 71: 709-712.

Mowry, K. and Melton, D. (1992) Vegetal messenger RNA localization directed by a 340-nt sequence element in *Xenopus* oocytes. Science 255: 991-994.

Nakamura, T., Takio, K., Eto, Y., Shibai, H., Titani, K. and Sugino, H. (1990) Activin-binding protein from rat ovary is follistatin. Science 247: 836-838.

Nieuwkoop, P. D. (1969a) The formation of mesoderm in urodelean amphibians. I. Induction by the endoderm. Roux Arch. EntwMech. Org. 162: 341-373.

Nieuwkoop, P. D. (1969b) The formation of the mesoderm in urodelean amphibians II. The origin of the dorso-ventral polarity of the mesoderm. Roux Arch. EntwMech. Org. 163: 298-315.

Nishimatsu, S., Takebayashi, K., Suzuki, A., Murakami, K. and Ueno, N. (1993) Immunodetection of *Xenopus* bone morphogenetic protein-4 in early embryos. Growth Factors 8: 173-6.

Peng, H. B. (1991) Solutions and protocols *in* "Methods in Cell Biology 36". B. K. Kay, H. B. Peng, eds., Academic Press, New York.

Rebagliati, M. R. and Dawid, I. B. (1993) Expression of activin transcripts in follicle cells and oocytes of *Xenopus laevis*. Dev. Biol. 159: 574-580.

Rebagliati, M. R., Weeks, D. L., Harvey, R. P. and Melton, D. A. (1985) Identification and cloning of localized maternal mRNAs from *Xenopus* eggs. Cell 42: 769-777.

Schulte-Merker, S., Smith, J. C. and Dale, L. (1994) Effects of truncated activin and FGF receptors and of follistatin on the inducing activities of BVg1 and activin: does activin play a role in mesoderm induction? EMBO J. 13: 3533-3541.

Sive, H. L. (1993) The frog prince-ss: A molecular formula for dorsal-ventral patterning in *Xenopus*. Genes Dev. 7: 1-12.

Slack, J. M. W. (1991a) "From Egg to Embryo: Regional Specification in Early Development," Cambridge University Press, Cambridge.

Slack, J. M. W. (1991b) The nature of the mesoderm-inducing signal in *Xenopus*: a transfilter induction study. Development 113: 661-669.

Slack, J. M. W. (1994) Inducing factors in *Xenopus* early embryos. Curr. Biol. 4: 116-126.

Slack, J. M. W., Darlington, B. G., Heath, J. K. and Godsave, S. F. (1987) Mesoderm induction in early *Xenopus* embryos by heparin-binding growth factors. Nature 326: 197-200.

Slack, J. M. W., Isaacs, H. V. and Darlington, B. G. (1988) Inductive effects of fibroblast growth factor and lithium ion on *Xenopus* blastula ectoderm. Development 103: 581-590.

Smith, J. C., Price, B. M. J., Green, J. B. A., Weigel, D. and Herrmann, B. G. (1991) Expression of a *Xenopus* homolog of *Brachyury (T)* in an immediate-early response to mesoderm induction. Cell 67: 79-87.

Smith, J. C., Price, B. M. J., Van Nimmen, K. and Huylebroeck, D. (1990) Identification of a potent *Xenopus* mesoderm-inducing factor as a homolog of activin A. Nature 345: 729-731.

Smith, W. B. and Harland, R. M. (1992) Expression cloning of noggin, a new dorsalizing factor localized to the Spemann organizer in *Xenopus* embryos. Cell 70: 829-840.

Smith, W. C., Knecht, A. K., Wu, M. and Harland, R.M. (1993) Secreted noggin protein mimics the Spemann organizer in dorsalizing *Xenopus* mesoderm. Nature 361: 547-549.

Smith, W. C. and Harland, R. M. (1991) Injected Xwnt-8 RNA acts early in *Xenopus* embryos to promote formation of a vegetal dorsalizing center. Cell 67: 753-765.

Sokol, S., Christian, J. L., Moon, R. T. and Melton, D. A. (1991) Injected wnt RNA induces a complete body axis in *Xenopus* embryos. Cell 67: 741-752.

Sokol, S., Wong, G. G. and Melton, D. A. (1990) A mouse macrophage factor induces head structures and organizes a body axis in *Xenopus*. Science 249: 561-564.

Sokol, S. Y. (1993) Mesoderm formation in *Xenopus* ectodermal explants overexpressing Xwnt8: evidence for a cooperating signal reaching the animal pole by gastrulation. Development 118: 1335-1342.

Sokol, S. Y. and Melton, D. A. (1992) Interaction of wnt and activin in dorsal mesoderm induction in *Xenopus*. Dev. Biol. 154: 1-8.

Stutz, F. and Sphor, G. (1986) Isolation and characterization of sarcomeric actin genes expressed in *Xenopus laevis* embryos. J. Mol. Biol. 187: 349-361.

Sudarwati, S. and Nieuwkoop, P. D. (1971) Mesoderm formation in the anuran *Xenopus laevis* (Daudin). Roux Arch. EntwMech. Org. 166: 189-204.

Suzuki, A., Thies, R. S., Yamaji, N., Song, J. J., Wozney, J. M., Murakami, K. and Ueno, N. (1994) A truncated BMP receptor affects dorsal-ventral patterning in the early *Xenopus* embryo. Proc. Natl. Acad. Sci. U.S.A. 91: 10255-10259.

Symes, K. and Smith, J. C. (1987) Gastrulation movements provide an early marker of mesoderm induction in *Xenopus laevis*. Development 101: 339-349.

Tannahill, D. and Melton, D. A. (1989) Localized synthesis of the Vg1 protein during early *Xenopus* development. Development 106: 775-785.

Tashiro, K., Yamada, R., Asano, M., Hasimoto, M., Muramatsu, M. and Shiokawa, K. (1991) Expression of mRNA for activin-binding protein (follistatin) during early embryonic development of *Xenopus laevis*. BBRC 174: 1022-1027.

Thomsen, G., Woolf, T., Whitman, M., Sokol, S., Vaughan, J., Vale, W. and Melton, D. A. (1990) Activins are expressed early in *Xenopus* embryogenesis and can induce axial mesoderm and anterior structures. Cell 63: 485-493.

Thomsen, G. H. and Melton, D. A. (1993) Processed Vg1 protein is an axial mesoderm inducer in *Xenopus*. Cell 74: 433-41.

Ueno, N., Shoda, A., Takebayashi, K., Suzuki, A., Nishimatsu, S.-I., Kikuchi, T., Wakimasu, M., Fujino, M. and Murakami, K. (1992) Identification of bone morphogenetic protein-2 in early *Xenopus laevis* embryos. Growth Factors 7: 233-240.

van den Eijnden-Van Raaij, A. J. M., van Zoelent, E. J. J., van Nimmen, K., Koster, C. H., Snoek, G. T., Durston, A. J. and Huylebroeck, D. (1990) Activin-like factor from a *Xenopus* cell line responsible for mesoderm induction. Nature 345: 732-734.

Vincent, J.-P. and Gerhart, J. C. (1987) Subcortical rotation in *Xenopus* eggs: an early step in embryonic axis specification. Dev. Biol. 123: 526-539.

Weeks, D. L. and Melton, D. A. (1987) A maternal mRNA localized to the vegetal hemisphere in *Xenopus* eggs codes for a growth factor related to TGF-β. Cell 51: 861-867.

Wilson, P. A. and Melton, D. A. (1994) Mesodermal patterning by an inducer gradient depends on secondary cell-cell communication. Curr. Biol. 4: 676-686.

Yisraeli, J. and Melton, D. A. (1988) The maternal mRNA Vg1 is correctly localized following injection into *Xenopus* oocytes. Nature 336: 592-595.

Yisraeli, J., Sokol, S. and Melton, D. A. (1990) A two step model for the localization of maternal mRNA in Xenopus oocytes: involvement of microtubules and microfilaments in the translocation and anchoring of Vg1 RNA. Development 108: 289-298.

TARGET SITE DETERMINATION FOR XENOPUS LAEVIS FORK HEAD RELATED TRANSCRIPTION FACTORS

Eckhard Kaufmann, Dorothée Müller,
Petra Dege, Heiko Rauer, Heike Rohm
and Walter Knöchel

Abteilung Biochemie
Universität Ulm
Albert Einstein Allee
D-89069 Ulm, Germany

INTRODUCTION

The correct temporal and spatial expression of developmentally controlled genes is achieved by the specific binding of trans-acting transcription factors (Beebee and Burke, 1992). These DNA-binding proteins are composed of various modules ensuring proper DNA-recognition, transcriptional regulation and, in some cases, protein oligomerization. Further, according to conserved sequence motifs these factors have been subdivided into several well known structural classes such as the helix-turn-helix, the helix-loop-helix, the zinc-finger and the leucine-zipper motif (Harrison, 1991; Knöchel 1994). Recently, proteins with the so called "fork head" motif were found to represent another group within this classification system (Weigel and Jäckle, 1990). All the members of this class contain a highly hydrophilic domain of about 110 amino acid residues, the fork-head domain. The amino acid sequence of the fork head domain does not resemble any of the other known structural motifs. The sequence similarity within this domain is very high, whereas the sequences outside this region are only distantly related or completely divergent among the various fork head proteins (Lai et al., 1993).

Structure determination by X-ray analysis at 0.25 nm resolution (Clark et al., 1993) of a co-crystal containing the fork head domain of the liver-specific transcription factor HNF-3γ and its DNA-target sequence, the TTR promoter, revealed a certain similarity to the eucaryotic class of helix-turn-helix proteins yet with some specific differences: a compact winged-helix or "butterfly" structure is achieved by the intimate association of a cluster of three α-helices located at the amino terminus with wing-like and β-sheet structures more towards the carboxyterminal end of the molecule. Besides several contacts with the sugar-phosphate backbone the major direct interactions between the nucleic acid and the protein are established by presenting the recognition helix three of the helical cluster to the major groove of the slightly bent DNA (Brennan, 1993; Lai et al., 1993).

This DNA-binding domain was originally described within the fork head protein of the fruitfly Drosophila melanogaster, where point mutations within the fork head domain give rise to homeotic transformations within the terminal regions of the embryo (Weigel et al., 1989). This domain has since been detected in a variety of different organisms ranging from yeast to primates (for a review see Lai et al., 1993). Moreover, each species contains a multitude of factors harbouring this motif. In particular, a multigene family comprising about one dozen genes coding for proteins with the fork head motif has been characterised in the frog Xenopus laevis (Dirksen and Jamrich, 1992; Knöchel et al., 1992; Ruiz i Altaba and Jessel, 1992, Bolce et al., 1993, Ruiz i Altaba et al., 1993; Lef et al., 1994). Amino acid sequences of the fork head domains found in these proteins deviate between 8 and 54 % from the progenitor of this class, the Drosophila fork head domain. The specific localisation of transcripts during early embryogenesis and, in particular, the induction of one gene (Xenopus fork head domain 1: XFD-1) by activin A gives strong evidence for the involvement of the fork head proteins in Xenopus development (Knöchel et al., 1992). However, the DNA-recognition sites of Xenopus fork head related proteins and their putative target genes are unknown. Therefore, we are presenting here an experimental approach towards the binding-site analysis of selected Xenopus fork head proteins.

According to the high level of sequence conservation within the DNA-binding domains, especially with respect to the residues contacting DNA bases, all members of the fork head family should share a similar structure of their DNA recognition sites. However, the wide variety of known DNA-recognition sequences found for some rodent fork head related proteins indicates that the structural similarity of the proteins is not reflected in a corresponding sequence similarity of their DNA binding sites. Since the amino acids for the direct contact with the DNA are highly conserved within the fork head domain (Clark et al., 1993) this variety must be due to other, less conserved regions of the domain; principally a region of twenty amino acid residues between helix two and the recognition helix three (Overdier et al., 1994). It seems rather that this variability enables specific template recognition, which is gained by more or less subtle structural changes within the protein's DNA-recognition sites.

We studied three Xenopus fork head proteins (XFD-1, 2 and 3 (Knöchel et al., 1992; Lef et al., 1994) as representatives for an increasing sequence divergence from the fork head domain of the Drosophila fork head protein (fkh) in the order: fkh -> XFD-1 -> XFD-3 -> XFD2 (Figure 1). In order to isolate specific binding sequences we embarked on two independent approaches: first, we expressed β-galactosidase fusion proteins in E. coli. These constructs enabled us by means of affinity chromatography to isolate restriction fragments harbouring specific recognition sites from a pool of enzymatically hydrolysed genomic Xenopus DNA (Payre and Vincent, 1991). Second, the fork head domain of all three proteins was expressed in E. coli and employed in a repetitive PCR-mediated binding assay on a manifold of degenerate deoxyoligonucleotides (Blackwell and Weintraub, 1990; Gogos et al., 1992).

Our results show that the proteins XFD-1 and XFD-3 share similar recognition sequences due to their high protein sequence similarity. In contrast, the target site for XFD-2 deviates remarkably from the other two proteins. We suggest that binding specificity might be achieved by subtle modulation of a conserved central sequence core by neighbouring sequences.

MATERIALS AND METHODS

Synthesis of a XFD-2 / β-galactosidase fusion protein

A Pst I/Sca I fragment of the corresponding XFD-2 cDNA (Knöchel et al., 1992; Lef et al., 1994) was blunt-ended by fill-in reaction with E.coli DNA-polymerase (Klenow fragment) and integrated into the Sma I site of the plasmid expression vector pAX (Markmeyer et al., 1990).

For expression of the recombinant protein the plasmid was transformed into TG1 cells in the presence of ampicillin. The cells were grown at 25 °C in 2xTY medium to an absorbency of 0.6 at 600 nm and subsequently induced with IPTG at a final concentration of 1 mM. The expression was continued for up to 5 hours, cells were harvested by centrifugation at 10 000 x g and resuspended in 1/20 of the original culture volume extraction buffer [80 mM KCl, 10 mM HEPES (pH 7.2), 10 mM MgCl$_2$, 1 mM EDTA, 1 mM DTT, 0.5 mM PMSF] and then disintegrated by several cycles of freezing/thawing and subsequent sonication with a microtip at 52 W for 30 sec. The lysate was cleared by centrifugation at 15 000 x g. The resulting supernatant was divided in aliquots and kept at - 70 °C.

Expression of the XFD-1, 2 and 3 fork head domains

Expression of the fork head protein domain proteins in E.coli was achieved by using a commercial T7 promoter (Studier and Moffat, 1986) based expression system (pRSET, invitrogen). This system introduces a histidine tag at the amino terminus and thus greatly facilitates the purification by Ni^{2+} chelate chromatography under denaturing conditions. Since the vector itself does not supply an extra stop codon, termination of translation was achieved by using a pRSET vector with an appropriate reading frame, such that a stop signal is generated in conjunction with the flanking sequences located at the carboxyl terminus of the respective insertion. A construct for the expression of the XFD-1 domain was made by inserting a blunt-ended Bam HI/Pst I fragment from the corresponding cDNA into the blunt-ended Bam HI site of the pRSET vector. The corresponding construct for XFD-2 was obtained by subcloning a Hae III/Pml I fragment from the XFD-2 cDNA into the blunt-ended Bam H I site of pRSET and the vector coding for the fork head domain of XFD-3 was generated by cloning a Sma I/Alu I fragment from the respective cDNA into the same vector. All reading frames were confirmed by DNA-sequencing. The constructs were transformed into BL21(DE3) cells and expression was induced with 1 mM IPTG and cells were incubated at 37 °C for 5 h. Details of the purification by affinity chromatography on a nickel-trinitrilo acetic acid matrix are described elsewhere (Kaufmann et al., 1994). Renaturation was achieved by extensive dialysis at 4 °C against [0.2 M ammonium acetate (pH 5.3), 1 mM DTT, 1 mM EDTA].

Isolation of specific DNA-target sequences by immunoaffinity chromatography

Genomic DNA from Xenopus laevis was prepared from erythrocytes (Kupiec et al., 1987) and subsequently cleaved with Sau 3A. DNA fragments were purified by phenol/chloroform extraction and ethanol precipitation. Affinity chromatography was modified from the procedure described by Payre and Vincent (1991) and performed on a Pharmacia-LKB HiLoad system: a 1 ml column was packed with protein A sepharose (Pharmacia), washed with 50 ml NET-2 buffer [50 mM Tris-HCl (pH 7.5), 150 mM NaCl, 0.05 % (w/v) NP-40] and loaded at a flow rate of 0.1 ml/min with 400 µl anti-β-galactosidase antibody (Sigma). The column was washed extensively with NET-2 and binding buffer [80 mM KCl, 10 mM HEPES (pH 7.2), 10 mM $MgCl_2$, 1 mM EDTA, 1 mM DTT, 0.5 mM PMSF] and subsequently loaded with 8 ml β-galactosidase fusion protein extract (see above) at a flow rate of 0.5 ml/min. After washing with 50 ml binding buffer the immobilised fusion protein was incubated for 50 min with restricted, genomic DNA. The column was washed again with binding buffer and then developed with a linear salt gradient in binding buffer ranging from 80 to 500 mM KCl. The elution of the DNA was recorded by UV-spectroscopy at 280 nm; fractions of 500 µl were collected. The DNA was purified by phenol/chloroform extraction and finally inserted into the Bam HI site of the plasmid pGEMEX (Promega) for DNA-sequencing.

PCR-supported binding site selection from degenerate deoxyoligonucleotides

The procedure as described by Blackwell et al. (1990) and Gogos et al. (1992) was modified as follows. Each of the affinity purified XFD-1, 2 and 3 fork head domain proteins (1.5 µg) was incubated in 30 µl binding buffer [20 mM Tris-HCl (pH 7.8), 50 mM KCl, 1mM EDTA, 1mM DTT, 10 % glycerol] in the presence of 1 µg poly (dIC) at 4 °C with ^{32}P-labelled degenerate deoxyoligonucleotides. The sequence of the 18-mer degenerate probe is:
5′ TCTAGAACTAGTGGATC (N)$_{18}$ CGATACCGTCGACCTCG 3′.
Labelling was achieved by annealing the degenerate deoxyoligonucleotide with primer I (5′ CGAGGTCGACGGTATCG 3′) and subsequent fill-in reaction with Klenow-fragment of E. coli DNA polymerase in the presence of ^{32}P-α-dCTP. 12 µl of the binding reaction were applied to a 7 % acrylamide gel in 0.5 x TBE at 4 °C, the gel was run at a constant voltage of 160 V. The DNA-protein complex was visualised by autoradiography and cut out from the wet gel. The DNA was eluted at 45 °C for 2 h in elution buffer [0.5 M ammonium acetate, 0.1 % SDS, 10 mM magnesium acetate], purified by extraction with phenol/chloroform and finally precipitated with ethanol. This purified DNA served as a template for the subsequent PCR-amplification step: 100 pmol each of primer I and primer II (5′ TCTAGAACTAGTGGATC 3′) were incubated with template DNA, 5 mM dNTP's and 2.5 units Taq polymerase in 50µl incubation buffer (Pharmacia) for 30 cycles (1 min, 94 °C; 40 s, 52 °C; 30 s, 72 °C). At this step it is mandatory to include a control without template DNA. The products of the amplification were purified on a 4 % low-melting gel and labelled as described for the next round of binding. On the average we performed five selection rounds with decreasing amounts of protein in order to select for high-affinity sites (Gogos et al., 1992). After the last cycle the DNA was purified and integrated into the Spe I and Sal I recognition sites of the plasmid vector pBluescript. About 50 templates selected for each of the three XFD domain proteins were sequenced on an ABI 373A sequenator.

Miscellaneous procedures

All DNA-standard procedures were done according to Sambrook et al. (1989). Deoxyoligonucleotides were synthesised on an ABI 391 synthesiser and purified by HPLC, with the exception of the degenerate templates for the binding site selection which were purified by denaturing gel electrophoresis.

All deoxyoligonucleotides were converted into the double-stranded form and labelled by ^{32}P-γ-dATP and bacteriophage T4 polynucleotide kinase. Gel retardation assays were done as described above. For footprinting reactions with DNAse I the deoxyoligonucleotides of interest were integrated into pBluescript and asymmetrically labelled after hydrolysis with Sal I and Xba I by a fill-in reaction with Klenow enzyme and ^{32}P-α-dCTP. The labelled fragment was purified by acrylamide electrophoresis as described (Maxam and Gilbert, 1980). Formation of protein-DNA complexes used for DNAse footprinting was performed under the same conditions as described above for the gel retardation experiment, but after complex formation the concentration of Mg^{2+} was raised and DNAse was added at room temperature (Galas and Schmitz, 1978).

RESULTS

The molecular weights of the Xenopus fork head related proteins XFD-1, 2 and 3 range between 41 and 45 kDa. The respective fork head domains themselves span a region of 111 amino acid residues (Dirksen and Jamrich, 1992; Knöchel et al., 1992; Ruiz i Altaba and Jessel, 1992; Lef et al., 1994). In order to study the in vitro binding properties of these XFD proteins we expressed the entire proteins as well as polypeptides comprising only the fork head domains in E. coli. The complete proteins were obtained as fusion constructs with β-galactosidase under the control of the lacZ promoter, whereas the fork head domains were produced in a T7 promoter based system. In the latter case the introduction of an amino-terminal histidine tag facilitated the purification by Ni^{2+} chelate chromatography under denaturing conditions (Hochuli, 1990).

Our first approach to isolate specific in vitro binding sites is based on a procedure described by Payre and Vincent (1991). Xenopus genomic DNA was isolated from erythrocytes and cleaved by hydrolysis with the four bp cutting enzyme Sau 3A, thereby generating fragments of an average size of 500 bp. Fragments harbouring specific binding sites for XFD-2 were selected by binding to the XFD-2/β-galactosidase fusion protein immobilised on a protein A sepharose/anti-β-galactosidase matrix. When the column was developed with a linear salt gradient, DNA fragments eluted within a concentration range of 230 to 330 mM salt. This DNA fraction represented about 0.3 % of input DNA. Nucleotide sequences of the retained fragments were determined and aligned in order to derive common sequence characteristics. The recognition sequence thereby determined for XFD-2 contains a conserved central sequence motif, 5' ATAAACA 3', whereas the flanking regions differ significantly (Figure 1).

Additionally to the target site selection by immunoaffinity chromatography we embarked on a second approach. The rationale of this strategy which had previously been employed for other target sequences (see for example: Blackwell and Weintraub, 1990) is the selection of potential recognition sites from a mixture of deoxyoligonucleotides degenerated at 12 to 26 positions. Bound DNA is retained as a protein-DNA complex and can be recovered by a gel-retardation assay. Since non-specifically binding fragments are co-isolated, the retained DNA is employed in another round of binding and selection. Thus, if successive rounds are performed under increasingly stringent conditions, highly specific target sites can be isolated. The DNA fragments of interest are present only in minute amounts on a molar scale and are thus amplified during each round by means of the polymerase chain reaction (PCR).

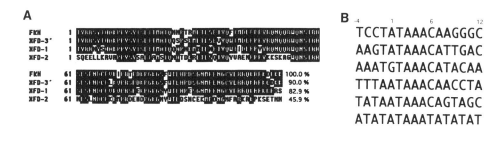

A   Alignment of the amino acid sequences of the fork head domains of XFD-1, 2 and 3 with the respective region of the <u>Drosophila</u> fork head protein. Positions of sequence identity are highlighted and calculated sequence similarities (in %) are indicated on the right.

Figure 1. Selection of DNA-recognition sequences for XFD-1, 2 and 3 by immunoaffinity chromatography and by PCR-mediated binding to a mixture of deoxyoligonucleotides degenerate at 18 positions.

A   Alignment of the amino acid sequences of the fork head domains of XFD-1, 2 and 3 with the respective region of the <u>Drosophila</u> fork head protein. Positions of sequence identity are highlighted and calculated sequence similarities (in %) are indicated on the right.

B   <u>Xenopus</u> genomic DNA fragments were selected by specific binding to immobilised XFD-2/β-galactosidase fusion protein (see Materials and Methods). Some arbitrarily selected sequences showing a frequent consensus motif are aligned with each other.

C   A deoxyoligonucleotides degenerate at 18 positions was incubated with the purified DNA-binding domains of XFD-1, 2 and 3 for five successive rounds of binding and amplification (see Materials and Methods). The two major DNA-sequence motifs isolated for each polypeptide are shown. All sequences (B, C) are aligned in order to achieve a maximum overlap within the core region (see Discussion). An arbitrary numbering is shown on top of the sequences.

These experiments were performed with the fork head domain of each of the three XFD proteins on an degenerate 18-mer deoxyoligonucleotide which is flanked by a defined 17 bp sequence on each side being complementary to the PCR primers. All three proteins gave rise to distinct complexes with the DNA. The protein concentration was reduced down to approximately 30 ng with increasing number of selections steps in order to enrich for high affinity binding sites (Gogos et al., 1992). Under these conditions more than 50 % of the DNA was retained as a protein complex after five rounds of selection and amplification. Again, the target sequences were determined and aligned for maximum homology. Figure 1 shows the two most frequent types of sequences being obtained for the XFD-1, 2 and 3 fork head domain. The binding sites for the fork head domains of XFD-1 and XFD-3 share a central 5' GT(^{A}/$_C$)AACA 3' core motif. The most prominent sequence for the XFD-1 domain is identical to the most frequent target for the XFD-3 fork head domain, thus reflecting the high degree of similarity between these two proteins within the DNA-binding domain. The core motif for the XFD-2 domain is 5' (^{A}/$_G$)T(^{A}/$_C$)AACA 3' which differs only in one position from that found for the other two protein targets (Figure 1).

Do the selected DNA-templates specifically recognise their cognate proteins? In order to answer this question we tested several DNA targets (18-mer plus flanking sequences) in a gel retardation assay and quantified the results by densitometry. For the XFD-2 fork head domain we found that the two selected templates (@14, 20) do bind with high specificity to the XFD-2 fork head domain only (Figure 2).

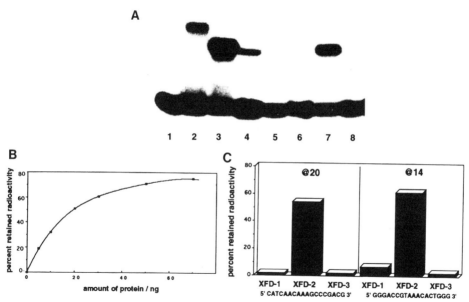

Figure 2. Two templates selected by the PCR-mediated isolation procedure bind to the fork head domain of XFD-2.

A Gel retardation assay with two targets (@14, @20) and the purified fork head domains of XFD-1, 2 and 3. Relative concentrations of the stock solutions of the three polypeptides applied were determined (XFD-1: 0.9 units; XFD-2: 0.2 units; XFD-3: 1.4 units; units referring to absorption at 278 nm). Lane 1: free target DNA @14; lanes 2, 3, 4: @14 plus XFD-1, 2 and 3 domain, respectively. Lanes 5, 6, 7, 8: as lanes 1 to 4, but with target DNA @20.

B Binding of target DNA @20 to the fork head domain of XFD-2 at different protein concentrations. Results of six bandshift experiments are graphically presented. The amount of DNA has been kept constant and is the same as in **A**.

C Quantification of bandshift experiments. The bars indicate the relative amount of radioactivity retained in the protein-DNA complex by the three polypeptides for each of the two target DNAs studied (sequences are shown at the bottom). In this calculation the different protein concentrations (see **A**) have been taken into account.

Since sequence @38 selected for the fork head domain of XFD-1 is identical to sequence @21 isolated for the respective domain of XFD-3, these two targets bind to the two protein domains with equal intensity. However, it is interesting to note that the other two DNA-sequences selected for the XFD-1 and 3 domains bind mutually to the respective proteins (XFD-1, 3), but also recognise the XFD-2 domain (Figure 3). The tight association of the PCR-mediated selected templates with their respective fork head domains was further corroborated by footprinting experiments with DNAse I on several DNA-recognition sites obtained (Figure 4). The regions encompassing the 18-bp target sequences (@14, 20) are completely protected from DNAse attack and apparently extend even beyond this region. Within this range we found in some cases, besides the usual protection, DNAse hypersensitive sites which might be attributed to DNA-bending (Clark et al., 1993).

Protein targets →		XFD-1	XFD-2	XFD-3
↓ DNA-recognition sequences				
ACCGAGTGTCAACACGTG	@38	+++	++	+++
CACAGTAAACAACGCGGC	@40	+++	++	+++
CATCAACAAAGCCCGACG	@20	-	+++	-
GGGACCGTAAACACTGGG	@14	-	+++	-
ACCGAGTGTCAACACGTG	@21	+++	++	+++
CCAAGGTAAACAGCTCGT	@31	+++	++	+++

Figure 3. DNA-binding preferences of XFD-1, 2 and 3 fork head domains.
Six DNA-fragments containing the PCR selected sequences including their flanking regions (see Results) were labelled and tested in bandshift experiments for their ability to bind to the fork head domain of the three proteins investigated.
The amount of retained radioactivity in the protein-DNA complex was determined by quantitative autoradiography. (+++), (++) and (-) indicate more than 50, 50-10 and less than 10 %, respectively. Values were normalised to protein concentration units.

We next performed gel retardation assays with 18-mers derived from PCR-selected targets sequences, which do not contain the flanking regions necessary for the PCR amplification (see Materials and Methods). As verified for the XFD-2 fork head domain specific binding site @20, we still observe a protein specific binding. However, the amount of binding seems to be markedly diminished if compared to the construct including the flanking regions. Densitometric analysis reveals that in the case of the target including the flanking regions, 58 % of the DNA is present in the protein-DNA complex as compared to only 10 % in the case of the deoxyoligonucleotide containing the binding site only (Figure 5). Nevertheless, such deoxyoligonucleotides are good candidates for mutagenesis experiments in order to search for nucleotide residues which might confer protein-specific recognition. These studies were performed on the two target sequences binding to the XFD-2 fork head domain with high specificity (@14, 20). Change of residue dA-(+5) in @20 (mut 20-1) completely abolishes binding to any of the three domains. On the other hand, transversion of dG-(+10) in @20 into dT (mut 20-2) or transition of dG-(+8) in @14 into dA (mut 14) does not effect the specificity for the XFD-2 fork head domain.

However, two other point mutations lead to a drastic change in binding specificity (Figure 6). If dC-(-1) in @20 is mutated to dA (mut 20-3), the DNA-sequence is preferentially recognised by the XFD-1 and 3 domains but only to a small extent by the XFD-2 domain. A change of dA-(+1) (mut 20-4) results in a high specificity only for the fork head domain of XFD-1 (Figure 6).

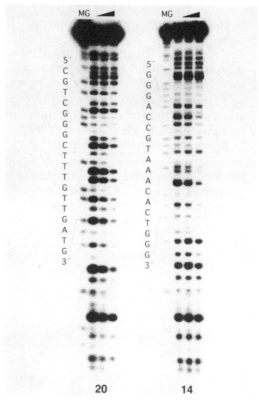

Figure 4. DNAse footprint analysis of XFD-2 target DNAs (@ 14, 20).
Xba I/Sal I fragments including the recognition site and flanking regions were ^{32}P-labelled on one strand, complexed with the fork head domain of XFD-2 and digested with 100 ng DNAse I for 45 s at room temperature. Shaded triangles on top of each footprint indicate increasing amounts of protein (0, 0.3 and 0.6 relative concentration units). MG: Maxam-Gilbert sequencing reaction specific for dG and dA residues, the corresponding sequence is shown on the left of each footprint.

DISCUSSION

A detailed in vitro analysis of the interaction between the Xenopus fork head proteins (XFD) and their cognate DNAs is pivotal to a further understanding of these factors on a molecular level. However, in contrast to many other members of the fork head protein family (Lai et al., 1991), DNA recognition sites for the XFD proteins are hitherto unknown. Thus, we embarked on an in vitro selection of target sites for XFD polypeptides by means of two independent experimental approaches. We studied three distinct XFD proteins with increasing divergence within the DNA-binding domain from that of the fork head protein of Drosophila. One of our experimental strategies is based on an immunoaffinity selection of binding sites from a pool of <u>Xenopus</u> genomic DNA restriction fragments by the XFD-2/β-galactosidase fusion protein (Payre and Vincent, 1991), whereas the second approach relies on the selection of potential targets out of a mixture of degenerate deoxyoligonucleotides by binding to XFD-1, 2 and 3 fork head domains. Since the proteins for the two experimental setups were expressed in <u>E. coli</u>, secondary modifications like phosphorylation or glycosylation do not seem to be required for proper DNA-recognition. The fork head domain of HNF-3γ for the diffraction study (Clark et al., 1993) has also been purified from expression in <u>E.coli</u>. When studying the DNA-binding behaviour of XFD-2 by means of two independent systems we used in one case a complete protein and in the other case the DNA-binding domain only. Thus, we could demonstrate that common sequence characteristics as determined by the two different methods are to be attributed to the DNA-binding fork-head domain which these two polypeptides have in common. It should be noted that the binding specificity for the immunoaffinity approach using crude protein is not diminished if compared to the experiment using the purified DNA-binding domains (Pollock and Treisman, 1990). The collection of <u>Xenopus</u> genomic fragments is naturally a complete representation of possible native binding sequences. The complexity of a degenerate deoxyoligonucleotide can be simply calculated (Keller and Maniatis, 1991) yielding a value of 3×10^{10} for an 18-mer. Thus, under normal experimental conditions (30 pmol of target DNA) the 18-fold degenerate template does represent all possible types of sequences, a conclusion which does not hold true in case of a longer degenerate matrix.

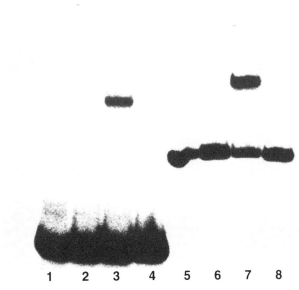

Figure 5. The binding of the XFD-2 fork head domain to its recognition site is apparently dependent upon the size of the target DNA.
A Gel retardation experiment showing the binding of the XFD-2 fork head domain to target site @20. The recognition site is a 18 bp deoxyoligonuleotide derived from the original target as being isolated from the PCR-mediated selection procedure. Lane 1: free DNA @20; lanes 2, 3 and 4: DNA @20 plus the fork head domain of XFD-1, 2 and 3, respectively.
B same binding experiment as in **A**, but the DNA target encompasses the original target site @20 which is flanked by two 14 bp primer regions. Lanes 5, 6, 7 and 8 correspond to lanes 1, 2, 3 and 4 in **A**. Equal amounts of protein were applied in **A** and **B**.

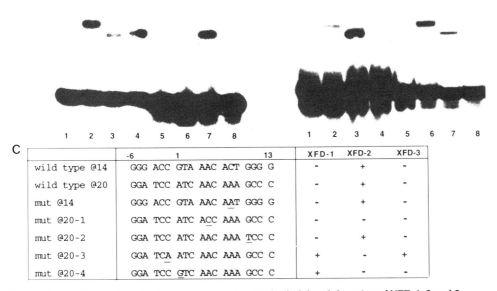

A				B	
1 2 3 4 5 6 7 8				1 2 3 4 5 6 7 8	

C

	-6 1 13	XFD-1	XFD-2	XFD-3
wild type @14	GGG ACC GTA AAC ACT GGG G	-	+	-
wild type @20	GGA TCC ATC AAC AAA GCC C	-	+	-
mut @14	GGG ACC GTA AAC AAT GGG G	-	+	-
mut @20-1	GGA TCC ATC ACC AAA GCC C	-	-	-
mut @20-2	GGA TCC ATC AAC AAA TCC C	-	+	-
mut @20-3	GGA TCA ATC AAC AAA GCC C	+	-	+
mut @20-4	GGA TCC GTC AAC AAA GCC C	+	-	-

Figure 6.   Binding of mutant recognition sites to the fork head domains of XFD-1, 2 and 3
A   XFD-2 binding site @14 was mutated by a single nucleotide exchange (mut 20-3). The binding of this mutant to the fork head domains of XFD-1, 2 and 3 is shown. Lane 1: free mut20-3 DNA; lanes 2, 3 and 4: mut 20-3 DNA plus the fork head domains of XFD-1, 2 and 3; lane 5: free @20 DNA; lanes 6, 7 and 8: @20 DNA plus the fork head domains of XFD-1, 2 and 3.
B   XFD-2 binding site @20 was mutated by a single nucleotide exchange (mut 20-4) Lane 1: free @20 DNA; lanes 2, 3 and 4: @20 DNA plus the fork head domains of XFD-1, 2 and 3; lane 5: free mut 20-4 DNA; lanes 6, 7 and 8: mut 20-4 DNA plus the fork head domains of XFD-1, 2 and 3.
C   The sequences of wild type and mutant deoxyoligonucleotides are shown. Mutated positions are underlined. An arbitrary numbering is shown on top of the sequences. The binding preference to the fork head domains of XFD-1, 2 and 3 is indicated on the right.

If the DNA-recognition sequences selected for the three protein domains are compared, a central consensus sequence is observed. Within this array of nucleotides no "scattering" but always one dominating nucleotide per position is found (Figure 1). In most positions within this core the most prominent residues selected by the two approaches are identical. Thus, as verified here for XFD-2 the two methods generate mutually supplementary data. Meanwhile, we are extending these studies by selecting recognition sites for XFD-1 and 3 by immunoaffinity chromatography, as demonstrated here for XFD-2, and are finding again an excellent agreement between the results derived by the two independent methods. Our data obtained by the PCR method yield a central core region of seven residues which correlates with other binding sites selected by this procedure (Blackwell et al., 1990; Pollock and Treisman, 1990; Keller and Maniatis, 1991; Gogos et al., 1992; Overdier et al., 1994). The similarity between this central region of XFD-1 and 3 targets versus XFD-2 targets corresponds well to the divergence of the respective amino acid sequences within the DNA-binding domain (Knöchel et al., 1992). Further, the difference between these two classes is pronounced by the isolation of two target sequences which do almost exclusively bind to the XFD-2 domain (Figures 2, 3). The central region is essential for DNA-binding, as shown by the drastic change in binding affinity, if an dA residue is converted into dC (Figure 6).

Outside of this core region a substantial sequence variety with respect to the selected residue is observed. Surely, this does not implicate that this region is unimportant in protein recognition: an eight bp deoxyoligonucleotide derived from the central region is incapable of binding to any of the proteins investigated (data not shown). Rather, it is likely that the specificity of the different proteins for their targets may be concealed in these side regions. The requirement of longer sequences does not imply that the nature of these 'spacer' sequences is important. One might speculate that specificity is not solely conferred by single residues but rather by an overall charge pattern which would explain the surprising variety of targets.

We undertook some mutagenesis experiments on two sequences with high binding specificity to the XFD-2 fork head domain (Figure 6). Mutations at position +8 and +10 of the XFD-2 domain targets (@14, 20) apparently do not change the binding preference, but alterations in position +1 and -1 lead to a change in specificity. These data indicate that specificity might be conferred by the 5' region of the central core mentioned above. A binding site with 18 residues seems to be sufficient for all three fork head domains investigated, but it might present a minimal requirement; the increased binding affinity of a target which includes flanking sequences (Figure 5) indicates that the region of DNA-protein interaction could be more extended than previously assumed (Clark et al., 1993; Overdier et al., 1994). This view is further supported by the results of DNAse I footprinting experiments on templates derived by the PCR-mediated approach; the protected region seems to encompass regions flanking the 18 bp recognition sequence (Figure 4). Experiments in order to substantiate this finding are currently in progress.

The amino acid residues which are responsible for direct base contacts between the TTR DNA target and the HNF-3γ fork head domain according to the crystallographic study of Clark et al. (1993) are conserved for all three XFD proteins investigated. Most significant differences between the XFD-1 and 3 fork head domains versus the XFD-2 domain arise within a region located between the carboxyl terminus of α-helix two and the amino terminus of α-helix three, which has been reported to be devoid of direct base contacts. In accordance with Overdier et al. (1994) we would suggest that changes within said sequence are transduced into subtle changes of the fork head domains overall structure, thereby leading to a corresponding modulation of the DNA recognition properties.

Obviously our in vitro studies select for DNA-binding sites from a thermodynamic point of view (Keller and Maniatis, 1992). It is feasible that the interaction in vivo might be different, possibly in such a manner that interactions which might be weak in vitro could be amplified by tertiary interactions with other proteins present in vivo. Thus, a DNA template with a lower in vitro binding affinity would be preferred in vivo. However, in many biological systems prior in vitro studies have greatly facilitated later in vivo studies and, thus, we gather our sequence studies might provide such a basis.

ACKNOWLEDGEMENTS

We gratefully acknowledge the excellent technical assistance of Bernd Guilliard. These studies were supported by grants from the Deutsche Forschungsgemeinschaft (Forschergruppe: "Regulation der zellulären Differenzierung") and from the Fonds der Chemischen Industrie.

REFERENCES

Beebee, T. and Burke, J., 1992, "Gene structure and transcription", IRL Press, Oxford.

Blackwell, T.K. and Weintraub, H., 1990, "Differences and similarities in DNA-binding preferences of MyoD and E2A protein complexes revealed by binding site selection.", *Science* 250:1104-1110.

Bolce, M.E., Hemmati-Brivanlou, A., and Harland, R.M., 1993, "XFKH-2, a Xenopus HNF-3 alpha homologue, exhibits both activin-inducible and autonomous phases of expression in early embryos.", *Dev. Biol.* 160:413-423.

Brennan, R.G., 1993, "The winged-helix DNA-binding motif: another helix-turn-helix takeoff", *Cell* 74:773-776.

Clark, K.L., Halay, E.D., Lai., E. and Burley, S.K., 1993, " Co-crystal structure of the HNF-3/fork head DNA-recognition motif resembles histone H5", *Nature* 364:412-420.

Dirksen, M.L. and Jamrich. M., 1992, "A novel, activin-inducible, blastopore lip-specific gene of Xenopus laevis contains a fork head DNA-binding domain.", *Genes Dev.* 6:599-608.

Galas, D. and Schmitz, A., 1978, "DNAse footprinting: a simple method for the detection of protein-DNA binding specificity.", *Nucl. Acids Res.* 5:3157-3170.

Gogos, J.A., Hsu, T., Bolton, J. and Kafatos, F.C., 1992, "Sequence discrimination by alternatively spliced isoforms of a DNA binding zinc finger domain.", *Science* 257:1951-1955.

Harrison ,S.C., 1991, "A structural taxonomy of DNA-binding domains", *Nature* 353:715-719.

Hochuli, E., 1990, Purification of recombinant proteins with metal chelate adsorbent, *in* "Genetic engineering, principle and methods.", J.K. Setlow, ed., Plenum Press, New York.

Kaufmann, E., Hoch, M. and Jäckle, H., 1994, "The interaction of DNA with the DNA-binding domain encoded by the Drosophila gene fork head", *Eur. J. Biochem.* 223:329-337.

Keller, A.D. and Maniatis, T., 1991, "Selection of sequences recognised by a DNA binding protein using a preparative southwestern blot.", *Nucl. Acids Res.* 19:4675-4680.

Knöchel, S., Lef, J., Clement, J., Klocke, B., Hille, S., Köster, M., and Knöchel, W., 1992, "Activin A induced expression of a fork head related gene in posterior chordamesoderm (notochord) of Xenopus laevis embryos.", *Mech. Dev.* 38:157-165.

Knöchel,W., 1994, "Transcription factors and induction in Xenopus laevis embryos.", Experientia (in the press).

Kupiec, J.J., Giron, M.L., Vilette, D., Jeltsch, J.M. and Emanoil-Ravier, R., 1987, "Isolation of high molecular weight DNA from eucaryotic cells by formamide treatment and dialysis.", *Anal. Biochem.* 164:53-59.

Lai, E., Prezioso, V.R., Tao, W.F., Chen, W.S. and Darnell Jr., J.E., 1991, "Hepatocyte nuclear factor 3 alpha belongs to a gene family in mammals that is homologous to the Drosophila homeotic gene fork head.", *Genes Dev.* 5:416-427.

Lai, E., Clark, K.L.,Burley, S.K. and Darnell Jr., J.E., 1993, "Hepatocyte nuclear factor 3/ fork head or "winged helix" proteins: a family of transcription factors of diverse biological function.", *Proc.Natl.Acad.Sci. USA* 90:10421-10423.

Lef, J., Clement,J.H., Oschwald, R., Köster, M. and Knöchel, W., 1994, "Spatial and temporal transcription patterns of the forkhead related XFD-2/XFD-2' genes in Xenopus laevis embryos.", *Mech. Dev.* 45:117-126.

Markmeyer, P., Rühlmann, A., Englisch, U., and Cramer, F., 1990, "The pAX plasmids: new gene-fusion vectors for sequencing, mutagenesis and expression in Escherichia coli.", *Gene* 93:129-134.

Maxam, A.M. and Gilbert, W., 1980, "Sequencing end-labelled DNA with base-specific chemical cleavages.", *Methods Enzymol.* 65:499-560.

Overdier, D.G., Porcella, A. and Costa, R.H., 1994, "The DNA-binding specificity of the hepatocyte nuclear factor 3/forkhead domain is influenced by amino acid residues adjacent to the recognition helix.", *Mol. Cell. Biol.* 14:2755-2766.

Payre, F. and Vincent, A., 1991, "Genomic targets of the serendipity β and δ zinc finger proteins and their respective DNA recognition sites.", *EMBO J.* 10:2533-2541.

Pollock, R. and Treisman, R., 1990, "A sensitive method for the determination of protein-DNA binding specificities.", *Nucl. Acids Res.* 18:6197-6204.

Ruiz i Altaba, A. and Jessell, T.M., 1992, "Pintallavis, a gene expressed in the organizer and midline cells of frog embryos: involvement in the development of the neural axis.", *Development* 116:81-93

Ruiz i Altaba, A., Cox, C., Jessell, T.M. and Klar, A., 1993, "Ectopic neural expression of a floor plate marker in frog embryos injected with the midline transcription factor Pintallavis.", *Proc. Natl. Acad. Sci. USA* 90:8268-8272.

Sambrook, J., Fritsch, E.F. and Maniatis, T., 1989, "Molecular cloning: a laboratory manual.", Cold Spring Harbour Press, New York.

Studier, F.W. and Moffat, B.A., 1986, "Use of bacteriophage T7 RNA polymerase to direct high-level expression of cloned genes.", *J. Mol. Biol.* 189:113-130.

Weigel, D. and Jäckle, H., 1990, "The fork head domain: a novel DNA binding motif of eucaryotic transcription factors?", *Cell* 63:455-456.

eFGF REGULATION OF MESODERM FORMATION AND PATTERNING DURING GASTRULA STAGES

M.E.Pownall and J.M.W. Slack

Developmental Biology Unit, ICRF, Zoology Department
Oxford University, South Parks Road, Oxford OX13PS

The current understanding of a role for FGF in the establishment of mesoderm during early Xenopus development is quite different from what was originally proposed. This paper will review recent findings, key players and discuss the new model that has evolved.

Mesoderm Induction

In Xenopus, the mesoderm forms in the marginal zone around the equator of the blastula embryo in response to signals emanating from the vegetal hemisphere. This induction begins before the onset of zygotic transcription indicating that factors involved in the initial stages of mesoderm induction must be present maternally. The animal pole cells do not normally contribute to mesoderm, however, if they are explanted and recombined with vegetal pole cells then their normal epidermal fate is redirected to forming mesoderm (Nieuwkoop, 1969). This observation by Nieuwkoop 25 years ago provided a powerful biological assay in which to test tissues and purified factors in culture for mesoderm inducing activity in order to elucidate the nature of the *in vivo* vegetal signals. (For reviews see Slack, 1994; Sive, 1993; Kimelman et al, 1992.) It was further found that if the ventral vegetal region is combined with the animal cap, the mesoderm formed has a ventral character, i.e. blood, mesenchyme and some muscle. When dorsal vegetal cells are recombined with the animal cap, the cells form mesoderm of a dorsal type, such as notochord and muscle. This implied that there are two vegetal signals specifying the induced mesoderm to adopt either a dorsal or a ventral character. From this animal cap assay the fibroblast growth factor (FGF) family, and some members of the transforming growth factor-β (TGFβ) family, were found to have potent mesoderm inducing activity. Based on

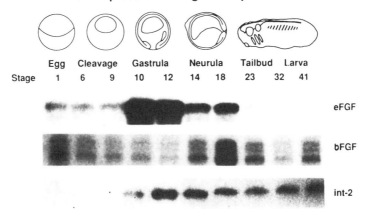

FGF Expression During Development

Figure 1. RNase protection showing temporal expression patterns of the Xenopus FGFs (Courtesy of Harry Isaacs and David Tannahill).

biological activity and the fact that bFGF was found to be present in the egg (Kimelman et al, 1988) it was fast becoming dogma that bFGF provided the ventro-vegetal signal and a TGFβ family member, activin, was the dorso-vegetal signal. Recently, however, the elucidation of gene expression patterns of the FGFs, along with results from experiments where the FGF signalling pathway is inhibited, has begun to argue for a very different role of FGFs in the formation of mesoderm.

eFGF

Three members of the FGF family have been cloned in Xenopus, bFGF (FGF-2) (Kimelman et al, 1988), int-2 (FGF-3) (Tannahill et al, 1992), and eFGF (FGF-4) (Isaacs et al, 1992) however, only bFGF and eFGF are expressed maternally (Figure1), and of these, only eFGF has a secretory signal sequence. This makes eFGF the best candidate, as a maternally expressed secreted factor, to be involved in induction of mesoderm in the blastula. eFGF also has a very potent inducing activity in the animal cap assay (Isaacs et al, 1994). However, eFGF is most highly expressed during gastrula stages (Figure 1). Furthermore, expression analysis of both bFGF and eFGF has indicated that in vivo, these FGFs are localized to the animal hemisphere during blastula stages, rather than to the vegetal hemisphere where the natural inducer is localized (Song and Slack, 1994; Isaacs, unpublished results). The presence of FGF members in the responding, rather than the signalling, tissue during blastula stages began to suggest that FGFs may be involved in events other than vegetal signalling as it was initially presumed. Furthermore, it is during gastrula stages that eFGF is expressed at its highest levels in a ring around the blastopore (Isaacs et al, 1992), suggesting that it has a role in events occuring later than the initial phases of mesoderm induction.

The strongest argument for a role for FGFs during early Xenopus development comes from the work of Amaya, et al. (1991) where the FGF signalling pathway is blocked. The FGF receptor is a membrane bound protein

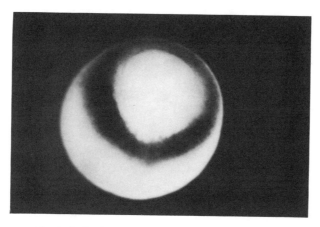

Figure 2. Whole mount *in situ* hybridization showing expression of Xbra around the blastopore in an early Xenopus Gastrula. (probe courtesy of Jim Smith; *in situ* courtesy of Abigail Tucker).

tyrosine kinase that is thought to dimerize in response to ligand binding. Therefore, a receptor lacking the intracellular kinase domain will act as a dominant negative FGF receptor by dimerizing with wild type receptor and rendering it inactive. Embryos injected with the truncated FGF receptor (XFD) show defects in posterior development while animal caps from these embryos can no longer be induced by eFGF (Amaya et al, 1991; Isaacs et al, 1994). XFD embryos also show disturbed gastrulation movements and the blastopores of these embryos fail to close. There are specific effects on the expression of certain genes, where *Xbra*, *XmyoD*, and *Xpo* are drastically downregulated by expression of the dominant negative FGF receptor, while other genes such as *goosecoid*, *Xwnt8*, and *Xsna* are not effected (Amaya et al, 1993; Isaacs et al, 1994). It is unclear for most of the affected genes whether they are direct targets for FGF signalling. There is good evidence, however, for a direct link between eFGF and the expression of the Xenopus *Brachyury* homologue, *Xbra*.

Xenopus *Brachyury*

Xbra expression has been shown to be an immediate early response to mesoderm induction by FGF (Smith et al, 1991), that is, transcription of *Xbra* is initiated in response to FGF in the absence of protein synthesis. Moreover, expression of *Xbra* in embryos expressing the dominant negative FGF receptor is downregulated at the start of gastrulation and continues to decline during gastrula and neurula stages (Amaya et al, 1993; Isaacs et al, 1994), arguing that FGF signalling is necessary for the activation of *Xbra* expression in the blastula and *Xbra* expression remains dependent on FGF after the initial stages of mesoderm induction. *Xbra* is expressed in a ring around the blastopore during gastrula stages (Figure 2), precisely co-localized to the expression of eFGF (Isaacs et al, 1992). Animal caps taken from *Xbra* injected embryos express eFGF, while caps treated with eFGF express *Xbra*. The tight correlation of eFGF and *Xbra* expression in culture, together with the co-expression of eFGF and *Xbra* around the blastopore in vivo, suggests these molecules may interact through an autoregulatory loop (see Isaacs et al, 1994). Furthermore, it has been recently shown that eFGF is sufficient to maintain the expression of *Xbra* in dissociated gastrula cells from the blastopore region (Isaacs et al, 1994).

The *T/Brachyury* gene, originally cloned in mouse (Hermann et al, 1990), is highly conserved among vertebrates and homologs have been found in Xenopus and zebrafish. In mouse, homozygous *T* mutants fail to develop a notochord and lack posterior structures, while in zebrafish, the *no tail* (*ntl*) mutants also lack a differentiated notochord and the posterior half of their bodies (Schulte-Merker et al, 1994). This common phenotype for an embryo lacking functional *Brachyury* shows a striking resemblance to embryos expressing the dominant negative FGF receptor where the anterior part of the body has developed normally but the posterior is deformed. The data suggest that both eFGF signalling and *Xbra* expression are aspects of a pathway common to all vertebrates, which is involved in the induction of posterior mesoderm.

Signal Transduction

FGF signalling in the embryo is transduced from the cell surface to the nucleus through a receptor mediated cascade of kinase activation, resulting in the modification of transcription factors that elicit specific gene expression controlling cell fate or behaviour. The intracellular aspects of the FGF signal transduction pathway involves the FGF receptor, which is a protein tyrosine kinase, the small GTP-binding protein, *ras*, which interacts with another kinase, *raf*, which leads to the activation of MAP kinase. Each of these components is specifically activated by FGF and not activin. Surprisingly, however, it has recently been shown that some of the aspects of activin induction in animal cap assays can be blocked by inhibiting the FGF signalling pathway at various points. (Cornell and Kimelman, 1994; Hartley et al, 1994; LaBonne and Whitman, 1994).

Animal caps from embryos expressing the dominant negative FGF receptor do not express *Xbra* in response to activin induction (Figure 3). This observation is consistent with the downregulation of *Xbra* in whole embryos injected with the dominant negative FGF receptor where activin signalling is not disturbed (Isaacs et al, 1994). Activin induction of *a-cardiac actin*, *XMyoD*, and *XNot* expression is also sensitive to the dominant negative FGF receptor, however, the expression of other genes like *goosecoid* and *Xwnt8* remain activin inducible. Animal cap cells expressing the dominant negative FGF

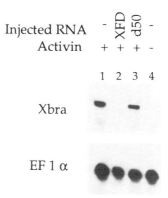

Figure 3. Xbra expression cannot be induced by activin in animal caps lacking a functional FGF signalling pathway (from Cornell and Kimelman, 1994). XFD is the dominant negative FGF receptor and d50 is a non-functional control. Expression is depicted by RNAase protection.

receptor still spread on fibronectin plates in response to activin (Cornell and Kimelman, 1994). Thus, some but not all of the aspects of activin induction are blocked by a defective FGF signalling pathway, and importantly, activin induction of *Xbra* requires FGF signalling. Furthermore, a dominant negative *ras* blocks both activin and FGF induction and a constitutively activated *ras* induces ventrolateral type mesoderm typical of FGF inductions (Whitman and Melton, 1992; La Bonne and Whitman, 1994). A dominant negative *raf* also blocks FGF as well as activin induction of *cardiac actin*. Interestingly, the dominant negative FGF receptor, the dominant negative *ras*, and the dominant negative *raf* all affect the posterior axial patterning of the Xenopus embryo with the same resultant phenotype. Moreover all of these constructs block some aspects of activin signalling.

The FGF signalling pathway is not directly activated in response to activin. Only FGF activates *ras, raf,* and MAP kinase. The FGF pathway is a prior, or perhaps coordinate requirement for a cell to respond to activin; that is, a functional FGF pathway is a competence factor. However, there are certain aspects of activin induction that are FGF independent, underscoring that FGF is only one component of the mesoderm-inductive response. Furthermore, embryos injected with the dominant negative activin receptor form no mesoderm in spite of the presence of a functional FGF signalling pathway. So it seems that FGF is not sufficient to induce mesoderm in the context of the whole embryo. Even so, the overall placement of FGF in the model for mesoderm induction has changed drastically, from a vegetally localized ventrolateral mesoderm inducer, to a factor involved in the response to mesoderm induction and in regulating gene expression during gastrulation.

Gastrulation

Although there is an early requirement for FGF in the expression of immediate early genes like *Xbra* and *Xpo* (Amaya et al, 1993; Isaacs et al, 1994), we believe, based on gene expression and activity, that an additional important role of eFGF is to maintain the expression of *Xbra* during gastrulation and the failure to maintain *Xbra* expression underlies the phenotype of embryos expressing the dominant negative FGF receptor. It has been proposed that in mouse embryos lacking *Brachyury* function, cells cannot move through the primitive streak and therefore cannot complete gastrulation (Beddington et al, 1992). The molecular mechanism of *Xbra* function is thought to be as a transcription factor regulating the expression of down stream genes(Kispert and Herrmann, 1993). Since there is evidence that *Xbra* is important in driving gastrulation movements (Wilson et al, 1993), these target genes could be extracellular matrix components or molecules involved in cell motility or cell adhesion. Both mouse *T* mutants and Xenopus embryos expressing the dominant negative FGF receptor fail to complete gastrulation.

The defects of XFD embryos are illustrated by the initial normal involution of the dorsal mesoderm, followed by the failure of the lateral and ventral mesoderm to involute correctly. The resultant tailbud stage embryos have normal heads and abnormal trunks and tails. The cells that involute first are part of the prechordal mesoderm and they do so by diverging and crawling along the blastocoel roof. The cells in the chordamesoderm, which involutes subsequent to the prechordal mesoderm, exhibit a different type of cell movement called convergent extension. It is the cells of chordamesoderm that express *Xbra*, and it is these cells that fail to involute properly in XFD embryos. The XFD embryos complete involution of the prechordal mesoderm and form normal heads, but do not complete the convergent extension of the chordamesoderm and thereby fail to finish gastrulation. It is speculated that the convergent extension cell behaviour is regulated by Xbra, and that *Xbra* expression requires a source of FGF signalling throughout gastrulation.

The new model

The case for FGFs playing a critical role in maintaining *Xbra* expression and the properties of the mesoderm necessary for gastrulation is well supported. (1) *Xbra* expression is abolished in embryos where FGF signalling is blocked (Amaya et al, 1993). (2) Activin cannot activate expression of *Xbra* in embryos where FGF signalling has been blocked (Cornell and Kimelman, 1994). (3) In animal caps taken from blastula, eFGF and Xbra can cross-activate expression of each other (Isaacs et al, 1994). (4) eFGF and Xbra are co-expressed in the blastopore through gastrula stages and much of XFD and Brachyury mutant phenotypes can be explained by defective cell movements of mesoderm in this region (Smith et al, 1991; Isaacs et al, 1992). (5) eFGF can maintain the expression of Xbra in gastrula mesoderm (Isaacs et al, 1994).

The current view of the role of FGF in the establishment of the mesoderm is depicted in Figure 4. In this model, the vegetally localized mesoderm inducing signal is not an FGF, but rather a member of the TGFβ family, perhaps Vg-1 (Dale et al, 1993; Thomsen and Melton,1993). During blastula stages, this activin-like molecule requires the presence of FGF in the responding tissue in order to activate the expression of *Xbra*, which in turn activates the zygotic transcription of eFGF.

In the marginal zone of the post-MBT blastula, *eFGF-Xbra* undergo a period of mutual autocatalytic activation that probably serves to amplify and spread the response to the primary inducing signal. During gastrula stages the expression of *eFGF* becomes independent of Xbra, however *Xbra* expression still requires eFGF. The expression of *Xbra* in blastopore is required to drive cell

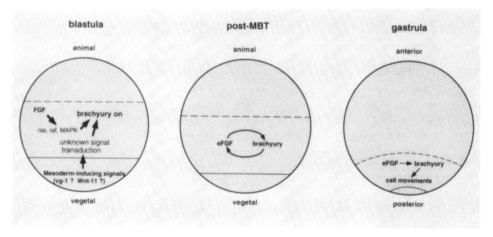

Figure 4. A model for the role of FGFs during the establishment of the mesoderm (from Isaacs et al, 1994).

movements during gastrulation. As mesoderm invaginates in from the blastopore region, the expression of *eFGF* is down regulated, and consequently *Xbra* expression is also extinguished, except in the notochord. Thus, in this model the requirement for eFGF function is both early, to permit the vegetal signal to act during blastula stages, as well as later, to maintain the expression of *Xbra* and those properties of the mesoderm necessary for gastrulation.

REFERENCES

Amaya, E., P.A. Stein, T.J. Musci, and M.W. Kirschner. 1993. FGF signalling in the early specification of mesoderm in Xenopus. Development 118: 477-487.

Amaya, E., T.J. Musci, and M.W. Kirschner. 1991. Expression of a dominant negative mutant of the FGF receptor disrupts mesoderm formation in Xenopus embryos. Cell 66: 257-270.

Beddington, R.S.P., P. Rashbass and V. Wilson. 1992. *Brachyury*- a gene affecting mouse gastrulation and early organogenesis. Development *suppl.* 157-171.

Cornell, R.A. and D. Kimelman. 1994. Activin-mediated mesoderm induction requires FGF. Development 120: 453-462.

Dale, L., G. Matthews, and A. Colman. 1993. Secretion and mesoderm inducing activity of the TGFb related domain of Xenopus Vg1. EMBO J. 12: 4471-4480.

Hemmati-Brivanlou, A. and D.A. Melton. 1992. A truncated activin receptor inhibits mesoderm induction and formation of axial structures in Xenopus embryos. Nature 359: 609-614.

Herrmann, B.G. and A. Kispert. 1994. The *T* genes in embryogenesis. TIGS10 (8): 280-286.

Herrmann, B.G., S. Labiet, A. Poustka, T.R. King and H. Lehrach. 1990. Cloning of the T gene required in mesoderm formation in the mouse. Nature 343: 617-622.

Isaacs, H.V., D. Tannahill and J.M.W. Slack. 1992. Expression of a novel FGF in the Xenopus embryo. A new candidate inducing factor for mesoderm formation and anteroposterior specification. Development 114: 711-720.

Isaacs, H.V., M.E. Pownall, and J.M.W. Slack. 1994. eFGF regulates Xbra expression during Xenopus gastrulation. EMBO J. (in press)

Kimelman, D., J.A. Abraham, T. Haaparanta, T.M. Palisi, and M. Kirschner. 1988. The presence of FGF in the frog egg: its role as a natural mesoderm inducer. Science 242: 1053-1056.

Kimelman, D., J.L. Christian, and R.T. Moon. 1991. Synergistic principles of development: overlapping patterning systems in Xenopus mesoderm induction. Development 116: 1-9.

Kispert, A. and B.G. Herrmann. 1993. The *Brachyury* gene encodes a novel DNA binding protein. EMBO J. 12: 3211-3220.

LaBonne, C. and M. Whitman. 1994. Mesoderm induction by activin requires FGF-mediated intracellular signals. Development 120: 463-472.

Nieuwkoop, P.D. 1969. The formation of the mesoderm in urodelean amphibians I. Induction by the endoderm. Wilhem Roux's Arch f Entw Mech Orgs. 162: 341-373.

Schulte-Merker, S., F.J.M. van Eeden, M.E. Halpern, C.B. Kimmel, C. Nusslein-Volhard. 1994. *no tail* (*ntl*) is the zebrafish homologue of the mouse T (*Brachyury*) gene. Development 120: 1009-1015.

Sive, H.L. 1993. The frog prince-ss: a molecular formula for dorsoventral patterning in Xenopus. Genes Dev 7: 1-12.

Slack, J.M.W. 1994. Inducing factors in Xenopus early embryos. Current Biology 4: 116-126.

Smith, J.C., B.M.J. Price, J.B.A. Green, D. Weigel, and B.G. Herrmann. 1991. Expression of a Xenopus homolog of Brachyury (T) is an immediate-early response to mesoderm induction. Cell 67: 79-87.

Song, J. and J.M.W. Slack. 1994. Spatial and temporal expression of bFGF (FGF-2) mRNA and protein in early Xenopus development. Mech. Dev. (in press)

Tannahill, D., H.V. Isaacs, M.J. Close, G. Peters and J.M.W. Slack. 1992. Developmental expression of the Xenopus int-2 (FGF-3) gene : activation by mesodermal and neural induction. Development 115: 695-702.

Thomsen, G.H. and D.A. Melton. 1993. Processed Vg1 protein is an axial mesoderm inducer in Xenopus. Cell 74: 433-441.

Whitman, M and D.A. Melton. 1992. Involvement of p21 ras in Xenopus mesoderm induction. Nature 357: 252-255.

Wilson, V., P. Rashbash and R.S.P. Beddington. 1993. Chimeric analysis of *T* (*Brachyury*) gene function. Development 117: 1321-1331.

THE ROLE OF INSULIN-LIKE GROWTH FACTOR II IN THE GROWTH AND DEVELOPMENT OF THE MAMMALIAN EMBRYO

Fotini Stylianopoulou

Laboratory of Biology-Biochemistry
Faculty of Nursing
University of Athens
P.O. Box 14224
Athens 11510, Greece

INTRODUCTION

An emerging body of evidence has established that mammalian embryogenesis is under the control of polypeptide molecules that influence not only growth, but differentiation and morphogenesis during mammalian development. Growth factors exert their regulatory functions by acting in an autocrine or paracrine mode. These embryonic polypeptide growth factors are usually identical to growth factors isolated from serum or tissues of adult animals. Insulin-like growth factor II (IGF-II) (for a review see Schofield, 1992) is one of the growth factors playing a significant role in the development of the mammalian embryo.

Structurally, IGF-II shares a 41% homology with the molecule of insulin, and a 65% homology with IGF-I, the classical somatomedin, which mediates the actions of growth hormone during postnatal growth and development. The mature IGF-II is a single-chain polypeptide, containing 67 amino acids. The tertiary structure of the molecule is stabilized by three disulfide bridges and consists of four domains. Domains A and B are similar to the respective domains of insulin, domain C is analogous to the connecting peptide of proinsulin, and domain D is not present in insulins (Daughaday and Rotwein, 1989).

The IGF-II gene exists as a single copy on human chromosome 11 (DePagter-Holthuizen et al., 1985), the syntenic rat chromosome 1 (Soares et al., 1986), and mouse chromosome 7 (Lalley and Chirgwin, 1984), downstream of the gene for insulin. The rat IGF-II gene (Sussenbach, 1989) consists of three alternatively used noncoding exons and three coding exons. It utilizes at least three promoters and multiple polyadenylation sites, and thus expresses multiple transcripts in a variety of tissues during the embryonic, fetal, and neonatal periods, without any apparent developmental or tissue specificity (Soares et al.,1986). IGF-II gene expression is high prenatally, and declines shortly after birth. In adult rats its expression is confined to the choroid plexus and the leptomeninges (Stylianopoulou et al., 1988b).

Organization of the Early Vertebrate Embryo
Edited by N. Zagris *et al.*, Plenum Press, New York, 1995

LOCALIZATION OF IGF-II GENE EXPRESSION IN THE MAMMALIAN EMBRYO

IGF-II gene expression has been studied, using in situ hybridization, throughout embryogenesis in the human, mouse and rat.

In the human embryonic conceptus, induction of IGF-II gene expression follows blastocyst implantation (Ohlsson et al., 1989). Following induction, high levels of IGF-II mRNA are localized in the placental cytotrophoblasts, where it is believed to play an important role in forming and maintaining the rapid proliferative phenotype of the trophoblastic shell.

Before implantation, the IGF-II gene is not expressed in the mouse conceptus either (Lee et al.,1990). After implantation, at embryonic day 5.5, a high level of IGF-II gene expression occurs in the extraembryonic ectoderm and the ectoplacental cone. In subsequent stages (E 6.5-7.0), IGF-II gene expression extends to the mural trophectoderm and the columnar visceral endoderm, as well as to the extraembryonic mesodermal cells (in the forming amniotic folds and exocoelom), as soon as they first appear. At the late primitive streak/neural plate stage (E 7.5), high levels of IGF-II mRNA are detected in all extraembryonic structures (the allantois, the amnion, the chorion and the visceral yolk sac) and for the first time, in the embryo proper, where it is localized in a restricted area of the embryonic mesoderm that does not include the primitive streak. At the stage of the neural folds (E 8.0), intense IGF-II transcription continues in all extraembryonic structures, with the exception of endodermal derivatives (parietal endoderm and the outer endodermal layer of the visceral yolk sac). In the embryo proper the highest level of IGF-II mRNA is found in the developing heart and the lateral (somatic and splanchnic) mesoderm. IGF-II mRNA is also detected, for the first time, in mesodermal derivatives, such as the head mesenchyme which probably includes neural crest cells. IGF-II mRNA is also found at this stage in the lining of the ventral and lateral walls of the foregut, which is a derivative of the definitive endoderm. As embryonic development progresses (E8.5), IGF-II mRNA levels increase in presomitic and somitic mesoderm, while it decreases in the ectoplacental cone (Lee et al., 1990). In day 8 mouse embryos the IGF-II precursor polypeptide has been localized immunocytochemically in the allantois, the amniotic mesoderm, the mesodermal layer of the visceral yolk sac, the head mesenchyme, the heart and the foregut, colocalizing with the mRNA (Lee et al., 1990).

In later stages of embryonic development (E10 - E16), in rat embryos, IGF-II transcripts are present in cranial mesenchyme and tissues of mesodermal origin, predominantly those derived from the lateral mesoderm. Abundant IGF-II mRNA is found in muscle and chondrocytes, prior to ossification. The liver and the bronchial epithelium are the only tissues derived from the endoderm in which the IGF-II gene is expressed. IGF-II transcripts are not detected in ectodermally derived tissues, at any stage of embryonic development, with the exception of the choroid plexus, the newly-forming pituitary rudiment, and to a lesser extent the auditory placode (Stylianopoulou et al., 1988a). Immunocytochemical detection of the IGF-II precursor polypeptide, in E14 mouse embryos, demonstrated that it is generally colocalized with the mRNA in mesodermally derived tissues (Lee et al., 1990).

The topographical pattern of IGF-II gene expression in human fetuses of 16-20 weeks gestation is very similar to that in the E10-E16 rat embryos mentioned above. Thus high levels of IGF-II mRNA are found in the dermis of the skin; the muscle perimysium and epimysium; the capsule and interlobular septa of the thymus; the epicardium and coronary vessel walls of the heart; the pleura, interlobular septa and pulmonary vessel walls of the lung; the hepatic perisinusoidal cells; the lamina propria, submucosa, and serosa of the stomach and the intestine; the retroperitoneal tissue of the pancreas; the adrenal capsule; the capsule and the calyces of the kidney; the perichondrium of costal cartilage; and the sclera of the eye (Han et al., 1987).

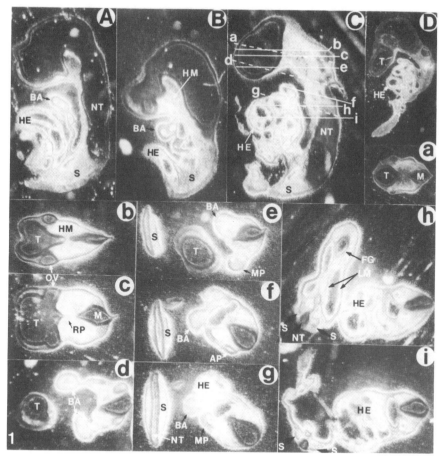

Figure 1. Pattern of IGF-II gene expression at embryonic day 10. In situ hybridization was employed to detect IGF-II mRNA. Areas rich in IGF-II mRNA appear white in the figure.

A-D:Oblique sagittal sections; a-i:Transverse sections at the levels indicated in C. BA:branchial arch; HE:heart; NT:neural tube; S:somites; HM:head mesenchyme; T:telencephalon; M:myelencephalon; OV:optic vesicle; RP:Rathke's pouch; MP:maxillary process; AP:auditory placode; FG:foregut; LM:lateral mesoderm lining the coelum. (From Stylianopoulou et al., 1988a).

TYPES OF IGF-II RECEPTORS AND THEIR LOCALIZATION IN THE DEVELOPING MAMMALIAN EMBRYO

IGF-II binds to two different types of cell membrane receptors : Type 1 (IGF1R) and type 2 (IGF2R) (Moxham and Jacobs, 1992), which are structurally unrelated. IGF1R is a transmembrane glycoprotein, similar to the insulin receptor. It is a heterotetramer composed of two α subunits 100-135 kDa, and two β subunits 90-95 kDa ($\alpha_2\beta_2$), linked by disulfide bridges (Ullrich et al., 1986). The ligand binding domain of the receptor is located extracellularly in a cysteine-rich region of the α subunits (Gustafson and Rutter, 1990), while the intracellular part of the receptor protein possesses tyrosine kinase activity. In addition to IGF-II, IGF1R also binds IGF-I, and with a 15-20-fold higher affinity than that for IGF-II (Germain-Lee et al.,1992).

The other type of IGF-II receptor, IGF2R, is a 220-300 kDa single-chain polypeptide, with no tyrosine kinase activity (Kornfeld, 1992). It bears no structural resemblance to

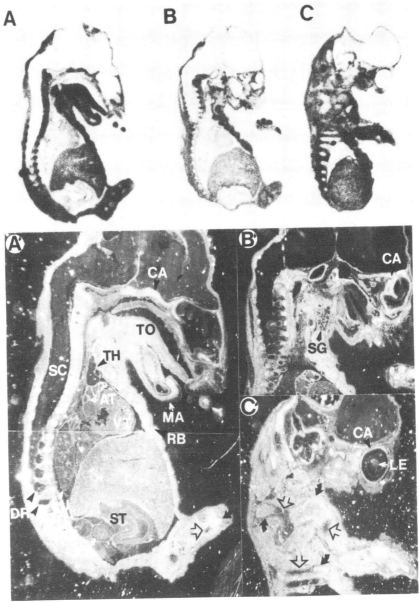

Figure 2. Pattern of IGF-II gene expression at embryonic day 16. A-C:Autoradiographs of sagittal (A) and parasagittal (B,C) sections after hybridization. A'-C': The same tissue sections as A-C, after exposure to photographic emulsion and subsequent developing (see also Figure 1). Closed and open arrows indicate different regions of the same cartilage before and after hypertrophy, respectively.
CA:cartilage; TO:tongue; TH:thymus; SC:spinal cord; AT:atrium; VT:ventricle of the heart; MA:mandible; RB:ribs; L:liver; ST:stomach; KI:kidney; AD:adrenal; LU:lung; DR:dorsal root ganglion; SG:submaxillary gland; LE:lens.(From Stylianopoulou et al., 1988a.)

IGF1R, or any other growth factor receptor. In contrast, it has been shown that it is nearly identical with the cation-independent mannose 6-phosphate receptor (CI-MPR), which is involved in sorting lysosomal enzymes from the Golgi apparatus and/or the plasma membrane to the lysosomes (Morgan et al., 1987). The two ligands (IGF-II and mannose 6-phosphate) bind to the receptor at different extracellular regions, since binding of one ligand does not inhibit the binding of the other (Nishimoto et al., 1991). The role of IGF2R in signal transduction remains unclear, although it has been suggested that it mediates extracellular signals via a G protein (Nishimoto et al., 1991). IGF2R does not appear to mediate the growth-promoting effects of IGF-II, since blocking IGF2R using antibodies, did not abolish the mitogenic effects of IGF-II in tissue culture (Kiess et al., 1987). On the other hand, analogous blocking of IGF1R did inhibit IGF-II function (Furlanetto et al., 1987). IGF2R is believed to participate in IGF-II turnover, by mediating the degradation of extracellular IGF-II after endocytosis (Oka et al., 1985).

Both types of receptors to which IGF-II binds are expressed in the developing mammalian embryo. The mRNA for IGF1R is widely distributed in E14 and E15 rat embryos (Bondy et al.,1990). It is localized in tissues derived from all three germinal layers, including the ectoderm, where, as mentioned before, IGF-II is never expressed. This pattern of distribution suggests both an autocrine and paracrine mode of action for IGF-II. In addition, IGF1R mediates the actions of IGF-I, which is also expressed (Bondy et al.,1990; Han et al.,1987) and plays a significant role during embryogenesis (Liu et al.,1993).

The developmental pattern of IGF-II/Mannose-6-phosphate receptor gene expression parallels that of IGF-II: in E8 mouse embryos the level of the receptor protein is low, but it can be detected immunohistochemically in some of the tissues that intensely express IGF-II, such as the allantois, the chorion, the heart, and to a lesser degree, the head mesenchyme and the foregut (Lee et al.,1990). In midgestation (E14) embryos IGF-II/Man.-6-P receptor immunoreactivity (Lee et al.,1990) colocalizes with IGF-II mRNA (Stylianopoulou et al. 1988a), giving intense staining in tissues like the heart, the tongue, and cartilage. Colocalization of IGF-II and IGF-II/Man.6-P receptor gene expression continues in later stage (E20) rat embryos, where the receptor mRNA is detected in the heart, limb/muscle, lung, intestine, kidney, liver, and to a lesser extent brain (Sklar et al.,1992). This pattern of distribution is consistent with the role of IGF2R in acting as a buffer system to control the levels of IGF-II available.

IMPRINTING OF THE IGF-II AND IGF2R GENES

Mice carrying one allele of the IGF-II gene disrupted by gene targeting and one normal allele are proportional dwarfs with a body weight 60% lower than that of their homozygous wild-type littermates (DeChiara et al., 1990), if they have inherited the disrupted allele from their father. If they inherit the disrupted allele from their mother, they are phenotypically normal (DeChiara et al., 1991). These results suggested that only the paternally derived allele is expressed. Indeed in situ hybridization and nuclease protection analyses showed that during embryogenesis only the paternal allele is expressed, while the maternal is silent in all tissues, with the exception of the choroid plexus and the leptomeninges, where both alleles are transcriptionally active. Thus the gene encoding IGF-II is paternally imprinted (DeChiara et al., 1991).

In mice, the T (brachyury) gene as well as the gene encoding IGF2R are located at the proximal region of chromosome 17 (Barlow et al., 1991). Two overlapping deletions in this region result in two different mutations, T Hairpin (T^{hp}) (Johnson,1974, 1975) and t^{lub2} (Winkling and Silver, 1984). Animals which inherit from their mother an allele carrying either of these mutations die in utero at about day 15 of gestation (Johnson, 1974, 1975; Winkling and Silver,1984). In contrast, transmission of either the T^{hp} or the t^{lub} alleles from the paternal germline results in viable heterozygous offspring. The lethal maternal phenotype is believed to be due to the expression of at least one imprinted genetic locus, which has been named Tme (T-associated maternal effect). The protein product of Tme is believed to be essential for normal embryogenesis, and to be transcribed from the maternal allele. Detailed study of the t^{lub2} locus (Barlow et al., 1991) has shown that of the five genes so far mapped to this region, only the gene coding for IGF2R is imprinted, since IGF2R mRNA is transcribed exclusively from the maternal allele (Barlow et al., 1991).

Thus the genes encoding IGF-II and its type 2 receptor (IGF2R/CI-MPR) are reciprocally imprinted: the former paternally, and the latter maternally.

ROLE OF THE IGFs (IGF-I AND IGF-II) AND THEIR RECEPTORS (IGF1R AND IGF2R) IN MAMMALIAN EMBRYOGENESIS

Construction of mice carrying targeted disruptions of the genes encoding the IGFs and their type I receptor (IGF1R), as well as the occurrence of natural mutants lacking the IGF2R gene have yielded important information regarding their roles in mammalian embryonic growth and differentiation.

As mentioned above, heterozygous animals carrying a paternally derived disrupted IGF-II gene are viable, proportionate dwarfs, with a body weight 60% that of wild type littermates (DeChiara et al., 1990). Homozygous animals lacking both IGF-II alleles are phenotypically identical to these heterozygous animals, since the maternal allele is not expressed (DeChiara et al., 1991).

Similarly, homozygous animals lacking both alleles of the IGF-I gene are dwarfs with body weight 60% of normal. It should be noted that, depending on genetic background, some of the IGF-I null mutants die shortly after birth (Liu et al., 1993). These results show that both IGF-I and IGF-II are important for normal embryonic growth, but that neither is indispensable. On the other hand, double null mutants lacking both IGF-I and IGF-II are severely retarded in growth (body weight 30% of normal), and invariably die at birth (Liu et al., 1993). Thus missing both IGFs is incompatible with life, and at least one of the two IGFs is necessary for neonatal survival.

In contrast to what happens when one of the IGFs is missing, lack of both alleles of the gene encoding IGF1R is incompatible with life. Animals homozygous for a disrupted IGF1R gene are even smaller than those lacking IGF-I (body weight 45% of normal), and they invariably die at birth of respiratory failure (Liu et al., 1993). A functional IGF1R is thus absolutely necessary for the growth, development and the survival of the mouse embryo. Histological examination of the IGF1R null mutant embryos revealed a generalized organ hypoplasia, including the muscles and epidermis; a delay in ossification, and neuropathological alterations (Liu et al., 1993). More specifically, the number of myocytes was decreased in the abdominal muscles, in the muscles of the neck and limbs and in the respiratory muscles, while the expression of muscle specific markers appeared normal. In the skin, the stratum spinosum of the epidermis was extremely thin in the mutants, consisting of fewer cells than in normal animals. In addition, a reduced number of hair follicles was observed in the skin of the mutant embryos. Bone development was qualitatively normal and only a time delay in the process of differentiation was noted. In the nervous system the alterations observed consisted of an increase in the cellular density of the mantle zone in the spinal cord and the brain stem. All histological aberrations observed could thus be attributed to a change in cell numbers in the respective tissues. This suggests that the signals transmitted through IGF1R primarily affect the rate of the cell cycle. The more severe phenotype of the mutants lacking IGF1R, compared to the mutants lacking IGF-I, suggests that IGF1R also mediates the growth- and differentiation-promoting effects of IGF-II.

Using body weight as an index, growth kinetics of mouse embryos lacking IGF-I, and/or IGF-II, and/or IGF1R were analyzed throughout embryonic development. This study demonstrated that between days E11.0 and 12.5 IGF1R mediates only IGF-II signals, while from E13.5 onwards IGF1R interacts with both IGF-I and IGF-II (Baker et al., 1993).

Double null mutants lacking both alleles at both the IGF-I and the IGF1R loci did not differ phenotypically from the IGF1R single null mutants, verifying the hypothesis that IGF-I interacts only with IGF1R. On the other hand, double null mutants lacking both IGF-II and IGF1R had a more severe growth deficit compared to the IGF1R single null mutants: Their body weight was 30% of normal, as compared to 45% (Liu et al., 1993). This indicates that IGF-II utilizes in addition to IGF1R, another, yet uncharacterized receptor in exerting its growth-and differentiation-promoting effects.

Mice carrying a maternal allele with the T^{hp} chromosomal deletion, which includes the IGF2R gene, die in utero at about day 15 of gestation (Johnson, 1974, 1975; Winkling and Silver, 1984). If these animals also carry a paternally derived disrupted allele of the

IGF-II gene, 70% of them survive to birth (Filson et al., 1993). These results suggest that the dominant lethal maternal effect (T*me*) could be due to an overabundance of IGF-II, that is not degraded, since IGF2R, which mediates its turnover, is missing. Thus the function of IGF2R in the developing embryo is probably related to the modulation of the amount of IGF-II available.

SUMMARY-CONCLUSIONS

IGF-II plays an important role in mammalian embryogenesis, acting mainly as a mitogenic factor. Its actions during embryonic development are mediated via type 1 IGF receptor (IGF1R), and via another as yet uncharacterized receptor (XR). The type 2 IGF receptor (IGF2R), which also serves as the cation independent mannose-6-phosphate receptor acts to modulate the amount of IGF-II available in the developing embryo. The genes for IGF-II and IGF2R are reciprocally imprinted: the former paternally and the latter maternally.

In addition to IGF-II, IGF-I also plays an important role during embryogenesis, acting through the IGF1R. Lack of either IGF-I or IGF-II can be tolerated by the embryo, although it results in a significant growth retardation. On the other hand, lack of either both IGFs, IGF1R or IGF2R is incompatible with life.

ACKNOWLEDGMENTS

The author wishes to express her deepest thanks to Dr Argiris Efstratiadis, for the opportunity to spend a most stimulating sabbatical in his laboratory and for providing continued information on the developments in the field of IGF-II research. I would also like to thank Ms Nafsica Violaki for her invaluable assistance in preparing the camera-ready manuscript.

REFERENCES

Baker, J., Liu, J.P., Robertson, E.J., and Efstratiadis, A. (1993). Role of insulin-like growth factors in embryonic and postnatal growth. Cell 75:73-82.

Barlow, D.P., Stoger, R., Herrmann, B.G., Saito, K., and Schweifer,N. (1991). The mouse insulin-like growth factor type-2 receptor is imprinted and closely linked to the Tme locus. Nature 349:84-87.

Bondy, C.A., Werner, H., Roberts, C.T.Jr., and LeRoith, D. (1990). Cellular pattern of insulin-like growth factor-I (IGF-I) and type I IGF receptor gene expression in early organogenesis: comparison with IGF-II gene expression. Mol.Endocrinol. 4:1386-1398.

Daughaday, W., and Rotwein, P. (1989). Insulin-like growth factors I and II: peptide, messenger ribonucleic acid and gene structures, serum, and tissue concentrations. Endocr.Rev. 10:68-91.

DeChiara, T.M., Efstratiadis, A., and Robertson, E.J. (1990). A growth-deficiency phenotype in heterozygous mice carrying an insulin-like growth factor II gene disrupted by targeting. Nature 345:78-80.

DeChiara, T.M., Robertson, E.J., and Efstratiadis, A. (1991). Parental imprinting of the mouse insulin-like growth factor II gene. Cell 64:849-859.

DePagter-Holthuizen, P., Hoppner, J.W.M., Jansen, M., Geurts van Kessel, A.H.M., van Ommen, G.J.B., and Sussenbach, J.S. (1985). Chromosomal localization and preliminary characterization of the human gene encoding insulin like growth factor II. Hum.Genet. 69:170-173.

Filson, A.J., Louvi, A., Efstratiadis, A., and Robertson, E.J. (1993). Rescue of the T-associated maternal effect in mice carrying null mutations in Igf-2 and Igf2r, two reciprocally imprinted genes. Development 118:731-736.

Furlanetto, R.W., DiCarlo, J.N., and Wisehart, C. (1987). The type II insulin-like growth factor receptor does not mediate deoxyribonucleic acid synthesis in human fibroblasts. J.Clin.Endocrinol.Metab. 64:1142-1149.

Germain-Lee, E.L., Janicot, M., Lammers, R., Ullrich, A., and Casella, S.J. (1992). Expression of the type I insulin-like growth factor receptor with low affinity for insulin-like growth factor II. Biochem.J. 281:413-417.

Gustafson, T.A., and Rutter, W.J. (1990). The cysteine-rich domains of the insulin and insulin-like growth factor I receptors are primary determinants of hormone binding specificity. J.Biol.Chem. 265:18663-18667.

Han, V.K.M., D'Ercole, A.J., and Lund, P.K. (1987). Cellular localization of somatomedin (insulin-like growth factor) messenger RNA in the human fetus. Science 236:193-197.

Johnson, D.R. (1974). Hairpin-tail: A case of post-reductional gene action in the mouse egg? Genetics 76:795-805.

Johnson, D.R. (1975). Further observations on the hairpin tail (T^{hp}) mutation in the mouse. Genet.Res. 24: 207-213.

Kiess, W., Haskell, J.F., Lee, I., Greenstein, I.A., Miller, B.E., Aarons, A.L., Rechler, M.M., and Nissley, S.P. (1987). An antibody that blocks insulin-like growth factor (IGF) binding to the type II IGF receptor is neither an agonist nor an inhibitor of IGF-stimulated biologic responses in L6 myoblasts. J.Biol.Chem. 262:12745-12751.

Kornfeld, S. (1992). Structure and function of the mannose 6-phosphate receptor/Insulin like growth factor II receptors. Ann.Rev.Biochem. 61:307-330.

Lalley, P.A., and Chirgwin, J.M. (1984). Mapping of the mouse insulin genes. Cytogenet.Cell Genet. 37:515-521.

Lee, J.E., Pintar, J., and Efstratiadis, A. (1990). Pattern of the insulin-like growth factor II gene expression during early mouse embryogenesis. Development 110:151-159.

Liu, J.P., Baker, J., Perkins, A.S., Robertson, E.J., and Efstratiadis, A. (1993). Mice carrying null mutations of the genes encoding insulin-like growth factor I (Igf-1) and type 1 IGF receptor (Igf1r). Cell 75:59-72.

Morgan, D.O., Edman, J.C., Standring, D.R., Fried, V.A., Smith, M.C., Roth, R.A., and Rutter, W.J. (1987). Insulin-like growth factor II receptor as a multifunctional binding protein. Nature 329:301-307.

Moxham, C. and Jacobs, S. (1992). Insulin-like growth factor receptors. In: The Insulin-like Growth Factors:Structure and Biological Functions, P.N.Schofield,ed., Oxford University Press, Oxford, England, pp.80-110.

Nishimoto, I., Murayama, Y., and Okamoto, T. (1991). Signal transduction mechanism of IGF-II/man-6-P receptor. In: Modern Concepts of Insulin-like Growth Factors, E.M.Spencer,ed., Elsevier, New York, pp.517-522.

Ohlsson, R., Larsson, E., Nilsson, O., Wahlstrom, T., and Sundstrom. P. (1989). Blastocyst implantation precedes induction of insulin-like growth factor II gene expression in human trophoblasts. Development 106:555-559.

Oka, Y., Rozek, L.M., and Czech, M.P. (1985). Direct demonstration of rapid insulin-like growth factor II receptor internalization and recycling in rat adipocytes: insulin stimulates ^{125}I insulin-like growth factor II degradation by modulating the IGF-II receptor recycling process. J.Biol.Chem. 260:9435-9442.

Schofield, P.N., ed. (1992). The Insulin-like Growth Factors: Structure and Biological Functions. Oxford University Press, Oxford, England.

Sklar, M.M., Thomas, C.L., Municchi, G., Roberts, C.T.Jr., LeRoith D., Kiess, W., and Nissley, P. (1992). Developmental expression of rat insulin-like growth factor-II/mannose 6-phosphate receptor messenger ribonucleic acid. Endocrinology 130:3484-3491.

Soares, M.B., Turken, A., Ishii, D., Mills, L., Episkopou, V., Cotter, S., Zeitlin, S., and Efstratiadis, A. (1986). Rat insulin-like growth factor II gene: a single gene with two promoters expressing a multitranscript family. J.Mol.Biol.192:737-752.

Stylianopoulou, F., Efstratiadis, A., Herbert, J., and Pintar, J. (1988a). Pattern of the insulin-like growth factor II gene expression during rat embyogenesis. Development 103:497-506.

Stylianopoulou, F., Herbert, J., Soares, M.B., and Efstratiadis, A. (1988b). Expression of the insulin-like growth factor II gene in the choroid plexus and the leptomeninges of the adult rat central nervous system. Proc.Natl.Acad.Sci.U.S.A. 85:141-145.

Sussenbach, J.S. (1989). The gene structure of the insulin-like growth factor family. Prog.Growth Factor Res.1:33-48.

Ullrich, A., Gray, A., Tam, A.W., Yang-Feng, T., Tsubokawa, M., Collins, C. Henzel, W., LeBon, T., Kathuria, S., Chen, E., Jacobs, S. Francke, U., Ramachandran, J., and Fujita-Yamaguchi, Y. (1986). Insulin-like growth factor I receptor primary structure: comparison with insulin receptor suggests structural determinants that define functional specificity. EMBO J. 5:2503-2512.

Winkling, H., and Silver, L.M. (1984). Characterization of a recombinant mouse T haplotype that expresses a dominant lethal maternal effect. Genetics 108:1013-1020.

EARLY AVIAN MORPHOGENESIS LEADING TO GASTRULATION IN THE WIDER CONTEXT OF VERTEBRATE EMBRYOGENESIS

Hefzibah Eyal-Giladi

Department of Cell and Animal Biology
The Hebrew University of Jerusalem
91904 Jerusalem, Israel

Gastrulation was regarded for many years as the first prominent morphogenetic event in the development of a vertebrate embryo. Consequently the earlier stages were relatively neglected, until some 30 years ago, when the pregastrulation stages of amphibians and birds became a focus of interest, mainly as far as mesoderm induction is concerned. The mammalian and especially the human embryo are the least studied, due to objective difficulties in performing experimental procedures on the stages concerned. Therefore the only way at present, is to draw conclusions based on a comparison of homologous areas of vertebrate embryos of different groups, at identical developmental stages. There seems to be a confusion in the comparative staging and the comparative terminology concerning early vertebrate embryos and I would therefore like to briefly review the homologous stages of amphibia, birds and mammals and the terms used to describe them. Especially nowadays, when genetic markers are used as criteria for developmental processes, as well as for evolutionary conclusions, one should be aware of what one may or may not compare. The gastrulae, as everybody agrees, are a good starting point for such a comparison from which one can go backwards to the early stages and trace the common denominators between the three embryonic types.

In contrast to the gastrula of the anamniotic amphibia, in which all cells contribute to the embryonic body, the gastrulae of amniotic birds and mammals have a different basic design: a flat, central embryonic disc, which is the formative part of the germ, surrounded by trophoblastic tissues, which are specialized devices developed by the germ to nourish the developing embryo. The trophoblastic tissues (see Fig. 1) do not contribute either to the embryo proper or to its extraembryonic parts. Normally when a comparison is made between the above mentioned gastrulae only the formative sections are compared. The same approach should be applied also to earlier stages of development. There also, a clear distinction should be made between formative and trophoblastic components.

The period between fertilization and gastrulation can be divided in all the three types into 1) cleavage, leading to the formation of 2) the morula (Fig. 1), a compact ball or disc of cells, after which comes 3) the blastula (Fig. 2), characterized by the formation of a

Organization of the Early Vertebrate Embryo
Edited by N. Zagris *et al.*, Plenum Press, New York, 1995

MORULA

a   AMPHIBIAN

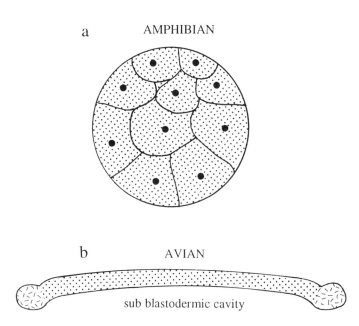

b   AVIAN

sub blastodermic cavity

c   MAMMALIAN

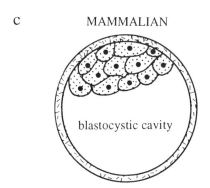

blastocystic cavity

Figure 1. A schematic comparison (sagital sections) of early morulae of amphibia (a); birds (b) and mammals (c).

formative cells

trophoblastic cells

BLASTULA
SAGITAL SECTIONS

a   AMPHIBIAN

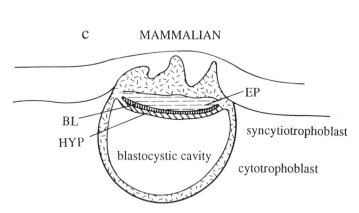

b   AVIAN

c   MAMMALIAN

Figure 2. A schematic comparison (sagital sections) of late blastulae of amphibia (a); birds (b) and mammals (c).
Abbreviations: AO - area opaca; BL - blastocoele; CD - central epiblastic disc; EC - ectoderm; EN - entoderm; EP - area pellucida; HYP - hypoblast; KS - Koller's sickle; PM - posterior marginal zone

blastocoele, which develops inside the formative tissue, and separates the ectoderm (epiblast) from the entoderm (hypoblast).

The ovulated eggs in all three types are basically radially symmetric (Nieuwkoop et al., 1985). Later on in each of the three above mentioned groups there is a change leading to bilateral symmetrization and the determination of the anterior and posterior ends. In amphibians the change in symmetry is very rapid following fertilization, so that the zygote is already bilateral symmetric, the plane of symmetry being determined by the point of sperm entry, opposite to which the dorsal lip of the blastopore will form (Ancel and Vintemberger, 1948). In birds the plane of bilateral symmetry is gravity dependent (Kochav and Eyal-Giladi, 1971) and is gradually established during the cleavage and morula stages by the spatial position of the blastodisc. Although there is probably some imprinting of bilaterality during the above stages, it is still very labile and can easily be overruled by changing the spatial position of the germ (Eyal-Giladi and Fabian, 1980). Normally in the avians, the cleavage and morula stages (I-X E.G&K, Eyal-Giladi & Kochav, 1976) are prelaying stages ("uterine"), during which the egg rotates inside the uterus of the mother (Clavert, 1962). As a result of this rotation, the germ acquires an oblique position where the germ's side facing the direction of the rotation is being forced down (Kochav and Eyal-Giladi, 1971). The other end which points upwards is gradually determined to become the posterior side of the future embryo, which is the side at which all the cytological (Raveh et al., 1976), physiological (Eyal-Giladi et al., 1979; Raddatz et al., 1987) and morphogenetic (Eyal-Giladi and Kochav, 1976; Kochav et al., 1980; Fabian and Eyal-Giladi, 1981) parameters of development are initially demonstrated, and from which they spread in an anterior direction. Until stage VII E.G&K, which should be considered as a mid-morula stage, both the plane of symmetry as well as the anterior-posterior polarity can be reversed by reversing the tilt of the blastoderm. At the above sensitive stages it is also possible to make the blastoderm "forget" its polarity by storing it for five days at 15°C. It is also possible to altogether prevent the imprinting of polarity during the sensitive stages by forcing the germ into a horizontal position (Eyal-Giladi et al., 1994). Prevention of bilateralization will leave the developing blastodisc as a radially symmetric system until the stage that it loses its ability (competence) for symmetrization. Very little is known about the way bilaterality is imposed upon the mammalian embryo. However, the picture emerging from a survey of the literature is the following: In the zygote there is no obvious cytoplasmic segregation (Eviskov et al., 1994) and until implantation the mammalian blastocyst (morula) is radially symmetrical. During implantation, the inner cell mass, which is the formative part of the blastocyst seems to be forced into an oblique position. The anterio-posterior axis of the developing embryo seems to be always in a consistent relationship and orientation to the uterus (in humans) or to the uterine horns (in the mouse) (Smith, 1985; Gardner et al., 1992), which comply with gravitation. It is therefore possible that also in mammals it is gravity that determines bilaterality, an idea that should be examined experimentally.

In the "uterine" avian morula there is an early morphological marker to bilateralization, which is expressed as an orderly posterior-anterior cell shedding process (Fabian and Eyal-Giladi, 1981), in which all the cells of the lower layers of the blastodisc (mid morula) lose contact with each other, a process occurring in a polar way from the future posterior to the future anterior side, and fall into the sub-blastodermic cavity where they disintegrate. The end result of this process is a stage X E.G&K blastoderm - the late morula (Fig. 3). At that stage the central part, the area pellucida, is an epithelium, one cell thick, which is surrounded by a thicker cellular belt, the peripheral area opaca, which is in contact with the yolk. At the completion of cell shedding the distinction between the formative part of the blastoderm (area pellucida) and the trophoblastic part (area opaca) becomes obvious. It is interesting to point out in this connection, that in case bilateralization is experimentally prevented, the process of cell shedding at stages equivalent to the "uterine stages" occurs,

114

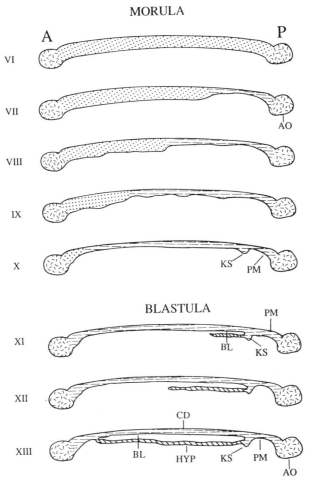

Figure 3. Sagital sections of chick blastoderms.
Morula stages VI - X E.G&K
Blastula stages XI - XIII E.G&K

Abbreviations: A - anterior; AO - area opaca; BL - blastocoele; CD - central epiblastic disc; EP - area pellucida; HYP - hypoblast; KS - Koller's sickle; P - posterior; PM - posterior marginal zone.

[⋯⋯⋯] Prior to cell shedding
[▭] After cell shedding

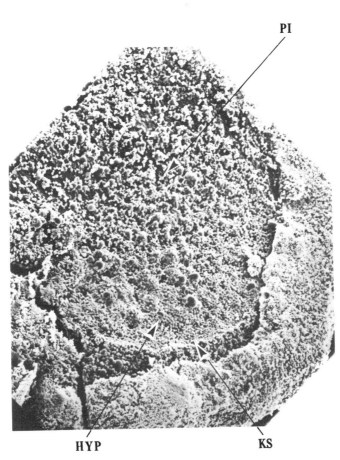

Figure 4. A scanning electron micrograph of a stage XI chick blastoderm. A clear difference is seen between the confluent hypoblastic sheet (HYP) growing from Koller's sickle (KS) and the more anterior hypoblastic cells from polyinvagination origin (PI).

not in a posterio-anterior direction as in normal development, but in a centripetal manner. In such blastoderms an area opaca and an area pellucida are formed, but as a consequence of the lack of any signal for symmetrization and axialization, only extraembryonic tissues, but no embryo will be formed.

In the chick, the egg is usually laid at stage X and needs to be incubated at 37°C in order to develop into a blastula. The stage X, late morula, though looking radial already, has distinct posterio-anterior developmental potencies, the morphological expression of which is the appearance of a sickle shaped thickening of the lower surface at the posterior side of the area pellucida. Koller's sickle marks the segregation of the area pellucida into a peripheral belt - the marginal zone - and a central disc with quite distinct and different developmental potencies.

The avian marginal zone[*] is the most potent region in the blastoderm on which any further axial development depends. Although it is circular, the avian marginal zone has a clear bilateral symmetry and the middle point of Koller's sickle marks the posterior side of the blastoderm and of the future embryo. During the incubation of a stage X blastoderm a single layered thin shelf of cells spreads from Koller's sickle in an anterior direction underneath the central disc, on the lower surface of which there are already patches of cells which have polyinvaginated from the epithelial disc (Fig. 4). The two distinct cell populations, one from Koller's sickle source and the other from polyinvagination origin, merge in a very characteristic pattern, to form the lower layer - the hypoblast[**]. At the end of the process a complete lower layer is formed underneath the central disc, thus gradually transforming the stage X, late morula, into a stage XIII blastula (Figs. 3 & 4). The narrow space between the now distinct epiblast and hypoblast is homologous to the amphibian blastocoele which forms inside the morula and separates the animal cap ectoderm, the epiblast, from the vegetal cap entoderm, the hypoblast.

In mammals, despite the difficulty in properly staging and systematically studying the early implantation stages, the situation seems to be very similar. The picture of how the hypoblast is formed in mammals and whether it expresses polarity is not clear. What is well known is that a lower layer (hypoblast) is formed from and underneath the inner cell mass and thus forms what has to be regarded as the blastula. Again the narrow space separating the epiblast from the hypoblast (Fig. 2C) is homologous to the amphibian and avian blastocoele. This definition calls attention to the misleading terminology used by mammalian embryologists, who usually call the blastocystic cavity a blastocoele. The blastocystic cavity is, in effect, from a comparative point of view, homologous to the sub-blastodermic cavity of

---

[*] Regretfully, the term marginal zone is used, in amphibians and in avians, for non-homologous embryonic sections. This term has been used for decades in many publications dealing with amphibian or avian morphogenesis, but with different connotations. I would like therefore to minimize the confusion by exactly defining what a marginal zone means in each group. In an amphibian blastula the marginal zone is the cellular belt separating the animal ectoderm from the vegetal entoderm. The belt is wider at the dorsal side which will form the axial mesoderm, while its ventral side is relatively narrow. During gastrulation the cells of the marginal zone involute into the embryo to form the mesoblast. In the avians the marginal zone is the peripheral belt of the area pellucida. It becomes distinct at stage X E.G&K (late morula) at the posterior side by the appearance of a thickening - Koller's sickle, which forms the border between the marginal zone and the central epiblastic disc. Cells of the marginal zone (including Koller's sickle) migrate anteriorly and participate in the formation of the lower layer - the hypoblast, which can schematically be compared to the entoderm of the amphibian blastula. In other words, while the amphibian marginal zone cells involute during gastrulation to form the mesoblast, the cells of the avian marginal zone migrate during the blastula stage to form the hypoblast (endoblast).

[**] The lower layer is referred to in this paper by the general term hypoblast because of its microscopic as well as macroscopic morphological uniformity. It contains, however, two cell populations from different developmental origins: one from marginal zone origin, which is inductive for primitive streak and another from polyinvaginated cells which are non-inductive.

the avians, which later becomes the yolk sac cavity. If this was only a matter of terminology it would have been tolerable, but the real problem is that by this terminology, what should be called a mammalian morula is called a blastula and thus might confuse the comparative studies on early induction and on the expression of molecular markers.

The blastula stages have been shown in amphibians to be the stages in which mesoderm induction starts, resulting somewhat later in gastrulation. Mesoderm induction is a result of the interaction between the inductive entoderm and the responsive competent ectoderm. If we adopt the above comparative staging we can demonstrate that a very similar situation exists in the avian blastoderm. The hypoblast (=entoderm) has been shown already by Waddington (1933) and later by Eyal-Giladi and Wolk (1970), Azar and Eyal-Giladi (1979, 1981, 1983), Mitrani et al. (1983) and Mitrani and Eyal-Giladi (1984) to be capable of inducing a primitive streak in the epiblastic central disc of the area pellucida. Azar and Eyal-Giladi (1979) have concluded from a series of deletion experiments on stage XIII (late blastula) chick blastoderms that the hypoblast probably derives its inductive capacity from the posterior section of the marginal zone. This idea was examined in a series of experiments on the marginal zone which were performed on stage X (late morula) and stage XII (mid blastula) blastoderms. The experiments which included transplantation of sections of the posterior marginal zone into other locations in the marginal zone of the same blastoderm, or blastoderms of different stages (Khaner and Eyal-Giladi, 1986, 1989; Eyal-Giladi and Khaner, 1989), have indeed indicated the crucial role of the marginal zone in the formation of the primitive streak. It was demonstrated that in the marginal zone belt there is a circular gradient of primitive streak inductive potential with a minimum at the anterior side and a maximum at the posterior side. The existence of a circular gradient of "embryo body forming capacity" with a gradual restriction to the posterior section of the marginal zone was already advocated by Spratt (1966). The experiments mentioned above have also shown that there is a temporal aspect concerning the inductive potential of the marginal zone which decreases from stage X onwards (Eyal-Giladi and Khaner, 1989). At stage XIII the posterior marginal zone still has enough potential which can be demonstrated, following the removal of the hypoblast, by the regeneration of a new inductive hypoblast (Azar and Eyal-Giladi, 1979). This potential is however too weak to be demonstrated in transplantation experiments. We have also shown that the region with the strongest inductive potential is the only one to express itself in a blastoderm, unless there is another one with an equal strength at a sufficient distance from it. The regulation in the system to form only one primitive streak is through an inhibitory signal emerging from the same center in the marginal zone where the processes of axis formation has been initiated. The inhibitory effect has been shown to be conveyed via the marginal zone belt and it probably does not pass through the central disc (Khaner and Eyal-Giladi, 1989).

Taking all the above indications together, we have formulated the following working hypothesis: The region with the highest inductive potential in the avian marginal zone, is the posterior section. It contributes cells that move via Koller's sickle into the forming hypoblast and are the ones that form the relatively coherent posterior sheet of cells that is seen to spread anteriorly. Only those cells possess the capacity to induce a primitive streak in the epiblast above them. A second cell population from polyinvagination origin that merges with and joins the cells from marginal zone origin to form the full hypoblast of the stage XIII blastula is not inductive for primitive streak. However, by being in the right place the polyinvaginated cells help to shape the advancing inductive cells and guide them into a postero-central position, which is important for the correct shaping of the primitive streak. To check whether this is what really happens, the following experiments have been done: From blastoderms at stages X and XI a section of the posterior marginal zone was cut out, with the anterior incision either anterior or posterior to Koller's sickle. The explant was incubated for 35-55 min in RDL (rhodamine-dextran-lysine) while the blastoderm was put on ice to avoid further development. After the labeling, the piece was put back into its previous position and the reconstituted blastoderm was further incubated. The position of the labeled cells after different periods of incubation was determined, and reconstructions were made (Eyal-Giladi et al., 1992). Koller's sickle and the marginal zone behind it were found to contribute all the centrally located cells of the growing hypoblast, a process that starts very intensively at stage X and slows down at the consecutive stages. The epiblastic cells of the central disc that are situated at stage X anterior to Koller's sickle, were shown to constitute the primitive streak first and later to move into the mesoblast.

The question whether the cells of the posterior marginal zone are already inductive while in the marginal zone, or whether they acquire their inductivity after their movement into the hypoblast has also been addressed. The central epiblastic disc of a stage XIII blastoderm has been shown to be incapable of forming a primitive streak when cultured on its own. It therefore can serve as a suitable system to check the inductivity of certain tissues and factors. When posterior marginal zone sections, either with or without Koller's sickle, were applied to the lower surface of epiblastic central discs, no primitive streaks developed from them. However, when equally sized hypoblast sections were applied instead, primitive streaks developed from most of the central discs. This means that the cells of the posterior marginal zone while in situ have the potential to become inductive, but that they cannot materialize it unless they move via Koller's sickle into the hypoblast.

Primitive streak induction depends on a diffusable substance from the hypoblast that affects the competent epiblastic cells. This has been proven by transfilter induction experiments (Eyal-Giladi and Wolk, 1970) as well as by the incubation of epiblastic central discs in a conditioned medium of hypoblasts. In both cases primitive streaks developed in the central discs. Contrary to that, incubation of epiblastic central discs in conditioned media of epiblast and marginal zone, in the same and even higher concentrations, did not have any primitive streak inductive effect (Eyal-Giladi et al., 1994). If we compare mesoderm induction in the avians with amphibians, we can see a clear parallel. In amphibia, Nieuwkoop's center localized in the posterior part of the entoderm is non-active in the morula and is activated in the early blastula to induce the dorsal mesodermal marginal zone in the ectoderm (Fukui et al., 1994). The equivalent of Nieuwkoop's center in the avians is the marginal zone, which in itself is not inductive during the morula stage and becomes so only during the formation of the blastula (stage X and on) when its cells move into the hypoblast (entoderm) of the forming blastula.

ABSTRACT

A comparison is made among the early stages of development in three vertebrate groups: Amphibia, Aves and Mammalia. The morula and blastula stages are defined in all three groups and the morphogenetic processes including axis determination and mesoderm induction are discussed. The term marginal zone is used for non-homologous regions in amphibians and birds, and causes much confusion. The marginal zone of amphibia is a region in the late blastula, which is already induced and determined to form the embryonic mesoderm, and is homologous to the avian primitive streak.

The avian marginal zone is a circular belt of epiblastic tissue in the periphery of the area pellucida. Its posterior section can first be identified at stage X E.G&K after the appearance of the thickening of Koller's sickle, which forms the most anterior section of the posterior marginal zone. A stage X E.G&K blastoderm can therefore be compared to a late amphibian morula, while stages X-XIII blastoderms exhibit the formation of the avian blastula. The hypoblast which grows from the posteriorly located Koller's sickle towards the anterior side of the area pellucida, can therefore be compared to the vegetal yolk mass of an amphibian blastula. The avian mesoderm, similarly to the amphibians, is induced during the late morula and the blastula stages. In the amphibians the induced mesoderm forms the marginal zone (Spemann's organizer) and in the birds it forms the anterior part of the primitive streak (the avian organizer). The inducer of Spemann's organizer is Nieuwkoop's center, while the inducer of the PS are the cells of the posterior section of the marginal zone (Koller's sickle included) that move into the growing hypoblast and interact with the epiblast. Experiments supporting the above conclusions are discussed.

Acknowledgement. I wish to express my thanks to Prof. P.D. Nieuwkoop for his valuable remarks on the manuscript.

REFERENCES

Ancel, P., and Vintemberger, P., 1948, Recherches sur le determinisme de la symetrie bilteral dans l'oeuf des amphibiens, *Bulletin Biologique de la France et de la Belgique*, Suppl. 31:1-182.

Azar, Y., and Eyal-Giladi, H., 1979, Marginal zone cells - the primitive streak inducing component of the primary hypoblast in the chick, *J. Embryol. exp. Morph.* 52:79-88.

Azar, Y., and Eyal-Giladi, H., 1981, Interaction of epiblast and hypoblast in the formation of the primitive streak and embryonic axis, as revealed by hypoblast rotation experiments, *J. Embryol. exp. Morph.* 61:133-144.

Azar, Y., and Eyal-Giladi, H., 1983, The retention of primary hypoblastic cells underneath the developing primitive streak allows for their prolonged inductive influence, *J. Embryol. exp. Morph.* 77:143-151.

Clavert, J., 1962, Symmetrization of the egg of vertebrates, *in*: "Advances in Morphogenesis," Vol. 2:27-60.

Evsikov, S.V., Morozova, L.M., and Solomko, A.P., 1994, Role of ooplasmic segregation in mammalian development, *Roux' Arch. Dev. Biol.* 203:199-204.

Eyal-Giladi, H., and Wolk, M., 1970, The inducing capacities of the primary hypoblast as revealed by transfilter induction studies, *Roux' Arch.* 165:226-241.

Eyal-Giladi, H., and Kochav, S., 1976, From cleavage to primitive streak formation: A complementary normal table, a new look at the first stages of the development of the chick. I. General morphology, *Develop. Biol.* 49:321-337.

Eyal-Giladi, H., Raveh, D., Feinstein, N., and Friedlander, M., 1979, Glycogen metabolism in the prelaid chick embryo, *J. Morph.* 161:23-30.

Eyal-Giladi, H., and Khaner, O., 1989, The chick's marginal zone and primitive streak formation. II. Quantification of the MZ's potencies - temporal and spatial aspects, *Develop. Biol.* 134:215-221.

Eyal-Giladi, H., Debby, A., and Harel, N., 1992, The posterior section of the chick's area pellucida and its involvement in hypoblast and primitive streak formation, *Development* 116:819-830.

Eyal-Giladi, H., Goldberg, M., Refael, H., and Avner, O., 1994, A direct approach to the study of the effect of gravity on axis formation in birds, *Ad. Space Res.* 14:(8)271-(8)279.

Eyal-Giladi, H., Lotan, T., Levin, T., Avner, O., and Hochman, J., 1994, Avian marginal zone cells function as primitive streak inducers only after their migration into the hypoblast, *Development* 120:2501-2509.

Fabian, B.C., and Eyal-Giladi, H., 1981, A SEM study of cell shedding during the formation of the area pellucida in the chick embryo, *J. Embryol. exp. Morph.* 64:11-22.

Fukui, A., Nakamura, T., Uchiyama, H., Sugino, K., Sugino, H., and Asashima, M., 1994, Identification of activins A, AB and B and follistatin proteins in Xenopus embryos, *Develop. Biol.* 163(1):279-281.

Gardner, R.L., Meredith, M.R., and Altman, G., 1992, Is the anterior-posterior axis of the fetus specified before implantation in the mouse? *J. Exp. Zool.* 264:437-443.

Khaner, O., and Eyal-Giladi, H., 1986, The embryo forming potencies of the posterior marginal zone in stage X through XII of the chick, *Develop. Biol.* 115:275-281.

Khaner, O., and Eyal-Giladi, H., 1989, The chick's marginal zone and primitive streak formation. I. Coordinative effect of induction and inhibition. *Develop. Biol.* 134:206-214.

Kochav, S., and Eyal-Giladi, H., 1971, Bilateral symmetry in the chick embryo - determination by gravity. *Science* 171:1027-1029.

Kochav, S., Ginsburg, M., and Eyal-Giladi, H., 1980, From cleavage to primitive streak formation: A complementary normal table and a new look at the first stages of the development of the chick. II. Microscopic anatomy and cell population dynamics, *Develop. Biol.* 79:296-308.

Mitrani, E., Shimoni, Y., and Eyal-Giladi, H., 1983, Nature of hypoblastic influence on the chick embryo epiblast, *J. Embryol. exp. Morph.* 77:143-151.

Mitrani, E., and Eyal-Giladi, H., 1984, Differentiation of dissociated-reconstituted epiblasts of the chick under the influence of normal hypoblast, *Differentiation* 26:107-111.

Nieuwkoop, P.D., Johnen, A.G., and Albers, B., 1985, The epigenetic nature of early chordate development, Cambridge Univ. Press.

Raveh, D., Friedlander, M., and Eyal-Giladi, H., 1976, Nucleolar ontogenesis in the uterine chick germ correlated with morphogenetic events, *Exp. Cell Res.* 100:195-203.

Smith, L.J., 1985, Embryonic axis orientation in the mouse and its correlation with blastocyst relationships to the uterus. II. Relationships from 4 1/2 to 9 1/2 days, *J. Embryol. exp. Morph.* 89:15-35.

Spratt, N.T., 1966, Some problems and principles of development, *Am. Zoologist* 6:9-19.

Waddington, C.H., 1933, Induction by the endoderm in birds, *Roux' Arch. Dev. Biol.* 128:502-521.

MAPPING OF GASTRULATION MOVEMENTS IN BIRDS

Lucien C.A. Vakaet[1] and Hilde Bortier[2]

[1] Laboratory of Experimental Cancerology, University Hospital,
De Pintelaan, 185, 9000 Gent, Belgium
[2] Department of Anatomy and Neurobiology, University of Salt lake City
Utah 84132, USA

INTRODUCTION

Mapping gastrulation movements in Birds is theoretically impossible : a map is a still image and shows at best a snapshot of gastrulation. A map cannot show movements even when it is armed with arrows : to that end movies are needed. To enable the registration and reproduction of the movements within avian blastoderms during the first 24 hours of incubation, a cinephotomicrographic device was set up. The results of these observations are preceded by a demonstration of the procedure for preparing the germs. This procedure is derived from that of New (1955). Its main characteristic is that it allows normal development in vitro for a 24 hour period in laid blastoderms that are cultivated with their ventral side up, towards the observer.

The first film, entitled "Movements of the layers of the chick blastoderm" is a cartoon. It is the synthesis of a schematic analysis of many shots of the development of normal blastoderms. It allows us to present the developmental stages we distinguish during avian gastrulation.

The second film, entitled "Experiments on the early chick blastoderm demonstrated with cinephotomicrography" is a selection of the most informative shots of developing normal and experimental chick embryos on which the images of the cartoon are based.

In the comments on both films we consistently use the following terminology :
- Anatomical orientation terms are used to indicate the presumptive axes of the embryo. So, we use dorsal-ventral and cranial-caudal, although for the latter antero-posterior is more often, but less accurately used.
- The layers of the blastoderm are named after their normal disposition within the egg : upper layer cells make part of the dorsal surface; deep layer cells touch the ventral surface and middle layer cells do not form part of the surface of the blastoderm.

MOVEMENTS OF THE LAYERS OF THE CHICK BLASTODERM

By the time a bird's egg is laid, it contains a young embryo that has already been developing for 24 hours within the body of the hen. During the next 24 hours, it will

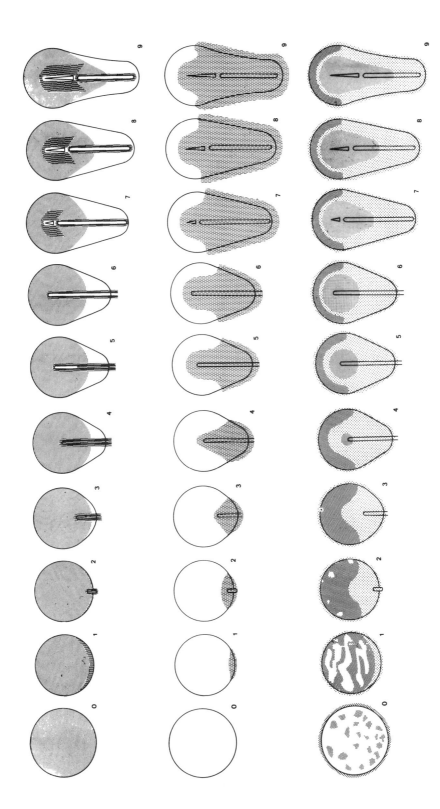

Figure 1. Synoptic representation of the stages of the gastrulation movements within the layers of the chick blastoderm. Lower row : deep layer; upper row : upper layer, middle row : middle layer. For comments. see text.

develop into an embryo in which the organs are about to begin to develop. The early development is illustrated schematically in Fig. 1.

The first shot is a fast-running sequence of diagrams that matches real development as closely as possible. We divide it into stages 0 through to 9, based on the general shape of the embryo during the first 24 hours of incubation. Although this subdivision into stages is artificial, it is indispensable since one cannot rely on the incubation time of the embryos to estimate their degree of development. The average incubation time (I.T. ± #) for each stage is given as a suggestion for the duration in hours of the optimal incubation time at 38°C to obtain blastoderms at a desired stage.

General Transformations

A stage 0 blastoderm (I.T. ± 0) is to be found in an unincubated freshly laid egg. Macroscopically, a less transparent area opaca forms the margin of the blastoderm. It surrounds the more translucent area pellucida. In a sagittal section, it appears that the area opaca is less transparent because of the presence of the germ wall, which in this stage represents all the layers of the area opaca. In the chick blastoderm, this germ wall may be larger and thicker in its caudal part. By contrast, the upper layer looks equally transparent all over the area pellucida.

During the first hours of incubation, the diameter of the blastoderm shrinks somewhat (H. Bortier and L. Vakaet, unpublished). The general shape of the area pellucida remains circular. At stage 1 (I.T. ± 5), an area centralis is situated within the area pellucida. The limit between both is most clearly seen caudally. The embryo will form within the area centralis.

Stage 2 (I.T. ± 7) is characterized by the primordium of the primitive streak. It appears as a linear condensation, perpendicular to the most caudal tangent to the area centralis. This tangent would cross the primitive streak at its mid-point. The primitive streak will develop markedly during the next few hours.

At stage 3 (I.T. ± 9) the primitive streak has not yet reached the centre of the original area centralis. At its caudal end, the primitive streak elongates into the area opaca. Concomitant with the overall elongation of the primitive streak, the border of the area pellucida gets a pear-shaped contour.

At stage 4 (I.T. ± 12) the cranial end of the primitive streak is in the centre of the area centralis. The area pellucida extends in a caudal direction, together with the caudal end of the elongating primitive streak that, therefore, does not extend farther into the area opaca than in the previous stages.

At stage 5 (I.T. ± 14) the primitive streak extends more cranially than the centre of the area centralis. From this moment on, its cranial end is named Hensen's node.

At stage 6 (I.T. ± 17) Hensen's node has attained its most cranial situation. The caudal end of the primitive streak still extends over a short distance into the area opaca. Together with the primitive streak, the area pellucida elongates in a caudal direction.

At stage 7 (I.T. ± 19) the head process is perceptible in the track of Hensen's node which has started its regression. The caudal end of the primitive streak, named Nodus Posterior, is now to be found within the area pellucida, that strongly gains in length in a caudal direction.

At stage 8 (I.T. ± 21) the head process has lengthened over the distance covered by the regressing Hensen's node. At the caudal end of the primitive streak, the Nodus posterior stays almost in place. Consequently, the primitive streak shortens.

Stage 9 (I.T. ± 24) is characterized by the head-fold. This is a tiny though clearcut crescent-shaped fold, cranial to the cranial end of the head process. Its appearance means to us the end of the early development, thus announcing the beginning of organogenesis.

The general transformations we have followed are the consequences of movements within the cell layers of the embryo. We now study the morphogenetic movements within each layer separately. We start with the deep layer, situated at the ventral aspect of the embryo, closest to the yolk and the best accessible layer of the blastoderm in New culture. After the deep layer, we illustrate the evolution of the upper layer. The middle layer is described last, as it is the least accessible.

Transformations in the deep layer

At stage 0, the deep layer consists of scattered cells, clinging to the ventral side of the upper layer. There may be more deep layer cells present in the caudal than in the cranial part of the area pellucida.

At stage 1, the deep layer cells form the endophyll layer, that presents many holes. Contrary to the scattered deep layer cells of stage 1, the endophyll is loosely connected with the upper layer. In the deep layer, movements start with a concentric ingrowth of cells from the germ wall. They form the marginal hypoblast. At the caudal limit of the area centralis, a crescent-like thickening of the deep layer appears, corresponding to the sickle generally named after Koller (1882) but that, as pointed out by Callebaut and Van Nueten (1994), should be named after Rauber (1876).

At stage 2, a growth of sickle endoblast (hypoblast) starts cranialward from Rauber's sickle and extends as a fountain-like movement into the area centralis, apparently pushing the endophyll forward. Concomitantly, the endophyll is fully closed. It is surrounded by hypoblast : caudally by the sickle endoblast and cranially and laterally by marginal hypoblast.

At stage 3, the endophyll is squeezed together cranialward by the migrating hypoblast, thus forming the endophyll wall or endophyllic crescent. This disposition is clearly visible on a sagittal section.

By the end of stage 4, a third component appears within the deep layer, namely the definitive endoblast. It originates ventrally from Hensen's node and the anterior half of the primitive streak by ingression as a continuous layer. The definitive endoblast is also surrounded by hypoblast.

During stage 5, the final placement of the endoblast continues and is finished by stage 6. The definitive endoblast forms an island with an oval border within the hypoblast. The hypoblast from now on encloses the endophyll wall and the endoblast. The definitive endoblast is, at its dorsal aspect, firmly attached to the middle layer, contrary to the hypoblast that stays at some distance from the other layers of the blastoderm. The endophyll wall (see Fig. 1) forms a semicircular irregular tube around the cranial periphery of the area centralis. On sections the endophyll tube is U-shaped and sticks by the extremities of the U to the upper layer. Its cavity is outlined mostly by endophyll and partly by the ventral aspect of the upper layer in the region that is called the proamnion.

During stage 7, stage 8 and stage 9, the caudal part of the endoblast follows the regression of the cranial part of the primitive streak. It gets a pear-shaped outline within the area pellucida.

Transformations in the upper layer

At stage 0, the upper layer has an even appearance.

By stage 1, the appearance of Rauber's sickle may be marked by a furrow on the dorsal aspect of the upper layer as the deepest of many small pits (H. Mehrbach, 1935) that give the upper layer of the area centralis the shagreen-like aspect mentioned by Gräper (1929). In a sagittal section, Rauber's sickle appears as a thickening of all germ layers. It may present a furrow in the dorsal surface of the upper layer.

Through stage 2, stage 3 and stage 4, the upper layer in the area centralis becomes a pseudostratified cylindrical epithelium. It is highly columnar and flat over the area centralis and over the primitive streak.

At stage 5, a longitudinal groove appears in the upper layer over the primitive streak. This groove is initiated cranially, where it forms the primitive pit, just behind Hensen's node and extends rapidly caudalwards. By stage 6 it reaches the caudal end of the primitive streak.

From stage 6 on, the primitive streak presents a groove over its full length. In a transverse section, the groove of the primitive streak seems formed by two closed lips built up by thickened upper layer. The deepest point of the groove descends to the level of the middle layer.

At stage 7, a thickening, outlined as an arrowhead, may be seen in the upper layer of the area centralis. This pseudo-stratified cylindrical layer is continuous with and morphologically inot distinguishable from that alongside the primitive streak. Cranial to Hensen's node, a median strip of upper layer is thinner over the head process (median hinge point, Schoenwolf et al., 1989). The thickened parts of the upper layer start moving in concert : the plate lateral to the median hinge point elongates while the cranial part of the primitive streak shortens during its regression. In this way, the upper layer of the area centralis gains in length.

At stage 8, the regression of the cranial half of the primitive streak and its overall shortening continue, while the upper layer condensation which lies cranially to the node elongates concomitantly.

At stage 9, this bilateral condensation appears to be the neural plate as it gets bordered cranially by the head-fold, immediately followed by the lifting up of the neural folds.

Transformations in the middle layer

At stage 0 some middle layer cells are already present at the margin of the area centralis, as suggested by the experiments of Settle (1954) and Zagris (1979).

At stage 1, middle layer cells are found central to the caudal border of the area centralis, on the track of the cranial rim of the hypoblast or sickle endoblast. (Bortier and Vakaet, unpublished scanning electromicroscopic observations).

At stage 2, the middle layer is spread out as an oval layer of cells in the caudal quarter of the area pellucida. An axial condensation within it corresponds to the early primitive streak.

At stage 3 the middle layer cells around the caudal half of the primitive streak extend somewhat outside the area pellucida. So far, the middle layer cells lateral to the cranial half of the primitive streak do not extend into the area opaca. This field therefore takes the form of a triangle with a blunt apex at the cranial end of the primitive streak.

As these processes continue during stage 4, the elongating primitive streak field looks more slender.

At stage 5, however, coinciding with the appearance of the groove in the upper layer over the primitive streak, an expansion starts in the middle layer field lateral to the cranial part of the primitive streak. This expansion is best perceptible cranial and lateral to Hensen's node along the cranial half of the primitive streak. This expansion of the middle layer coincides with the final placement of the definitive endoblast within the deep layer.

At stage 6, the expansion of the middle layer continues so that gradually the middle layer outside the area pellucida is found more cranially.

At stage 7, the head process elongates in concert with the regressing Hensen's node and the cranial half of the primitive streak.

At stage 8, the regression of Hensen's node and the shortening of the cranial half of the streak continue and the extra-embryonic middle layer gradually extends forward. At stage 9 the foremost borders of the middle layer's wings reach the level of the endophyll wall. The most cranial middle layer cells do not extend cranially beyond the head-fold. That head-fold is formed just behind the caudal border of an area that is free of middle layer : the pro-amnion.

Synthesis

The last part of the cartoon is a synoptic synthesis of the evolution of the three layers.It demonstrates that, whereas the movements of the three layers can be represented and understood separately, this is not possible when they are shown at the same time.

EXPERIMENTS ON THE EARLY CHICK BLASTODERM DEMONSTRATED WITH CINEPHOTOMICROGRAPHY

The first shots show the development of chick blastoderms marked with iron oxide particles on the ventral aspect of the deep layer. These marks allow to follow the fountain-like movement arising at the posterior limit of the area centralis, where Rauber's sickle vanishes in the process. This movement sweeps the endophyll cells together anteriorly, so as to form the endophyll wall. Caudal to the forming endophyll wall the primitive streak appears and elongates up to cranial to the central point of the area centralis. At St7 the streak starts its regression that will be followed after St9 by the initial phases of organogenesis.

The final placement of the definitive endoblast is demonstrated in two shots of chick blastoderms in which the deep layer had been experimentally removed from a St4 and a St5 blastoderm respectively. In the St4 blastoderm, iron oxide markers were deposited on the ventral aspect of the denuded middle and upper layer cells. No markers were used in the St5 blastoderm, as yolk grains suspended in saline covered the ventral aspect of the blastoderm. The final placement of the definitive endoblast starts in both embryos, ventrally to Hensens' node and progressively ventrally to the anterior half of the primitive streak. From there, a continuous layer extends in the normal way. This occurs in the absence of any deep layer. It is not a regeneration as no definitive endoblast had been ablated. Moreover, the further development of the embryos was normal.

Evocation of secondary axes is demonstrated by rotation of the deep layer at St5 after Waddington (1933), by the grafting of a Hensen's node in the endophyll wall (after Gallera (1958) and the grafting of a Nodus posterior in the endophyll wall of blastoderms at St4.

- The rotation of the deep layer evoked an ephemeral secondary primitive streak in the host. The significance of the experiment is not clear, as we do not know, for lack of available markers, whether or not middle layer cells were rotated together with the deep layer.

- Grafting a Hensen's node evoked in the host a secondary neural plate of the same age as that of the graft.

- Grafting of a Nodus posterior in the endophyll wall initially led to the disappearance of the graft together with the formation of a clear area. In this area a secondary primitive streak arises in the host and elongates normally. Although this primitive streak evanesced during further development, its ephemeral existence influenced the host embryo : the half neural plate next to the graft was atrophic. This might be caused by recruitment through the secondary streak of material originally due to become neural plate.

PROSPECTS

We made no comments here on the direction of the movements of the different Anlage Fields within the embryonal layers nor on the nature of their movements. More recent results of marking experiments as well as of investigations into the mechanisms and dynamics of the morphogenetic movements will be shown and discussed in the lecture by H. Bortier.

REFERENCES

Callebaut, M., Van Nueten, E., 1994, Rauber's (Koller's) Sickle : The Early Gastrulation Organizer of the Avian Blastoderm, European Journal of Morphology, 32, 35-48.

Gallera, J., 1965, Excision et transplantation des diffé rentes ré gions de la ligne primitive chez le poulet, C. R. Ass. Anat, 125, 632-639.

Grä per, L., 1929, Die Primitiventwicklung des Hü hnchens nach stereomikrographischen Untersuchungen, kontrolliert durch vitale Farbmarkierung und verglichen mit der Entwicklung anderer Wirbeltiere, Roux' Archiv, 116, 382-429.

Koller, C., 1882, Untersuchungen ü ber die Blä tterbildung im Hü hnerkeim, Archiv fü r Mikroskopische Anatomie, 20, 174-211.

Merbach, H., 1935, Beobachtungen an der Keimscheibe des Hü hnchens vor dem Erscheinen des Primitivstreifens, Zeitschrift fü r Anatomie und Entwicklungsgeschichte, 104, 635-652.

New, D.A.T., 1955, A new technique for the cultivation of the chick embryo in vitro. J.Embrol. Exp. Morph., 7, 146-164.

Rauber, A., 1877,Über die Stellung des Hü hnchens im Entwicklungsplan, Leipzig.

Schoenwolf, G., Bortier, H., Vakaet, L., 1989, Fate mapping the avian Neural Plate with Quail-Chick Chimeras : Origin of Prospective Median Wedge Cells, J. Exp. Zool., 249, 271-278.

Settle, G.W., 1954, Localization of the erythrocyte-forming areas in the early chick blastoderm cultivated in vitro, Contrib. Embryol. n° 241 Carn.Inst. 50, 223-227.

Vakaet L., 1970, Cinephotomicrographic investigations of gastrulation in the chick blastoderm, Arch. Biol (Liè ge), 81/3, 387-426.

Waddington, C.H., 1933, Induction by the Endoderm in Birds, Philos. Trans. Roy. Soc., ser. B, 221, 179-230.

Zagris N., 1979, Differentiation capacity of unincubated chick blastoderm in culture, J. Emb. Exp. Morph. 50, 47-55.

MORPHOGENETIC MOVEMENTS IN THE AVIAN BLASTODERM: MECHANISM AND DYNAMICS

Hilde Bortier[1] and Lucien C.A. Vakaet[2]

[1] Department of Anatomy and Neurobiology, University of Salt lake City
Utah 84132, USA
[2] Laboratory of Experimental Cancerology, University Hospital, De
Pintelaan, 185, 9000 Gent, Belgium

INTRODUCTION

Life is movement : living organisms are recognized by their movements. Contrary to mechanical movements, biological movements are unpredictable and they look autonomous. Biological movements are physiological or morphogenetic. Physiological movements are shape changes that are reversible and repetitive, such as limb movements, respiration, heartbeat. Morphogenetic or developmental movements are irreversible and non-repetitive shape changes. "Hic fugit irreparabile tempus" : nothing happens twice in morphogenesis. Both physiological and morphogenetic movements occur together during the lifetime of an organism, but morphogenetic movements are more prominent during development. L. Wolpert used to say that "not birth, not marriage, but gastrulation is the most critical period in the lifetime of an individual".

Morphogenesis is temporo-spatial. It occurs in a 3-dimensional organism that irreversibly changes shape in time. Morphogenesis is thus 4-dimensional : impossible to understand. We can, at best, reconstruct 3-dimensional structures in our mind. Therefore histology is the least difficult of the morphological sciences, as it is 2-dimensional. Anatomy is 3-dimensional and thus much more difficult. Embryology is 4-dimensional and therefore impossible to understand.

We study morphogenetic movements in chick and quail blastoderms. Avian blastoderms have an advantage over amphibian and mammalian blastoderms as the area pellucida is almost flat during gastrulation. Morphogenetic movements are too slow to be observed directly by the eye, therefore time-lapse videography and videomicrography are used to demonstrate these movements.

NORMAL DEVELOPMENTS IN VITRO

The shell covers the yolk and the egg white (Fig. 1). The yolk is fixed to the shell by the chalazae. The vitelline membrane surrounds the yolk and the blastoderm. The plasma membrane separates the blastoderm from the yolk (Fig.2).

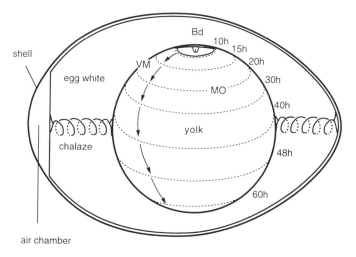

Fig. 1 Chick egg. VM : vitelline membrane, Bd : blastoderm, MO : margin of overgrowth, arrows : direction of epiboly, indication of time : hours of incubation

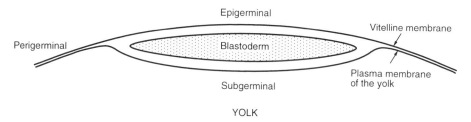

Fig. 2 Position of the avian blastoderm on the yolk

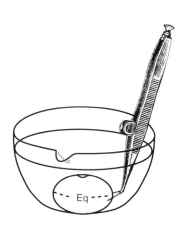

Fig. 3 Incision in the vitelline membrane. Eq : equator

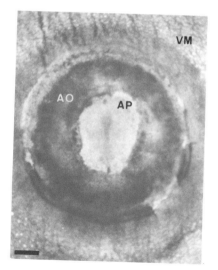

Fig. 4 Chick blastoderm St 5V observed from the ventral side. AO : area opaca,
AP : area pellucida, VM : vitelline membrane, bar = 1 mm

Avian blastoderms can be cultured in vitro since New (1955) described how the vitelline membrane can be incised at the equator and lifted from the yolk together with the blastoderm (Fig. 3).

During gastrulation a chick blastoderm St 5V of Vakaet (1970) has an area pellucida about 2 mm long and 1.5 mm wide. The overall thickness of the area pellucida is about 0.1 mm. The vitelline membrane, the area opaca, the area pellucida and the primitive streak are recognized by the difference in transparency (Fig. 4).

A section of a chick blastoderm at St 5V shows the three flat layers of the area pellucida (Fig. 5). The upper layer cells are part of the dorsal surface, the deep layer cells are part of the ventral surface and the middle layer cells have no external surface. This definition is used to avoid contamination of the terms epiblast, endoblast and mesoblast. For details concerning the development of the different layers we refer to Vakaet (1970) and Vakaet and Bortier (in press).

For time-lapse videography a camera was mounted on a stereomicroscope (Bortier, 1991; Bortier and Vakaet, 1992a). One video-image was recorded every 30 seconds, yielding an acceleration of 750x at normal projection speed (25 images/second). During development extraembryonic structures are derived from the area opaca. Intraembryonic structures are derived from the area pellucida. From the deep layer only the definitive endoblast develops into splanchnic structures. All the other somatic structures are derived from the upper layer by two types of morphogenetic movements : either movements in the plane of the upper layer or ingression perpendicular to the primitive streak.

MORPHOGENETIC MOVEMENTS

We define ingression as de-epithelialization of upper layer cells and migration of these cells between upper layer and deep layer. Ingression is visible on sections by the presence of bottle-like cells with long necks at the primitive streak and by cells leaving the upper layer, suggesting migration. Descendants of ingressing cells will colonize a presumptive region of the blastoderm. There is a similarity in the behaviour of ingressing cells and that of invasive cancer cells. Invasion also starts with de-epithelialization, followed by migration and later by colonization of other regions of the organism. The difference between both is essentially that ingressing cells are committed

133

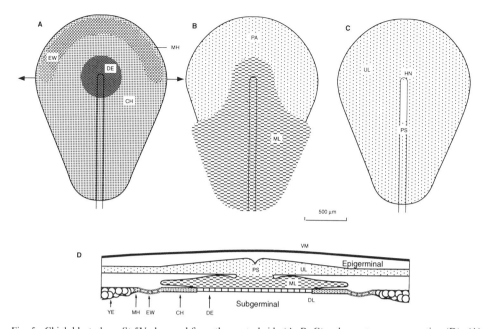

Fig. 5 Chick blastoderm St 5V observed from the ventral side (A, B, C) and on a transverse section (D). **(A)** Deep layer, **(B)** after resection of the deep layer, **(C)** upper layer after resection of the deep layer and the middle layer, **(D)** section in A (horizontal arrows). CH: central hypoblast, DE: definitive endoblast, DL: deep layer, EW: endophyll wall, HN: Hensen's node, MH: marginal hypoblast, ML: middle layer, PA: proamnion, PS: primitive streak, UL: upper layer, YE: yolk endoderm, VM: vitelline membrane

to a region and are programmed to stop the process, while invading cells do not have these restraints.

The morphogenetic movements in the plane of the upper layer have been studied in various ways. The best way is to use the nucleolar marker technique of N. Le Douarin (1973). Quail-chick xenografts can be recognized after fixation and Feulgen-Rossenbeck staining based on the difference between chick and quail nuclei. We used a combination of xenografting and time-lapse videography to follow the morphogenetic movements during gastrulation and neurulation (Bortier 1992 a, b). The xenografts were made by aspiration-punching of quail and chick upper layer fragments with a Pasteur pipette (Fig. 6).

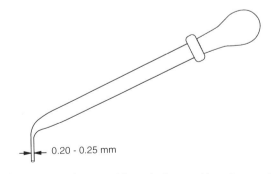

0.20 - 0.25 mm

Fig. 6 Pasteur pipette used for aspiration-punching of xenografts

The xenografts were made isochronically (at a similar stage), isotopically (in a similar region) and isotropically (in the same dorsoventral polarity). After healing, the more transparant quail grafts were followed with time-lapse videography due to the fact that the quail cells contain less yolk than the chick host and stay together. The quail grafts could be followed as long as they moved in the upper layer. This means that one xenograft experiment followed with time-lapse videography represents innumerable experiments on every point of the track of the graft. Time-lapse videography shows grafts that take part in the development of the neural tube; others stay in the epiblast; others ingress through the primitive streak and develop into mesoblastic structures.

Compared to the fate map of Spratt (1952), we found that the presumptive neural tissue in the upper layer of the St 4-6V chick blastoderm lies anterior to the primitive streak (Fig. 7). The transverse caudal limit of the presumptive neural plate is the cranial limit of the presumptive mesoblast. The presumptive parts of the neural plate are separated by transverse lines, forming bands that narrow during gastrulation and extend during neurulation. The presumptive notochord, somites and lateral plate are disposed from medial to lateral in parallel bands consisting of cells that converge from anterior to posterior towards the anterior half of the primitive streak.

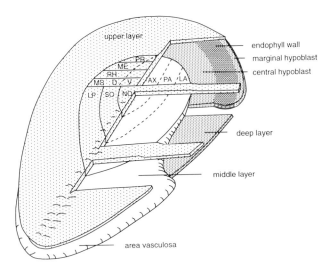

Fig. 7 Borders of the neural plate and the intraembryonic mesoblast in the chick blastoderm at St 6V. PR : prosencephalon, ME : mesencephalon, RH : rhombencephalon, MS : medulla spinalis, LP : lateral plate, SO : somites, NO : notochord, AX : axial, PA : paraxial, LA : lateral

CLOSURE OF EXPERIMENTAL EXCISION-WOUNDS

From the observed morphogenetic movements questions arise about the mechanism and dynamics that guide cells during these movements : how do the cells move in the upper layer, what is the driving motor, what guides the movements ?

We observed that punched-out wounds in the upper layer healed without a graft. We thought that study of the closure of these wounds could give an answer to our questions. Therefore we studied the closure of experimental wounds in the bare upper layer of the proamnion (Bortier et al. 1993). Time-lapse videography and video-micrography show the closure of these wounds.

We found that wound closure is not realized by one of the classically fundamental mechanisms. One such fundamental mechanism is an increase in mitotic figures. Counts of metaphases plus anaphases around wound submarginal zones and mirror image areas at the opposite side of the primitive streak of the blastoderm were significantly not different. S-phases counted after autoradiography of tritiated thymidine marked nuclei around wound submarginal zones and mirror image areas showed no difference either. Even if there were a difference, it would be hard to tell, as mitotic figures are high in gastrular and neural development. A second fundamental mechanism could be a contraction in the wound rim. One would then expect an even wound rim during closure and we did not see that with time-lapse videography. Wound rims are irregular during closure and the shape changes from circular to splitlike and again to circular. A continuous ring of microfilament bundles has not been found around the wounds. Even if microfilament bundles were present, this would not prove that contraction is occurring. A third fundamental mechanism could be migration on a substrate. As there are no other cell layers at the wound submarginal zones, one can only think of the basal lamina as a candidate substrate. With TEM and SEM we found that there is no basal lamina within the wounds. With immunohistochemistry for laminin we found no laminin inside the wounds. The structure of the epithelium of the wound submarginal region remained unchanged throughout closure : the cells did not acquire a fibroblast-like shape, as cells migrating on a substrate should. So, if none of these three fundamental mechanisms explains wound closure, how do these wounds close? We suggest that wounds close by movements of cells of the submarginal zone towards the wound gliding on their own basal lamina.

MECHANISM AND DYNAMICS OF MORPHOGENETIC MOVEMENTS

But what is the driving motor for the movements of the cells? To answer this question we considered some peculiar properties of the upper layer. The cells of this embryonic layer divide every 7 hours, adding cells to the upper layer. The upper layer is a pseudostratified cylindrical epithelium, where most of the nuclei are positioned halfway between the dorsal and the ventral side (Fig. 8). Contrary to the classic pseudostratified cylindrical epithelia (Krstic, 1984) the mitotic figures lie at the dorsal side of the upper layer by "interkinetic migration" (Sauer, 1936). Starting from the S-phase the nuclei move towards the dorsal side of the upper layer. During anaphase B the daughter nuclei move apart. In late telophase this leads to the formation of intercellular bridges or connecting cords.

We suggest that the cells of the upper layer move on their basal lamina driven by mitotic pressure which we define as the combination of mitotic activity of the upper layer (adding cells) and the pressure (force) that move the daughter cells apart. Daughter cells can be found one to four cells from each other as is demonstrated with SEM.

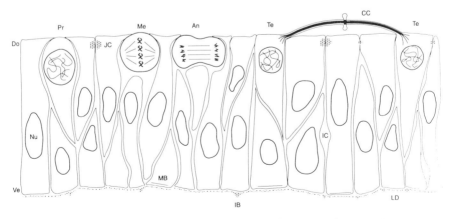

Fig. 8 Interkinetic migration in the chick upper layer. Do : dorsal, Ve : ventral, Nu : nucleus, Pr : prophase, Me : metaphase, An : anaphase, Te : telophase, CC : connecting cord, JC : junctional complex, MB : microfilament bundle, IB : interstitial body, LD : lamina densa of the basal lamina

The direction of the movements of cells is towards a sink in mitotic pressure. In wound closure the direction is towards the wound; in gastrulation it is towards the primitive streak. At the primitive streak cells ingress and disappear, creating a sink in mitotic pressure. This would inevitably lead to the formation of holes in the upper layer if morphogenetic movements did not fill the gap. Cells in the upper layer are indeed attracted to the primitive streak, as is to be seen in the film of chick experiments where induction of a secondary primitive streak in the proamnion aspires part of the neural lip of the host (Vakaet and Bortier, in press). Mitotic pressure could also explain bending of the neural plate (Schoenwolf et al. 1989). At the primitive streak cells ingress and disappear. This is not possible at the neural plate as there is a continuous laminin layer under the upper layer at neurulation (Bortier et al. 1989). The only way out for the upper layer is to bend.

REFERENCES

Bortier, H., De Bruyne, G., Espeel, M., and Vakaet, L., 1989, Immunohistochemistry of laminin in early chicken and quail blastoderms, *Anat.Embryol.* 180:65-69.

Bortier, H., 1991, Chicken organogenesis, *in*: "A dozen eggs: Time-lapse microscopy of normal development", 28-29, R. Fink, ed., Sinauer Associates, Sunderland.

Bortier, H., and Vakaet, L.C.A., 1992a, Mesoblast anlage fields in the upper layer of the chicken blastoderm at stage 5V, *in*: "Formation and Differentiation of Early Embryonic Mesoderm", NATO ASI Series, Series A: Life Sciences, Vol. 231:1-7, R. Bellairs, E.J. Sanders, J.W. Lash, eds., Plenum Press, New York.

Bortier, H., and Vakaet, L.C.A., 1992b, Fate mapping the neural plate and the intraembryonic mesoblast in the upper layer of the chicken blastoderm with xenografting and time-lapse videography, *in*: "Gastrulation", Development 1992 Suppl. 93-97, C. Stern and P.W. Ingham, eds.

Bortier, H., Vandevelde, S., and Vakaet, L.C.A., 1993, Mechanism of closure of experimental excision-wounds in the bare upper layer of the chick blastoderm, *Int. J. Dev. Biol.* 37:459-466.

Krstic, R.V., 1984, Illustrated Encyclopedia of Human Histology, Springer-Verlag, Berlin, Heidelberg, New York, Tokyo.

Le Douarin, N. M., 1973, A Feulgen-positive nucleolus, *Exp. Cell Res.* 77:459-468.

New, D.A.T., 1955, A new technique for the cultivation of the chick embryo in vitro, *J. Embryol. Exp. Morphol.* 3/4:326-331.

Sauer, F.C., 1936, The interkinetic migration of embryonic epithelial nuclei, *J. Morphol.* 60: 1-11.

Schoenwolf, G.C., Bortier, H., Vakaet, L., 1989, Fate mapping the avian neural plate with quail/chick chimeras: origin of prospective median wedge cells, *J. Exp. Zool.* 249: 271-278.

Spratt , N.T., 1952, Localization of the prospective neural plate in the early chick blastoderm, *J. Exp. Zool.* 120: 109-130.

Vakaet, L., 1970, Cinephotomicrographic investigations of gastrulation in the chick blastoderm, *Arch. Biol. (Liége)* 81/3: 387-426.

Vakaet, L.C.A., and Bortier, H., in press, Mapping of gastrulation movements in birds, *in*: "Organization of the early vertebrate embryo", NATO ASI Series, Series A: Life Sciences, N. Zagris et al. , eds., Plenum Press, New York.

HENSEN'S NODE: THE AMNIOTE EQUIVALENT
OF SPEMANN'S ORGANIZER

Claudio D. Stern

Department of Genetics & Development
College of Physicians and Surgeons of Columbia University
701 West 168th Street
New York, NY 10032
U.S.A.

INTRODUCTION

The dorsal lip of the blastopore of the gastrulating amphibian embryo is a unique region which, when transplanted to an ectopic position in a host embryo, induces the formation of a secondary embryonic axis, organised along the rostrocaudal, dorsoventral and mediolateral axes (Spemann & Mangold, 1924). Cells of the ectoderm of the host adjacent to the graft undergo a change in fate, from epidermal to neural. This property of the amphibian dorsal lip is now widely referred to as "Spemann's organizer". In the early 1930's, C.H. Waddington demonstrated that amniote embryos (ducks, chicks and rabbits) could also be made to generate a second nervous system by transplanting the tip of the primitive streak, a region known as Hensen's node (Waddington, 1932; 1933).

In amphibians, the dorsal lip is characterised not only by its ability to induce a second axis but also by the fates of its cells during subsequent development and by the localised expression of several molecular markers. Here, some of our present knowledge on these properties of Hensen's node will be surveyed.

CELL FATES IN THE ORGANIZER

Recently, it has been possible to map the descendants of cells in Hensen's node of the gastrulating chick embryo, using carbocyanine dyes (DiI, DiO) to mark small groups of cells or lysinated rhodamine-dextran to follow the progeny of single cells (Selleck & Stern 1991). The study confirmed earlier findings (e.g. Spratt, 1955; Rosenquist, 1966; 1983) that the node contains cells whose progeny contribute to several specific tissues: notochord/head process, prechordal plate, somites, definitive (gut) endoderm and the ventral portion of the neural tube. These tissues are similar to those derived from the dorsal lip of the blastopore of amphibian embryos (see Hadorn, 1970; Purcell & Keller, 1993) and the node of mammalian embryos (see Beddington, 1994; Thery et al. 1994).

One unexpected finding from the study of Selleck & Stern (1991), however, was that the node contains cells that contribute only to the *medial* halves of the somites. Their lateral halves were found to come from progenitors located further posteriorly in the primitive streak. At the time, there was no obvious reason to expect this on functional or embryological grounds. But an independent study by Ordahl & Le Douarin (1992) reported that the medial halves of somites normally contribute only to the axial musculature, while their lateral halves contribute to muscles of the body wall and limbs.

Prospective notochord cells are situated in the anterior, median quadrant of the node (Selleck & Stern, 1991). Progenitors of the gut endoderm and of the floor plate of the neural tube are also present in the node, but more widely distributed. The intermediate region of the node, situated between the medial somite precursors and the prospective notochord area, contains cells that contribute to **both** notochord and somites, as revealed by analysis of the descendants of single, marked cells (Selleck & Stern 1991). These are therefore **multipotent precursor cells**. It was suggested (Selleck & Stern 1992a, b; Stern et al. 1992) that these multipotent precursors give rise to **committed** progenitors of medial half somite and notochord, situated in the adjacent regions. These findings beg the question: do the pluripotent precursor cells in the intermediate region of the node have *stem cell* properties, that is, are they able to renew themselves? Such a mechanism would ensure the maintenance of a progenitor population of constant size, which would be responsible for generating the large number of cells required to make notochord and somites spanning the entire length of the embryonic axis.

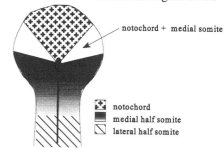

Fig. 1 Fate map of Hensen's node.

Further Evidence for a Resident Population of Stem Cells in Hensen's Node

When chick embryos are given a single, short heat shock, discrete anomalies of somite development are seen at regular intervals along the axis of the embryo, appearing every 6-8 segments (Primmett et al. 1988). The anomalies consist of an abnormal (either large or small) number of cells being allocated to the defective somites. Because somite formation occurs sequentially, this suggests that heat shock affects some *repeated, cyclic* process. A pair of somites forms every 100 minutes or so (Menkes & Sandor 1969), therefore the time interval between these anomalies corresponds to groups of cells that are about 10 hours apart. An obvious candidate for the repeated process is the cell division cycle. We measured it by ^{3}H-thymidine pulse and chase, which confirmed that presumptive somite cells divide every 10 hours (Primmett et al. 1989). These findings suggest that cells that segment at the same time as each other divide relatively synchronously. This is consistent with measurements of the mitotic index of presumptive somite cells in the segmental plate (Stern & Bellairs 1984): a large peak of cells in the M-phase of the cycle is seen just before segmentation (at the anterior tip of the segmental plate), another in the middle and a third one at the posterior end of the plates. Since the segmental plate contains 13 presumptive somites (Jacobson & Meier 1986), these peaks of mitosis are separated by about 7 somites.

Thus, the cell cycle is probably involved somehow in punctuating the continuous stream of cells into a discrete pattern of somites (Stern et al. 1988; Primmett et al. 1989). Further evidence for this comes from the single cell lineage analysis of presumptive somite cells at earlier stages of development, in Hensen's node (Selleck & Stern 1991; 1992b and see above). When a single cell in the presumptive medial half somite region of the node is marked by intracellular injection of lysinated rhodamine-dextran (LRD), its progeny appears clustered in small groups, separated by about 5-8 somites. Interestingly, when the injected cell is in a region that contributes to the notochord, repeated clusters of labelled cells are also seen, but these are much closer together: 1.5-2 somite-lengths apart. This suggests that the notochord precursors divide about 2.5 times faster than the somite progenitors, and predicts a doubling time of about 3.5-4 hours.

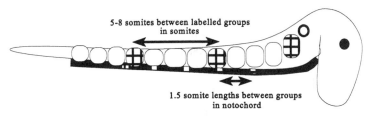

Fig. 2 Distribution of descendants of a single marked
Hensen's node progenitor cell.

Since, as discussed above, there are cells in the intermediate region of the node that contribute to both notochord and somite, then it is expected that they will divide also at the faster rate of 3.5-4 hours, and that their descendants slow down to a cycle time of about 10 hours if they become committed to a somite fate. The above findings can therefore be interpreted according to the model illustrated below (see also Stern et al. 1992). This model suggests that: the intermediate region of the node contains multipotent stem cells (MSC), whose progeny can give rise to founder cells for notochord (FC_n) and founder cells for the medial halves of the somites (FC_s). These founder cells also have stem cell properties: at each cell division of a founder cell, one daughter would leave the node region whilst the other daughter would remain in the node until the next division, the process repeating itself thereafter.

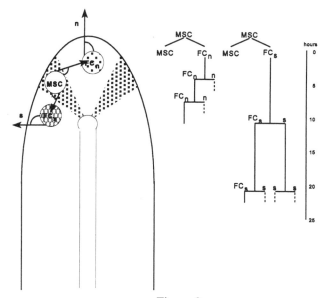

Figure 3

SOME CELLS MAY ACQUIRE MESODERMAL FATES
AT RELATIVELY LATE STAGES OF DEVELOPMENT

The proposed existence of *pluripotent precursor cells* in Hensen's node is important because it indicates that individual cells at the *end* of the primitive streak stage still contribute progeny to both ectodermal (floor plate, neural tube) and mesendodermal (somites, notochord/head process, gut endoderm) derivatives (Selleck & Stern 1991). How do cells decide between these fates? From studies in amphibians, it is generally believed that responsiveness to mesodermal induction is lost at the *beginning* of gastrulation (see Green & Smith, 1991 for review). Hensen's node of the amniote embryo is therefore either different from *Xenopus*, or the appropriate region of the frog (dorsal lip of the blastopore) has not been adequately explored.

In amphibians, there is good evidence for the involvement of the peptide growth factor activin in mesodermal induction (see Green & Smith, 1991). Recent results (Yu, Umesono, Carveth, Evans, Vale, Kintner and Stern, in preparation) suggest that activin could also play a role in allocating cells to mesodermal fates during the later phases of gastrulation in the amniote embryo. We (Yu et al., in preparation) have cloned two activin receptors, homologous to the amphibian ActRIIA and IIB. Chick transcripts of ActRIIB are first expressed when the primitive streak appears, and ActRIIA is expressed even later, in Hensen's node of the full-length streak; neither is detected before the start of gastrulation. Moreover, when different tissues are treated with activin and allowed to differentiate in culture, the primitive streak is found to be the most responsive tissue, giving rise to all mesodermal derivatives, including the most axial/dorsal types, in a concentration-dependent manner, as has been found in the frog blastula. These results argue that the role of activin in normal development may extend to rather late stages of mesoderm formation. Consistent with this proposal is the finding (Mitrani et al. 1990) that transcripts of both activin-A and -B can be detected by Northern analysis of chick embryos at the primitive streak stage. Thus, Hensen's node appears to be a site where "mesoderm induction" takes place at advanced primitive streak stages.

A MOLECULAR MARKER FOR THE ORGANIZER:
THE HOMEOBOX GENE *GOOSECOID*

The evidence that the amniote equivalent of the amphibian dorsal lip is Hensen's node came initially from Waddington's transplantation experiments, which demonstrated the inducing properties of the node. Now there are also several molecular markers expressed in the dorsal lip of *Xenopus* and in the node of chick and mouse embryos. One of these is the homeobox gene *goosecoid* (Blumberg et al. 1991; Blum et al. 1992; Izpisúa-Belmonte et al. 1993). Its name is derived from the two *Drosophila* genes with which it shares sequence homology: the segmentation gene *gooseberry* and the maternal gene *bicoid*. With the latter, *goosecoid* shares a Lysine residue at position-50 of the homeodomain, which confers both genes with their DNA binding sequence specificity. *Goosecoid* expression appears to be involved in defining the organizer property, because ectopic expression in the ventral side of a frog embryo can elicit the formation of a partial ectopic axis, including notochord, a second neural tube and ectopic ear vesicles (see Izpisúa-Belmonte et al. 1993). Furthermore, it has recently been shown that overexpression of *goosecoid* stimulates specific patterns of cell movement (Niehrs et al. 1993), suggesting that it is involved in defining the behaviours of specific cell groups during gastrulation. And activin, which stimulates the cell movements of convergence and extension that accompany gastrulation, induces the expression of *goosecoid* (Blumberg et al. 1991; Smith et al. 1991; Blum et al. 1992).

During primitive streak stages in the chick embryo, *goosecoid* mRNA is strongly expressed in Hensen's node, from where it disappears as the head process starts to form (stage 4[+] of Hamburger & Hamilton, 1955). But transcripts are first detectable much earlier, from stage XI of Eyal-Giladi & Kochav (1976), before primitive streak formation. At this stage, there are only a few cells expressing the gene, situated in the sparse middle layer that forms at the posterior end of the blastodisc between epiblast and hypoblast, closely associated with Koller's sickle. Do these early-expressing cells contribute to the later population of *goosecoid*-expressing cells in Hensen's node?

EMBRYONIC ORIGIN OF THE ORGANIZER: MULTIPLE SOURCES OF CELLS

To answer the above question, the carbocyanine dye DiI was used to label small groups of these early middle layer cells in stage XI chick embryos; the embryos were then incubated to stage 4, when *goosecoid* expression is confined to the node, and then the embryos were fixed. After this, the dye was photoconverted into an insoluble product that could withstand the subsequent *in situ* hybridization to reveal *goosecoid* mRNA (Izpisúa-Belmonte et al. 1993). The results of this experiment showed that the early middle layer cells do indeed contribute to the node at the definitive streak stage, and are subsequently found mainly in the definitive (gut) endoderm. However, the number of cells found initially in the middle layer is not sufficient to account for all the cells that express the gene at the later stage, suggesting that recruitment of cells into the node region also occurs during gastrulation.

To examine whether regions containing *goosecoid*-expressing cells can induce other neighbouring cells also to express the gene, quail middle layer cells and adjacent tissues, or the tip of a quail primitive streak, were transplanted to an ectopic position in a young chick embryo. The chimaera was incubated again to the definitive streak stage, and processed by *in situ* hybridization with a chick-specific *goosecoid* probe (Izpisúa-Belmonte et al. 1993). In addition to the normal expression of *goosecoid* in the node of the host embryo, there was also a second, ectopic site of expression, containing chick (host) cells that expressed the gene. This result suggests that during primitive streak formation, some cells that do not express *goosecoid* initially could be recruited by the elongating primitive streak to form part of the node.

This conclusion is supported further by recent fate maps of the pre-primitive streak embryo (Hatada & Stern, 1994). Just prior to primitive streak formation (stages XIII-XIV), the regions contributing to the notochord/head process, somites and gut endoderm (all derived from the node) are situated in the epiblast, near the middle of the blastoderm. Therefore, the node is largely derived from epiblast cells which are in this position before the tip of the primitive streak reaches it, but also contains a small number of cells derived from the middle layer of the posterior margin of the blastoderm before streak formation, and which may migrate along with the tip of the elongating primitive streak.

"SPEMANN'S ORGANIZER" AND HENSEN'S NODE: NEURAL INDUCTION

By transplanting Hensen's node between rabbit, chick and duck embryos, Waddington (1932, 1933) showed that at least some inducing signals from the node can act across species. In more recent experiments, it was shown that a chick Hensen's node is able to induce and to pattern even amphibian animal caps when sandwiched between them (Kintner and Dodd, 1991). Similarly, Blum et al. (1992) showed that when the tip of the mouse embryo, which contains the organizer region (Beddington, 1994) is grafted into the blastocoel cavity of a *Xenopus* embryo, a second axis is generated in the frog host.

One difference between amphibians and avian embryos is that the latter also contain a region of epiblast/ectoderm that lies **outside** of the normal fate map, and which only contributes cells to the **extra**embryonic membranes. This region is called the *area opaca*, and, importantly, it has been shown (Gallera, 1966; Storey et al. 1992) that it can respond to a graft of Hensen's node by producing a complete nervous system, expressing regional markers.

Therefore, the *area opaca* epiblast can be considered as a collection of cells which, from the neural point of view, are "naïve".

In avian embryos, it is possible to use the carbohydrate epitope L5 (Streit et al. 1990; Roberts et al. 1991) as a marker for early events during neural induction. As soon as the neural plate forms, immunoreactivity with the L5 antibody is restricted to this region. This specificity persists to later stages, when the neural tube has formed. However, prior to the appearance of a visible neural plate, during gastrulation, the epitope recognised by this antibody is distributed more widely, starting from the early primitive streak stage (Roberts et al. 1991). At this stage, the antibody decorates the region that is competent to respond to a graft of Hensen's node. To investigate whether the L5 epitope is involved in the response of competent epiblast to such a graft, Roberts et al. (1991) grafted a Hensen's node together with hybridoma cells secreting the antibody. They found that in the presence of the hybridoma cells, neural induction is inhibited. Thus, the L5 antigen, or the molecules carrying it, may be required by epiblast cells in the response to signals involved in neural induction. Moreover, the gradual restriction of immunoreactivity to more and more central regions of the embryo mimics a similar restriction in competence of the epiblast to respond to a graft of Hensen's node (Storey et al. 1992). Therefore the L5 epitope may be a marker of neural competence.

Hepatocyte Growth Factor Scatter Factor and its Role in Neural Induction

When human or mouse cells secreting the peptide Hepatocyte Growth Factor (also known as Scatter Factor, or HGF/SF; Stoker and Perryman, 1985) are grafted into an early chick embryo, an ectopic neural plate is sometimes seen to form in the vicinity of the graft (Stern et al. 1990). This led to the suggestion that HGF/SF may have neural inducing properties. In culture, treatment of extraembryonic (*area opaca*) epiblast explants with recombinant human or mouse HGF/SF sometimes causes the differentiation of neurones (Streit et al., in preparation).

Such explants, after overnight culture in the absence of factors, lose their expression of the L5 epitope. However, when cultured with HGF/SF concentrations as low as 100 pM, strong L5 immunoreactivity is seen throughout the explant after 24 hours (Streit et al., in preparation). This finding suggests that HGF/SF may act by either enhancing or maintaining the competence of the epiblast to respond to neural inducing signals.

Since L5 immunoreactivity and competence gradually become condensed to progressively more central regions of the epiblast during development, it seems likely that Hensen's node produces a factor that delays the loss of competence from adjacent regions of the epiblast in a concentration-dependent way. Could this factor be HGF/SF? Consistent with this possibility, *in situ* hybridization experiments reveal that the node is the only region of the primitive streak stage embryo that expresses mRNA encoding this factor (Streit et al., in preparation). Further experiments will be required to test this hypothesis more directly, but these considerations suggest that one of the functions of Hensen's node may be the maintenance of the competence of neighbouring epiblast to respond to neural inducing signals.

Multiple Steps in Neural Induction:
Neuralization and Regionalization are Separable Events

If the function of HGF/SF in early neural development is the induction or maintenance of neural competence in the epiblast, this suggests that neural induction comprises many different stages and/or signals that cooperate to produce a complex response: the development of a patterned nervous system. Further evidence that neural induction is complex and that it involves several steps comes from experiments from Martínez et al. (1990) and others, who showed that transplantation of metencephalic neuroepithelium into the diencephalon of late (stage 10) embryos elicits the ectopic expression of a metencephalic marker, *engrailed-2* and subsequent development of ectopic cerebellar traits in the host neuroepithelium. By stage 10, the neural tube is already closed in this region. This suggests that many hours after the node has lost its ability for neural induction, the induced neuroepithelium is still responsive to signals that can provide rostrocaudal positional information ("regionalization"; Storey et al. 1992).

Interspecies transplantation of the organizer region provides further support for this notion. Avian nodes lose their ability to induce an ectopic nervous system between stages 4 and 5 when transplanted to the *area opaca* of a competent host embryo (stage 3+) (Storey et al. 1992). However, when a chick node is sandwiched between two *Xenopus* animal cap explants, it retains its ability to induce the expression of frog neural and region-specific markers until much later stages (at least stage 7) (Kintner & Dodd, 1991). This finding suggests that the response of *Xenopus* animal caps leading to the expression of these markers is less complex than that of the chick epiblast, and therefore that the animal cap ectoderm may already have received some neural inducing signals by the stages at wich it is isolated in these experiments (mid-gastrula; stage 10). It also suggests that the node continues to emit signals, perhaps concerned with regionalization or with the maintenance of the neural phenotype, long after it has lost its ability to induce a new nervous system from truly naïve (extraembryonic) epiblast. Interestingly, the stage-dependence seen in the interspecies combinations with *Xenopus* animal caps is mimicked by a similar stage-dependence in the ability of the chick node to cause duplication of digits when transplanted to the anterior margin of a chick wing bud (Hornbruch & Wolpert, 1986; Stocker & Carlson, 1990): nodes as old as stages 8-9 still retain this ability.

All these speculations seem to point to the conclusion that Hensen's node is the source of several different signals. Some of these are restricted to early stages, such as the ability to maintain expression of L5, the competence of the epiblast to respond to later signals, the expression of HGF/SF and the ability to induce a new ectopic neural plate in extraembryonic epiblast. Others, like the process of "regionalization", continue to much later stages, such as the ability of the node to induce expression of neural and regional markers in frog animal caps and to cause duplication of digits after transplantation to the anterior margin of a chick limb bud.

SUMMARY AND CONCLUSIONS

Hensen's node is probably the most important region of the amniote embryo during gastrulation. Not only is it the only region of the embryo at this stage that can induce the formation of a fully patterned secondary axis, but it is also the source of the most axial tissues of the embryo: notochord/head process, the medial halves of the somites and the gut endoderm. Several genes, including the homeobox gene *goosecoid*, are expressed specifically in this region. This gene has been a useful marker for studies of the embryonic origin of the cells that make up the node, which suggested that two or more distinct populations of cells contribute to its formation. One peptide factor expressed in the node is the kringle-containing, plasminogen-related Hepatocyte Growth Factor/Scatter Factor (HGF/SF). This factor induces the expression of a marker for neural competence; its role may be the maintenance of responsiveness of epiblast cells to neural inducing signals. This and other findings suggest that the role of Hensen's node in neural induction is to emit several different signals, including some that maintain neural competence, others that effect the change of fate from epidermal to neural, and perhaps others that maintain the neural fate.

REFERENCES

Beddington, R.S.P. (1994) Induction of a second axis by the mouse node. *Development* **120**: 613-620.

Blum, M., Gaunt, S.J., Cho, K., Steinbeisser, H., Blumberg, B., Bittner, D. & De Robertis, E.M. (1992) Gastrulation in the mouse: the role of the homeobox gene *goosecoid*. *Cell* **69**: 1097-1106.

Blumberg, B., Wright, C.V.E., De Robertis, E.M. & Cho, K.W. (1991) Organizer-specific homeobox genes in Xenopus laevis embryos. *Science* **253**: 194-196.

Eyal-Giladi, H. & Kochav, S. (1976) From cleavage to primitive streak formation: A complementary normal table and a new look at the first stages of the development of the chick. *Devl Biol.* **49**: 321-337.

Gallera, J. (1966) Le pouvoir inducteur de la chorde et du mésoblaste parachordal chez les oiseaux en fonction du facteur "temps". *Acta Anat.* (Basel) **63**: 388-397.

Green, J.B.A. & Smith, J.C. (1991) Growth factors as morphogens: do gradients and thresholds establish body plan? *Trends Genet.* **7**: 245-250.

Hadorn, E. (1970) `Experimentelle Entwicklungsforschung'. Berlin: Springer Verlag.

Hamburger, V. & Hamilton, H.L. (1951) A series of normal stages in the development of the chick. *J. Morph.* **88**: 49-92.

Hatada, Y. & Stern, C.D. (1994) A fate map of the epiblast of the early chick embryo. *Development* (in press).

Hornbruch, A. & Wolpert, L. (1986) Positional signalling by Hensen's node when grafted to the chick limb bud. *J. Embryol. exp. Morph.* **94**: 257-265.

Izpisúa-Belmonte, J.C., De Robertis, E.M., Storey, K.G. & Stern, C.D. (1993) The homeobox gene *goosecoid* and the origin of the organizer cells in the early chick blastoderm. *Cell* **74**: 645-659.

Jacobson, A. & Meier, S. (1986) Somitomeres: the primordial body segments. **In:** *Somites in Developing Embryos*. (ed. Bellairs, R., Ede, D.A. & Lash, J.W.). Plenum Press. pp. 1-16.

Kintner, C.R. & Dodd, J. (1991) Hensen's node induces neural tissue in Xenopus ectoderm. Implications for the action of the organizer in neural induction. *Development* **113**: 1495-1506.

Martínez, S., Wassef, M. & Alvarado-Mallart, R.-M. (1991) Induction of a mesencephalic phenotype in the 2-day-old chick prosencephalon is preceded by the early expression of the homeobox gene en. *Neuron* **6**: 971-981.

Menkes, B. & Sandor, S. (1969) Researches on the development of axial organs. *Rev. Roum. Embryol. Cytol.* **6**: 65-88.

Mitrani, E., Ziv, T., Thomsen, G., Shimoni, Y., Melton, D.A. & Bril, A. (1990) Activin can induce the formation of axial structures and is expressed in the hypoblast of the chick. *Cell* **63**: 495-501.

Niehrs, C., Keller, R., Cho, K.W.Y. & De Robertis, E.M. (1993) The homeobox gene goosecoid controls cell migration in Xenopus embryos. *Cell* **72:** 491-503.

Ordahl, C.P. & Le Douarin, N.M. (1992) Two myogenic lineages within the developing somite. *Development* **114**: 339-353.

Primmett, D.R.N., Stern, C.D. & Keynes, R.J. (1988) Heat-shock causes repeated segmental anomalies in the chick embryo. *Development* **104**: 331-339.

Primmett, D.R.N., Norris, W.E., Carlson, G.J., Keynes, R.J. & Stern, C.D. (1989) Periodic segmental anomalies induced by heat-shock in the chick embryo are associated with the cell cycle. *Development* **105:** 119-130.

Purcell, S.M. & Keller, R.A. (1993) A different type of amphibian mesoderm morphogenesis in Ceratophrys ornata. *Development* **117**: 307-317.

Roberts, C., Platt, N., Streit, A., Schachner, M. & Stern, C.D. (1991) The L5 epitope: an early maker for neural induction in the chick embryo and its involvement in inductive interactions. *Development* **112**: 959-970.

Rosenquist, G.C. (1966) A radioautographic study of labeled grafts in the chick blastoderm. Development from primitive streak stages to stage 12. *Contrib. Embryol. Carnegie Inst. Washington* **38**: 71-110.

Rosenquist, G.C. (1983) The chorda center in Hensen's node of the chick embryo. *Anat. Rec.* **207**: 349-355.

Selleck, M.A.J. & Stern, C.D. (1991) Fate mapping and cell lineage analysis of Hensen's node in the chick embryo. *Development* **112**: 615-626.

Selleck, M.A.J. & Stern, C.D. (1992a Commitment of mesoderm cells in Hensen's node of the chick embryo to notochord and somite. *Development* **114**: 403-415.

Selleck, M.A.J. & Stern, C.D. (1992b Evidence for stem cells in the mesoderm of Hensen's node and their role in embryonic pattern formation. **In:** *Formation and differentiation of early embryonic mesoderm.* (ed. Lash, J.W., Bellairs, R. and Sanders, E.J.). New York: Plenum Press. pp. 23-31.

Smith, J.C., Price, B.M.J., Green, J.B.A., Weigel, D. & Herrmann, B.G. (1991) Expression of a Xenopus homolog of Brachyury (T) is an immediate-early response to mesoderm induction. *Cell* **67**: 79-87.

Spemann, H. & Mangold, H. (1924) Über Induktion von Embryonanlagen durch Implantation artfremder Organisatoren. *Wilh. Roux Arch. EntwMech. Organ.* **100**: 599-638.

Spratt, N.T. Jr. (1955) Analysis of the organizer center in the early chick embryo. I. Localization of prospective notochord and somite cells. *J. exp. Zool.* **128**: 121-163.

Stern, C.D. & Bellairs, R. (1984) Mitotic activity during somite segmentation in the chick embryo. *Anat. Embryol.* **169**: 97-102.

Stern, C.D., Fraser, S.E., Keynes, R.J. & Primmett, D.R.N. (1988) A cell lineage analysis of segmentation in the chick embryo. *Development* **104 Suppl.:** 231-244.

Stern, C.D., Ireland, G.W., Herrick, S.E., Gherardi, E., Gray, J., Perryman, M. & Stoker, M. (1990) Epithelial scatter factor and development of the chick embryonic axis. *Development* **110**: 1271-1284.

Stern, C.D., Hatada, Y., Selleck, M.A.J. & Storey, K.G. (1992) Relationships between mesoderm induction and the embryonic axes in chick and frog embryos. *Development* **1992 Suppl.:** 151-156.

Stocker, K.M. & Carlson, B.M. (1990) Hensen's node, but not other biological signallers, can induce supernumerary digits in the developing chick limb bud. *Roux's Arch. Dev. Biol.* **198**: 371-381.

Storey, K.G., Crossley, J.M., De Robertis, E.M., Norris, W.E. & Stern, C.D. (1992) Neural induction and regionalisation in the chick embryo. *Development* **114**: 729-741.

Streit, A., Faissner, A., Gehrig, B. & Schachner, M. (1990) isolation and biochemical characterization of a neural proteoglycan expressing the L5 carbohydrate epitope. *J. Neurochem.* **55**: 1494-1506.

Streit, A., Thery, C. & Stern, C.D. (1994) Of mice and frogs. *Trends Genet.* **10**: 181-183.

Waddington, C.H. (1932) Experiments on the development of chick and duck embryos, cultivated in vitro. *Phil. Trans. R. Soc. Lond. B* **221**: 179-230.

Waddington, C.H. (1933) Induction by the primitive streak and its derivatives in the chick. *J. Exp. Biol.* **10**: 38-46.

THE EXTRACELLULAR MATRIX IN DEVELOPMENT

Albert E. Chung

Department of Biological Sciences
University of Pittsburgh
Pittsburgh, Pennsylvania 15260
U. S. A.

INTRODUCTION

The extracellular matrix is required for embryogenesis and development in complex organisms. It provides a scaffolding for the organization of cells into tissues, a substrate for cell attachment and adhesion, a pathway for cell movement and migration, neuronal guidance, and maintenance of differentiated functions. The multitude of functions derives from the diverse nature of the chemical and physical properties of the constituent molecules that comprise the extracellular matrix, as well as their propensity to form macromolecular complexes with distinctive properties. The extracellular matrix is elaborated by the co-operative biosynthetic activities or one or more cell types. The biosynthesis of the molecules in the extracellular matrix is spatially and temporally regulated to fulfill the physiological requirements of the developing organism. Concomitant with the generation of the extracellular matrix, cells express plasma membrane integrin receptors which bind to cognate epitopes in the nascent matrix. The engagement of these receptors transmits signals to the cells which modify their behavior in the appropriate manner to promote migration, polarization, specific gene expression, adhesion, proliferation, or shape modulation.

The focus of this discussion will be on the basal lamina which is a specialized form of the extracellular matrix that is in intimate contact with the plasma membrane of epithelial, endothelial, muscle, nerve and fat cells. The major components of the basal lamina are collagen IV, laminin, entactin/nidogen, and the proteoglycan, perlecan (Timpl, 1989). Fibronectin (Ruoslahti, 1988) and tenascin (Chiquet-Ehrismann et al., 1986) are sometimes also associated with the basal lamina. These molecules contain amino acid sequences that are recognized by one or more members of the integrin family of receptors. The amino acid recognition sequences that are accessible will depend on the organization of the macromolecular complex that is present during development. The combination of extracellular matrix assembly and integrin receptor expression is therefore an important aspect of development. The biochemical properties of laminin and entactin that implicate them as important molecules in development will be summarized as well as promising directions for elucidating the *in vivo* roles of individual extracellular matrix molecules.

Organization of the Early Vertebrate Embryo
Edited by N. Zagris *et al.*, Plenum Press, New York, 1995

LAMININ

Structure

Laminin is now recognized to be a family of molecules, each member of which consists of three polypeptide chains linked by disulfide bonds to form a cross-shaped structure with globular termini (Engel, 1992). The prototypical form of laminin isolated from mouse parietal endoderm cells (Chung et al., 1979) and EHS tumors (Timpl et al., 1979, Engel et al., 1981), as illustrated in Figure 1, is composed of three glycosylated polypeptide chains, a 400 kDa A or alpha-1 chain, a 210 kDa or beta-1 chain, and a 210 kDa B2 or gamma-1 chain. Isoforms of the alpha (Ohno et al., 1986; Leivo and Engvall, 1988; Ehrig et al., 1990), beta (Hunter et al., 1989; Gerecke et al.,1994) and gamma (Kallunki et al., 1992) chains of laminin have been identified and classified according to their sequence identity with the prototype chains whose sequences have been determined for murine (Sasaki et al. 1987; Sasaki and Yamada, 1987; Sasaki et al., 1988; Durkin et al., 1988), *Drosophila* (Montell and Goodman, 1988; Chi and Hui, 1989; Montell and Goodman, 1989; Garrison et al., 1991; Kusche-Gullberg et al., 1992; Henchcliffe et al., 1993) and human laminin (Pikkarainen et al., 1987; Pikkarainen et al., 1988; Nissinen et al., 1991). The domain structure of laminin shown in Figure 1 was derived from physical and biochemical data. The isoforms represent variations of this basic theme in which one or more domains have been deleted. The rigid long arm, consisting largely of domains I and II, has a coiled-coil structure formed from the three chains. This region is rich in alpha-helices in which the amino acids are arranged in heptad repeats with hydrophobic amino acids located in the first and fourth positions of the repeat, favoring the formation of coiled-coil structures. The rod-like segments of the short arms, domains III and V, consist of EGF-like cysteine homology repeats, each containing 50-60 amino acids with 8 cysteine residues in highly conserved positions. The globular structures seen in the electron microscope are domains IV and VI. The terminal globule at the end of the long arm, derived exclusively from the carboxyl end of the A chain, contains five homology repeats shown in the diagram as a foot-like structure and is referred to as the G domain. Domains III and IV in the alpha 1 chain are split into subdomains, a and b, based on their amino acid sequences. The beta 1 chain contains a characteristic 40 amino acid loop between domains I and II, referred to as domain alpha. In mouse laminin, the three short arms are derived individually from the N-terminal regions of the alpha-1, beta-1 and gamma-1 chains.

The laminin molecule is extensively glycosylated with 15-30 % of its weight consisting of carbohydrate. There are 46 potential N-glycosylation sites on the alpha-1 chain and 13 and 14 on the beta-1 and beta-2 chains respectively. Many of these sites are located on the long arm of the molecule. The carbohydrate structure is extremely complex containing bi- and triantennary chains, the blood group I structure, and repeating sequences of 3 Gal beta-1,4GlcNacbeta-1 units (Knibbs et al., 1989). Although the basic structure of the intact molecule is conserved, there are substantial variations in the amino acid sequences among species, the differences being most pronounced at the carboxyl ends of the chains.

Variant forms of laminin are composed of one each of the alpha, beta and gamma isoform chains. These variant laminin molecules are usually restricted to specific tissues and may be expressed transiently (Leivo and Engvall, 1988; Engvall et al., 1990) suggesting that they play distinctive roles in development and organogenesis. At present there is no evidence for the presence of molecules with less than three chains.

Biological activities

During development cells attach to the substratum, migrate in a directional pattern, change shape, associate to form tissues and organs and express specific genetic programs. The evidence for the role of laminin in these activities is substantial but the mechanisms are not yet completely understood.

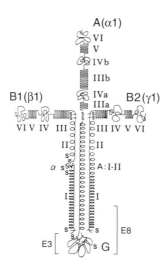

Figure 1. Domain structure of laminin. (Sasaki et al., 1988, with permission).

The first demonstrated biological activities of laminin were the enhancement of cell attachment to collagen substrates and the promotion of neurite extension (reviewed by Kleinman et al., 1985). These observations led to the investigation of the role of laminin in development and its mechanism of action. The expression of the laminin genes is regulated in early embryogenesis. The β1 chain of laminin can be detected in the mouse oocyte, and by the four cell stage of the preimplantation embryo both the beta-1 and gamma-1 chains are present, all three chains are coordinately synthesized beginning with the eight cell embryo (Cooper and MacQueen, 1983). Subsequent to implantation, laminin is localized to all morphologically distinct basement membranes.

The participation of laminin in morphogenesis and tissue development has been explored in mouse, *Drosophila*, chick and amphibians. In the mouse, the kidney and the brain and to a lesser extent the eye have served as models for development. The major evidence for the role of laminin is the correlation of morphological changes with the appearance of laminin subunits and their messenger RNAs. The distribution of laminin in the kidney of mouse embryos and in an *in vitro* transfilter model system revealed a temporal and spatial correlation with the early stages of tubule formation (Ekblom et al., 1980; Klein et al., 1988; Laurie et al., 1989; Ekblom et al., 1990). Laminin appeared in punctate deposits in areas of formation of pretubular aggregates and was later integrated into the tubular basement membrane. In the *in vitro* transfilter model system the appearance of laminin preceded morphogenesis when the nephrogenic mesenchyme was exposed to inducer tissue. These results suggested that laminin was involved in the aggregation of induced cells in the early stages of tubule formation. In an interesting set of experiments Holm et al. (1988), using organ cultures of 11-day embryonic metanephric kidneys grafted on chicken chorioallantoic membranes, were able to demonstrate with mouse specific antibodies that the beta-1 and gamma-1 chains were synthesized prior to the alpha-1 chain, which appeared only with tubular differentiation. Klein et al. (1988) in a more detailed analysis of the transfilter induction of metanephric mesenchyme obtained similar results. It was proposed that the alpha-1 chain was required for development of epithelial cell polarity, presumably by allowing the assembly of the appropriate laminin isoform. The development of epithelial polarity was inhibited by antibodies directed against the terminal domains of the long arm of laminin suggesting that this region was responsible for interacting with the induced cells to generate polarity. The mechanism by which the polarity is established is not known but the alpha-6 beta-1 and alpha-3 beta-1 integrin receptors for the E8 domain of laminin are probably involved.

Laminin coated substrates promote the dramatic elongation of neurites from both peripheral and central neurons (reviewed by Sanes, 1989). In an *in vitro* system, using time-lapse video microscopy, it was shown that laminin coated pathways were able to guide growth cones of peripheral and central nervous system-derived neurites (Hammarback et al., 1988). The guidance was correlated with the adhesive properties of the substratum. However, the role of laminin in neuronal guidance *in vivo* remains unclear. The spatial and temporal distribution of laminin antigens in the developing brain suggest that it is important (Letourneau et al., 1988). Laminin antibodies were found to stain the external limiting membrane of the neural tube at Day 8 of mouse embryogenesis and at Day 9, punctate staining could be seen on the surfaces of neuroepithelial cells in the marginal zone through which the axons of the ventral longitudinal pathway extend. This staining persisted through E14 but was reduced at E16. The results indicated that laminin deposition preceded neurite extension and could provide cues for growth cone extension. The factors which are involved in guidance of retinal axons to the optic tectum have recently been reviewed (Hynes and Lander, 1992). Retinal axons are adhesive to laminin as the axons grow towards their target, the optic tectum, but then lose their adhesiveness. The loss of adhesiveness was probably due to the down regulation of laminin integrin receptors of the beta-1 family on the retinal ganglion cell surface (de Curtis et al., 1991). This suggests that an important regulatory aspect of neuronal guidance is the signal transduction mediated through laminin and its receptors. Laminin has been shown to have both adhesive and anti-adhesive properties for neurons (Calof and Lander, 1991). Presumably, these properties are influenced by interacting components in the local environment which ultimately determine the direction and targeting of neurons.

Laminin appears to play an important role in lens and eye morphogenesis (Dong and Chung, 1991). The temporal expression of the genes for the alpha-1, beta-1, and gamma-1 chains was examined in the mouse embryo by immuno-staining and *in situ* hybridization from Day 9.5 of embryogenesis. Laminin beta-1, and gamma-1 transcripts were detected in the lens placode and neuroectoderm at the 25 somite stage. As the lens pit formed there was a marked increase in the level of message in the lens cells and in the pigment epithelium at the posterior aspect of the optic cup. In the pigment epithelium laminin synthesis preceded the formation of pigment granules. At E15.5 new patterns of laminin expression became apparent. Punctate groups of cells in the vitreous aspect of the retina and closely apposed to the limiting membrane expressed the transcripts. Corneal epithelial cells also synthesized laminin at this stage (Figure 2). Transcripts could be detected in the ganglion cell layer at E18.5 and persisted through day 7 postnatally.

Surprisingly, transcripts for the α1 laminin gene could not be readily detected in the ganglion cell layer, suggesting that a distinct laminin isoform may be involved in the guidance of retinal axons.

Several recent developments have provided new insights into the *in vivo* function of laminin and its isoforms. Merosin is the predominant isoform of laminin in the basal lamina of striated muscle and peripheral nerve. In merosin, the alpha-1 chain of the prototypic laminin is replaced with the alpha-2 chain. The other chains in merosin are either the β1 or β2 chain and the γ1 chain. Recent studies have provided strong evidence for the role of merosin in muscle function and development (Sunada et al., 1994; Xu et al., 1994) The *dy* mutation in mice results in a severe neuromuscular disease resembling human muscular dystrophy. In normal striated muscle the linkage between the extracellular matrix and the subsarcolemmal actin is provided by a complex consisting of dystrophin and several dystrophin-associated proteins (Sunada et al., 1994). Four of the dystrophin-associated proteins 35 DAG, 43 DAG (beta-dystroglycan), 50 DAG (adhalin) and a 25 kDa protein are transmembrane proteins. This complex is associated in the cytoplasmic compartment with dystrophin which binds to actin, and outside the cell to a 156-kDa extracellular protein alpha-dystroglycan (156 DAG) which binds specifically to merosin. In the *dy/dy* mouse skeletal muscle and peripheral nerve had greatly reduced amounts of the alpha-2 chain of laminin (Figure 3). The other two chains of laminin were at normal levels. The basal lamina of striated muscle was also shown to be morphologically defective. Furthermore, the gene for the alpha-2 chain mapped to mouse chromosome 10 in the

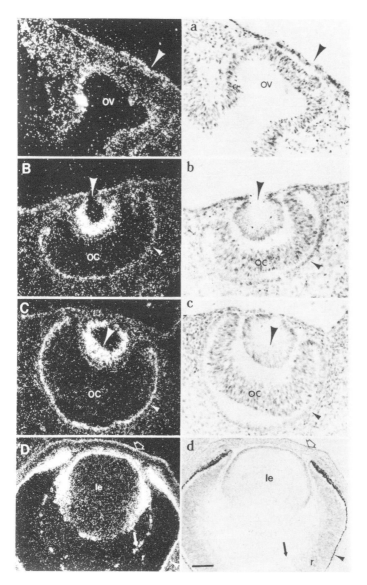

Figure 2. Laminin β1transcripts in the embryonic mouse eye. Panels A-D are darkfield pictures from E9.5-10 (A); E10.5 (B); E10.5-11 (C); E15.5 (D) embryos. Panels a-d are the corresponding brightfield images. Panels Aa, Bb, Cc bar=80 micrometers; panels Dd bar =160 micrometers. Open arrows identify the basal lamina of the corneal epithelium. OC, optic cup; OV, optic vesicle; le, lens; r, retina. (Dong and Chung, 1991, with permission).

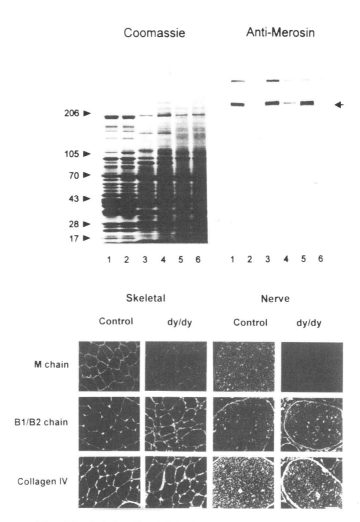

Figure 3. Absence of the alpha-2 chain of laminin in the *dy/dy* mouse. Upper panels, lanes 2, 4, 6, are extracts from skeletal muscle, cardiac muscle, and sciatic nerve of *dy/dy* mice; lanes 1, 3, and 5 are corresponding extracts from control animals. Lower panels, skeletal and nerve basal lamina stained with antibodies against merosin, B1/B2 chains of laminin and collagen IV. (Sunada et al., 1994, with permission).

region of the locus for the *dy/dy* mutation. These data suggest that the primary defect in the *dy* mutation is a deficiency in the alpha-2 chain. This defect is believed to disrupt the communication between the extracellular matrix and the cytoskeleton of the muscle or nerve resulting in necrosis and peripheral neuropathy.

The laminin isoform known as kalinin consists of the alpha-3, beta-3 and gamma-2 chains which are truncated forms of the prototypic laminin chains. Kalinin is localized to the lamina lucida of the basal lamina and is a component of the anchoring filaments that link the basal lamina with the underlying connective tissue network. The integrity of the basal lamina in the skin and mucous membranes of patients with the autosomal recessive disorder known as epidermolysis bullosa is severely compromised leading to blistering and erosion of the skin. The disease is sometimes fatal. It has recently been shown that the gamma-2 chain of laminin is abnormal in one form of the disease known as junctional epidermolysis bullosa (Pulkkinen et al., 1994: Aberdam et al., 1994). In one of the patients a point mutation in the splice junction of the gene, which resulted in the deletion of an exon in the EGF

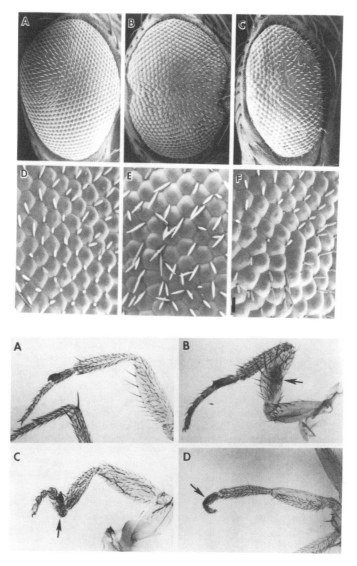

Figure 4. Defects in *Drosophila* eye and limbs in partial alpha-3 laminin deficiency. Upper panels, scanning electron micrographs of wild type and *lamA* mutant adult eyes; A, normal eye, B-F mutant eyes. Bar, 50 micrometers, A-C, 12 micometers, D-F. Lower panel, photomicrographs of mutant leg phenotypes in *lamA* mutant adults. Panels A-C cuticle of male foreleg, and panel D third leg, panel A, normal, panels C-D, mutant.. (Henchcliffe et al., 1993, with permission).

repeats, was identified as the cause of the defect. In another patient a deletion-insertion mutation, which resulted in premature termination of the protein, was identified as the abnormality. From these studies it is apparent that kalinin is essential for maintenance of tissue integrity. The defective kalinin probably could not form proper associations with other macromolecules that are necessary for the intact fuctional structure of the basal lamina.

    Drosophila laminin is similar to mammalian laminin and retains the typical heterotrimeric cross-shaped structure, with slight differences in the short arm projected by the alpha chain. Goodman and his colleagues have generated a series of complete or partial loss of function mutant alleles of the *lamA* gene in *Drosophila* (Henchcliffe et al., 1993). The phenotypes of these mutants indicate that the alpha chain of laminin, and therefore the

Table 1. Biologically active peptides in laminin[1]

Peptide	Chain	Activity
IKVAV	alpha-1	Cell attachment, neurite outgrowth, cell migration.
CCFALRGDNPQ	alpha-1	Cell attachment, differentiation of human endothelial cells
SINNR	alpha-1	Cell attachment of type II alveolar and HT 1080 cells
KQNCLSSRASFR GCVRNLRLSR	alpha-1	Heparin binding
YIGSR	beta-1	Cell attachment, chemotaxis, receptor binding, cell differentiation
LRE	beta-2	Attachment of ciliary ganglion neurons
RNIAEIIKDI	gamma-1	Neurite outgrowth

[1]For additional active peptides see Yamada and Kleinman, 1992 and Matter and Laurie, 1994.

entire laminin molecule, serves an important role in the development of the organism. Complete loss of function resulted in late embryonic lethality, whereas in escaper adults with partial loss of function deformities in the eyes, legs (Figure 4) and wing structure were observed. Changes in the number and organization of photoreceptor cells in the ommatidia were seen in the more severe cases, suggesting the possibility of changes in cell fate or effects on cell differentiation. Although detailed structure function analyses are yet to be done, this system offers the exciting opportunity to dissect the domain functions of the alpha chain. The *unc* mutations in *Caenorhabditis elegans* whose phenotypes include muscle dysfunction are characterized by alterations in the extracellular matrix molecules that include collagen and laminin isoforms.

It is apparent from the above examples that laminin and its isoforms are necessary for normal development and maintenance of tissue integrity. The details of the mechanism of how laminin influences these diverse activities *in vivo* have yet to be established. However, clues can be obtained from the information gained by *in vitro* studies on the intact molecules and fragments derived from them.

Analysis of laminin function

Laminin peptides

The determination of the primary structures of many of the laminin chains and the synthesis of laminin sequences both chemically and by expression vectors have provided opportunities to relate the biological activities ascribed to laminin with specific amino acid sequences (Yamada and Kleinman, 1992). In addition, proteolytic fragments have been useful for assigning activities to particular domains of the intact molecule. Although the results obtained from studies with these reagents cannot be directly extrapolated to intact organisms, they do provide clues for the *in vivo* activity of laminin. These experiments have revealed sequences or domains that are implicated in cell attachment, neurite outgrowth, chemotaxis, association with other macromolecules, assembly of both the laminin molecule and the extracellular matrix, morphogenesis, and gene expression. There is clearly an overlap of these activities since cell attachment may play a role in morphogenesis and gene expression. Table I gives a partial list of biologically active laminin peptides. Several of these support cell attachment, others promote neurite extension and another group interacts with proteoglycans. Access to these sequences *in vivo* will depend on the organization of the extracellular matrix and perhaps in other cases the action of proteolytic enzymes that may be present during tissue remodeling. Peptides containing the SIKVAV sequence have been studied extensively (Yamada and Kleinman, 1992). The SIKVAV sequence is derived from the α chain of laminin. It has been reported to promote cell attachment, morphogenesis, neurite outgrowth, angiogenesis, chemotaxis of neutrophils and endothelial cells, and to associate with laminin-binding non-integrin proteins. The mechanism by which this peptide elicits these activities is not known, furthermore, it is uncertain if the sequence is accessible to cells in an intact basal lamina.

The proteolytic fragment of laminin referred to as E8 contains the terminal portion of the rod-like structure of the long arm and its associated large globular domain. This fragment retains some of the biological activities of laminin such as the promotion of neurite outgrowth and cell adhesion. There is convincing evidence that the integrin receptors alpha-1 beta-6 and alpha-3 beta-1 bind to this region of laminin (Sonnenberg et al., 1990; Gehlsen et al., 1992) which could provide a pathway for laminin to influence cell behavior.

Signal transduction

The intracellular regulatory pathways through which laminin influences cellular activity is largely unknown. However, the discovery and characterization of the integrin receptors that engage sequences in the extracellular matrix molecules have provided a conceptual framework for futher experiments. The integrin receptors are transmembrane heterodimeric proteins, each of which is composed of an alpha and a beta subunit linked by non-covalent interactions (Hynes, 1987). The alpha and beta subunits comprise two separate families of proteins in which the members are related in structure and function. The receptors participate in cell-cell as well as cell-extracellular matrix associations. Each cell has its own repertoire of integrin receptors that can recognize cognate ligands. The nature of the cellular response depends on the specific ligand-receptor combination that is activated (Hynes, 1992). A given integrin receptor has the potential to recognize sequences on several different extracellular matrix molecules, for example, the integrin receptor alpha-3 beta-1 binds to laminin, collagen and fibronectin. Similarly, each extracellular matrix molecule is recognized by a panel of integrin receptors, which for laminin include alpha-1 beta-1, alpha-3 beta-1, alpha-6 beta-1, and alpha-4 beta 6 (Mecham, 1991). The complexity of the role of the extracellular matrix in morphogenesis and differentiation can be appreciated if one considers the plasticity of the extracellular matrix and integrin receptor expression during development.

The elucidation of the intracellular pathways by which signals from the extracellular matrix are transduced is a major challenge. Although extensive studies have not been reported for laminin, results obtained with matrigel (a mixture of laminin, entactin, collagen IV and perlecan) and fibronectin have provided clues. When activated, the cytoplasmic domains of integrin receptors bind to alpha-actinin and talin, components of the cytoskeleton and focal contacts (reviewed by Burridge et al., 1988; Hynes, 1992). Regulation of this interaction can trigger reorganization of the cytoskeleton and the attendant consequences on cell motility, shape and adhesiveness. The spreading of fibroblasts on fibronectin and RGD containing peptides derived from fibronectin has been shown to trigger the phosphorylation of a focal adhesion protein tyrosine kinase pp125 FAK (reviewed by Burridge et al, 1992). Since the integrin itself is not a protein kinase, the phosphorylation of this protein indicates that engagement of the integrin promotes the assembly or activation of cytoplasmic membrane and other proteins that are involved in phosphorylation and signal transduction. The role of pp125 FAK in the signal transduction pathway is not known at present but provides a hint that signals from the extracellular matrix may enter common regulatory pathways used by growth factors and hormones that utilize protein phosphorylation mechanisms.

Model systems have provided additional information. The mammary gland cell cultures studied by Bissell and collaborators (reviewed by Lin and Bissell, 1993) have been especially informative. Primary cultures of mammary epithelial cells from pregnant mice undergo terminal differentiation to form alveolar structures and to produce milk proteins in response to lactogenic hormones and an extracellular matrix. Studies on the extracellular matrix requirement revealed that laminin, but not type IV collagen or fibronectin, induced beta-casein production. An integrin mediated signal transduction pathway was indicated since antibodies against the beta-1 integrin subunit blocked the induction. These investigators developed a cell line which mimicked the primary cultures. Transfection of these cells with a beta-casein promoter-reporter construction led to the discovery of a regulatory element in the beta-casein promoter. This 161 base pair enhancer element, BCE-1 (bovine casein element one), responded to prolactin and the extracellluar matrix when transfected into mammary cells, but not CHO or MDCK cells, suggesting that it is tissue specific. This exciting result describes for the first time an extracellular matrix response element. The pathway between the initial event of integrin engagement and activation of the gene remains to be defined and promises to provide a very fruitful line of investigation.

Other potential pathways for the initiation of signal transduction by engagement of the integrin receptors include the organization and activation of transport systems, changes in calcium fluxes, cooperation with growth factor receptors, and modulation of enzymes implicated in the phospholipase C pathway (reviewed by Damsky and Werb, 1992; Adams and Watt, 1993).

Laminin assembly

Laminin has many isoforms, some of which appear to have specific functions. Merosin, as mentioned previously, is concentrated in the basal lamina of peripheral nerves and striated muscle and the absence of the alpha-2 merosin chain leads to the dystrophic phenotype in the *dy/dy* mouse. The S-laminin or beta-2 chain also has limited tissue distribution and is concentrated in the basal lamina of synapses as well as in kidney glomerulus (Engvall et al., 1990). The inactivation of the beta-2 gene in the mouse results in proteinuria and early death (personal communication). In addition, junctional epidermolysis bullosa is the result of an abnormal gamma-2 laminin chain. The laminin genes are located on several chromosomes and the chains are independently regulated. Although detailed analyses have not been carried out it is not unlikely that cells switch laminin isoforms that are assembled into the basal lamina during development in order to fulfill specific requirements for motility, differentiation or organogenesis. Very little is known about the mechanisms which regulate the expression of the laminin genes, the selection of the set of chains to be assembled, and the deposition of a particular trimeric isoform into the basal lamina.

The PC12 pheochromocytoma cell line has been used as a model system to explore the role of the beta-1 chain in the deposition of laminin into the extracellular matrix. These cells deposit fibronectin but not laminin into the extracellular matrix, however, they express the genes for the beta-2 and gamma-1 chains and not those for the alpha-1 or beta-1 genes (Reing et al., 1992). Transfection of the cells with the full length cDNA for the beta-1 chain under the control of the human beta-actin promoter and selection for stable transformants resulted in a new phenotype. The transformants synthesized and deposited laminin into the extracellular matrix and appeared to down-regulate fibronectin deposition (Figure 5) but not its transcripts. The morphology of the cells was dramatically altered from a substrate adherent flat shape to rounded clusters of tighly adherent cells that were resistant to dissociation with trypsin. These experiments suggest that in PC12 cells the beta-1 chain plays an important role in the assembly and deposition of laminin. The laminin matrix somewhat surprisingly did not contain a high molecular weight alpha-type chain. The presence of beta-2 transcripts did not lead to productive laminin deposition. These experiments illustrate the complexity of analyzing the mechanisms which regulate laminin deposition in the extracellular matrix and the regulation of the type of matrix that is elaborated by PC12 and other cells by analogy.

The advent of recombinant laminin proteins and peptides has stimulated attempts to unravel the determinants neccessary for laminin assembly. These experiments have focused on the carboxyl terminal region of the molecule in the rod domain. It is possible to reconstitute a segment of the triple helical domain with peptides derived from each of the constituent chains (Utani et al., 1994). It appears that amino acid determinants preclude the formation of homodimeric molecules and *in vitro* the assembly proceeds through

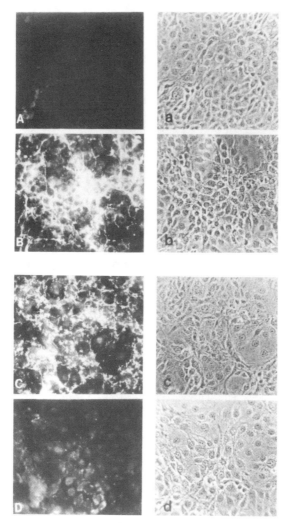

Figure 5. Indirect immunofluorescence of PC12 cells transfected with laminin beta-1 full length cDNA.. Panels A and C, control untransfected cells; panels B and D transfected. Cells in panels A and B were stained with anti-laminin antiserum and cells in panels C and D with anti-fibronectin antiserum. Panels a-d are phase contrast micrographs corresponding to panels A-D. (Reing et al., 1992, with permission).

dimerization of the beta-1 and gamma-1 chains followed by association with the alpha-1 chain. The situation *in vivo* may, however, be more complicated since glycosylation of the laminin chains and the role of chaperonins have to be considered. The mechanisms that might be involved in the selection of chains for assembly *in vivo* when there is a choice remain to be determined.

ENTACTIN

Structure

The general features of entactin are shown in Figure 6. As depicted, the prototype molecule has, at the amino terminus, two globular domains (G1 and G2) separated by a

Figure 6. Domain structure of entactin.

short protease sensitive link which is connected by a rod (E), composed of a thyroglobulin-like and four EGF-like cysteine-rich homology repeats, to a terminal globular domain (G3) at the carboxyl end (reviewed by Chung and Durkin, 1990; Timpl and Aumailley, 1993). The carboxyl segment of the molecule has significant sequence identity with the low density lipoprotein receptor and EGF precursor. The first EGF-homology repeat in the connecting rod has an RGD sequence recognized by the alpha-v beta-3 integrin receptor (Dong et al., unpublished). Calcium ions bind to sequences in the hydrophilic connecting segment between the two amino terminal globular domains as well as in the rod-like segment. The complete amino acid sequences of mouse (Durkin et al., 1988; Mann et al., 1989), human (Nagayoshi et al., 1989) and ascidian (Nakae et al., 1993) entactin have been determined. The ascidian entactin differs from the mammalian molecule in the organization and number of cysteine-rich homology repeats in the rod domain. Thus far, no isoforms of entactin are known and no alternate splicing of its message has been reported.

Biological activities

The biological functions of entactin have been inferred from studies both *in vivo* and *in vitro* . It is well established that entactin is widely distributed in the basal lamina with the implication that it plays a structural role. The temporal appearance of entactin and its message also suggests that it participates in the early phases of development. Entactin is present in the hatched blastocyst stage (Wu et al., 1983) and perhaps even earlier in the mouse embryo (Dziadek and Timpl, 1985). It has further been reported that entactin promotes the outgrowth of trophoblast cells and may therefore be important in the implantation of the embryo (Yelian et al., 1993). The appearance of entactin and its messenger RNA (Figure 7) in the embryonic mouse lens and eye indicates its importance in the development of the eye (Dong and Chung, 1991). In addition, entactin appears at the early stages of chick embryogenesis in the epiblast and hypoblast at the blastula stage and its presence appears to be correlated with directional migration of cells (Zagris et al.,1993). The early synthesis of entactin in liver regeneration and in experimental fibrosis (Schwoegler et al., 1994) support the idea that entactin has an important function in these processes as well. These and other experiments on the distribution of entactin in differenet tissues do not reveal the molecular basis for the biological role of entactin. However, some clues have been obtained from *in vitro* experiments.

Entactin and its structural domains have been synthesized in several expression systems (Tsao et al., 1990; Fox et al., 1991, Hsieh et al., 1994). The recombinant molecules have been employed in different experimental contexts which have revealed a number of potential functions for entactin. Association with other extracellular matrix macromolecules and promotion of cell attachment are two of the major biological activities attributable to entactin. Other activities include chemotaxis (Senior et al., 1992) long term maintenance of cultured regenerated skeletal myotubes (Funanage et al., 1992) and epithelial morphogenesis (Ekblom et al., 1994).

Entactin was first identified in a macromolecular complex containing laminin that was isolated from mouse parietal endoderm cells (Chung et al., 1977). Rotary shadowing of a laminin-entactin complex revealed a 1:1 stoichiometric complex (Paulsson et al., 1986). Reconstitution with recombinant fragments of entactin and laminin have shown that the carboxyl globular domain of entactin binds to a cysteine-rich homology repeat in the gamma-1 chain of laminin (Fox et al., 1991; Mayer et al., 1993). It has further been shown that the G2 domain of entactin (towards the N-terminal end) binds to collagen IV (Fox et al, 1991), the protein core of perlecan (Reinhardt et al, 1993), the N-terminal domain of fibronectin (Hsieh et al. 1994), and fibrinogen (Wu and Chung, 1991). The implications of

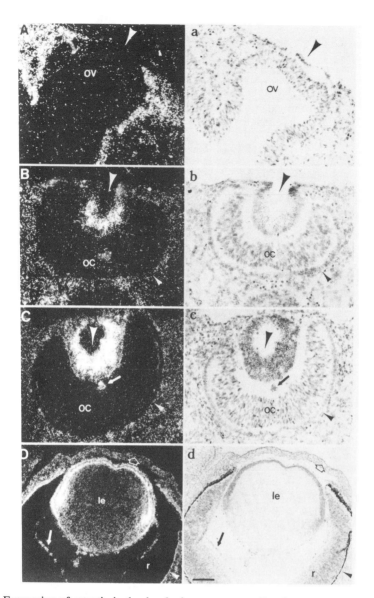

Figure 7. Expression of entactin in the developing mouse eye. Panels A-D are darkfield pictures from E9.5-10 (A); E10.5 (B); E10.5-11 (C); E15.5 (D) embryos. Panels a-d are the corresponding brightfield mages. OC, optic cup; OV, optic vesicle; le, lens; r, retina. Small arrows point to hyalocytes. In panels A and a arrows locate the lens placode; in B and b, the lens pit; and in C and c, the lens vesicle. The small arrowheads locate the pigment epithelium, and in panel D and d the open arrows point to the corneal stroma. Panels Aa, Bb, Cc bar=80 micrometrs; panels Dd bar =160 micrometers. (Dong and Chung, 1991, with permission).

the binding properties of entactin are twofold, (a) it may be essential as a bridging molecule for the assembly of the basal lamina in some tissues and (b) it may modulate the biological activities of the macromolecules to which it binds. Regulation of both the structure and biological activities of the extracellular matrix is clearly important during development when cell-extracellular matrix interactions are continuously changing as cells migrate and establish new associations. Competition for binding to the G2 domain of entactin may provide a mechanism for modifying the extracellular matrix. There is substantial evidence that entactin promotes cell adhesion through the RGD sequence in the first EGF-repeat in the connecting rod. The intact entactin molecule, either isolated from the extracellular matrix or synthesized from recombinant vectors, supports the attachment of several cell types in a concentration dependent manner. The attachment can be blocked by peptides containing the RGD sequence and synthetic peptides which contain the RGD sequence and flanking amino acids derived from entactin retain the cell attachment activity. In addition, recombinant entactin fusion proteins containing the E-domain in which the RGD sequence was mutated to RAD or RGE or completely deleted do not support cell attachment (Dong et al. , unpublished). The alpha-v beta-3 integrin has been identified as a potential receptor for entactin since attachment of melanoma M21 cells to the RGD containing peptides derived from entactin is blocked by antibodies against this integrin. However, attachment of human neutrophils to the RGD entactin-derived peptides is mediated through a different integrin receptor, the leukocyte response integrin (LRI) (Senior et al., 1992). A second potential binding site in entactin has recently been identified (Dong et al., unpublished). It was discovered that although antibodies against alpha-v beta-3 integrin could completely abolish the attachment of melanoma cells to the E-domain of entactin, attachment to the intact molecule could not be completely blocked. Furthermore, recombinant entactin, in which the RGD sequence had been deleted, could promote cell attachment approximately half as well as the unmodified entactin. This cell attachment site has been located to a 39 amino acid peptide in the EGF repeat adjacent to the amino terminal side of the G2 domain. Preliminary data suggest that the receptor for this peptide belongs to the beta-1 family of integrins, possibly alpha-3 beta-1. This receptor has previously been reported to bind to entactin (Dedhar et al., 1992).

The recognition of entactin by more than one plasma membrane receptor is reminiscent of other extracellular matrix molecules such as laminin and fibronectin. The multiplicity of receptors and binding sites provides a mechanism for stimulating different types of cellular responses when cells bind to the extracellular matrix. The interaction of human neutrophils with entactin is interesting in this context. In unstimulated neutrophils entactin promotes chemotaxis and serves as an attachment substrate through an RGD-dependent receptor (Gresham et al., 1993). In activated neutrophils however, entactin promotes phagocytosis which is inhibited by antibodies against the alpha-3 beta-1 receptor. Thus two distinct types of responses are observed depending on the receptor-ligand combination. It has further been shown that the G2 domain of entactin contains the ligand for the alpha-3 beta-1 receptor. It has also been observed that human melanoma M21 cells utilize the RGD sequence for attachment and spreading, the G2 site promotes cell attachment but not spreading. It seems that the two sites affect the cytoskeleton in distinctly different ways.

FUTURE DIRECTIONS

The importance of the extracellular matrix in development is well established. It serves not only as a scaffold for cell movement and tissue organization but also as a regulator of cell behavior. The mechanisms by which the extracellular matrix influences cell behavior in the embryo are largely unknown at present . The basal lamina which is in direct contact with the plasma membrane is heterogeneous both in its composition and the organization of its constituent molecules. However, much remains to be learned about the regulation of the assembly of the individual components into the functional structure of the matrix which determines the density, accessibilty, and types of integrin and other recognition sites. The extracellular matrix can be modified by varying the constituent molecules, and in the case of laminin itself, insertion of one of the many isoforms could have marked effects on the properties of the matrix. The timing of the synthesis of molecules such as entactin could have a major effect on the type of molecules that are inserted into the matrix and thereby its properties. The notion that the extracellular matrix is a homogeneous passive structure is no longer tenable and it must be considered to be as specialized and varied as the cells with which it interacts. Although at present it is not possible to probe the extracellular matrix at the level of resolution that distinguishes differences in organization new genetic approaches may be fruitful.

The substantial body of information that is available on the effects of individual components on isolated cells cannot always be extrapolated to the developing organisms where the physical and humoral environment are very different. Neither can the temporal and spatial expression of the genes be accepted as an indication of indispensable function *in vivo*. For example, tenascin has been shown to have both adhesive and anti-adhesive properties on isolated cells and the expression of its gene was shown to be tightly regulated during development (Chiquet-Ehrismann et al., 1986). However, inactivation of the gene had no obvious effects on mouse development (Saga et al., 1992). On the other hand, inactivation of the fibronectin gene resulted in embryonic lethality confirming its indispensible function. However, fibronectin was not essential for the early migration of cells in the embryo since development proceeded to an early blastula stage (George et al., 1993). The defects appear to arise from defective mesodermal migration, proliferation or differentiation in the absence of fibronectin. These results indicated that other molecules were involved in providing the substrate and cues for the earliest morphogenetic movements. The inactivation of the beta-2 laminin gene in the mouse has resulted in interesting phenotypes. The animals had abnormal kidney function, neuro-muscular development and life span (personal communication). This direct evidence, *in vivo*, supports the thesis that different isoforms of laminin have specialized tissue functions. Further analysis of this interesting mutation promises to yield many new and exciting conclusions. Studies with mutant animals in which specific genes have been inactivated offer the best promise for a thorough understanding of the complex functions of the extracellular matrix and its constituent molecules.

Cells respond to the extracellular matrix through plasma membrane receptors which include integrin and non-integrin receptors. A complete analysis of the role of the extracellular matrix in development must therefore include these receptors, their regulation and the pathways for signal transduction. Remarkable progress has been made in identifying the receptors and in studies on their mechanism of action. Among the more significant advances have been (a) the inactivation of the alpha-5 integrin subunits in mice with the resulting embryonic lethal phenotype (Yang et al., 1993), (b) analysis of the intracellular domains of the integrin receptors by mutations to determine the structural requirements for specific signal transduction (Hynes, 1992), and (c) recognition of intracellular and extracellular factors which regulate integrin activity (Hynes, 1992). The challenge for the future is to understand how the signals generated by the extracellular matrix, and mediated by cell surface receptors, become integrated with other signals and translated into specific responses such as directed cell movement and regulation of gene activity.

REFERENCES

Aberdam, D., Galliano, M.-F., Vailly, J., Pulkkinen, L., Bonifas, J., Christiano, A.M., Tryggvason, K., Uitto, J., Epstein Jr., E.H., Ortonne, J.-P., and Meneguzzi, G., 1994, Herlitz's junctional epidermolysis bullosa is linked to mutations in the gene (LAMC2) for the gamma-2 subunit of nicein/kalinin (Laminin-5), *Nature Genet.* 6:299.

Adams, J.C., and Watt, F.M., 1993, Regulation of development and differentiation by the extracellular matrix, *Dev.* 117:1183.

Burridge, K., Fath, K., Kelly, T., Nuckolls, G., and Turner, C., 1988, Focal adhesions: transmembrane junct.

Burridge, K., Petch, L.A., and Romer, L.H., 1992, Signals from focal adhesions, *Current Biol.* 2:537.

Calof, A.L., and Lander, A.D., 1991, Relationship between neuronal migration and cell-substratum adhesion: laminin and merosin promote olfactory neuronal migration but are anti-adhesive, *J. Cell Biol.* 115:779.

Chi, H.C., and Hui, C.F., 1989, Primary structure of the Drosophila laminin B2 chain and comparison with human, mouse, and laminin B1 and B2 chains, *J. Biol. Chem.* 254:1543.

Chiquet-Ehrismann, R., Mackie, E.J., Pearson, C.A., Sakakura, T. , 1986, Tenascin: an extracellular matrix protein involved in tissue interactions during fetal development and oncogenesis, *Cell* 47:131.

Chiquet-Ehrismann, R., Mackie, E.J., Pearson,C.A., and Sakakura, T., 1986, Tenascin: an extracellular matrix protein involved in tissue interactions during fetal development and oncogenesis, *Cell* 47:131.

Chung, A.E. and Durkin, M.E., 1990, Entactin: structure and function, *Am. J. Respir. Cell Mol. Biol.* 3:275.

Chung, A.E., Freeman, I.L., and Braginski, J.E., 1977, A novel extracellular membrane elaborated by a mouse embryonal carcinoma derived cell line, *Bichem. Biophys. Res. Comm.* 79:859.

Chung, A.E., Jaffe, R., Freeman, I.L., Vergnes, J.-P., Braginski, J.E., and Carlin, B., 1979, Properties of a basement membrane-related glycoprotein synthesized in culture by a mouse embryonal carcinoma derived cell line, *Cell* 16:277.

Cooper, A.R., and MacQueen, H.A. 1983, Subunits of laminin are differentially synthesized in mouse eggs and early embryos, *Dev. Biol.* 96:467.

Damsky, C.H., and Werb, Z., 1992, Signal transduction by integrin receptors for extracellular matrix: cooperative processing of extracellular information, *Curr. Opin. Cell Biol.* 4:772.

de Curtis, I., Quaranta, V., Tamura, R.N., and Reichardt, L.F., 1991, Laminin receptors in the retina: sequence analysis of the chick integrin alpha-6 subunit: Evidence for transcriptional and post-transcriptional regulation, *J. Cell Biol.* 113:405.

Dedhar, S., Jewell, K., Rojiani, M., Gray, V., 1992, The receptor for the basement membrane glycoprotein entactin is the integrin alpha-3 beta-1, *J. Biol. Chem.* 267:18908.

Dong, L.J., and Chung, A.E., 1991, The expression of the genes for entactin, laminin A, laminin B1 and laminin B2 in murine lens morphogenesis and eye development, *Differentiation* 48:157.

Durkin, M.E., Bartos, B.B., Liu, S.-H., Phillips, S.L., and Chung, A.E., 1988, The primary structure of the mouse laminin B2 chain and comparison with laminin B1, *Biochemistry* 27:5198.

Durkin, M.E., Chakravarti, S., Bartos, B.B., Liu, S.-H., Friedman, R.L., and Chung, A.E., 1988, Amino acid sequence and domain structure of entactin. Momology with epidermal growth factor precursor and low density lipoprotein receptor, *J. Cell Biol.* 107:2749.

Dziadek, M., Timpl, R., Expression of nidogen and laminin in basement membranes during mouse embryogenesis and in teratocarcinoma cells, 1985, *Dev. Biol.* 111:372.

Ehrig, K., Leivo, I., Argraves, W.S., Ruoslahti, E., and Engvall, E., 1990, Merosin, a tissue-specific basement membrane protein, is a laminin-like protein, *Proc. Natl. Acad. Sci. USA* 87:3264.

Ekblom, M., Klein, G., Mugrauer, G., Fecker, L., Deutzman, R., Timpl, R., and Ekblom, P., 1990, Transient and locally restricted expression of laminin A chain mRNA by developing epithelial cells during kidney organogenesis, *Cell* 60:337.

Ekblom, P., Alitalo, L., Vaheri, A., Timpl, R., and Saxen, L., 1980, Induction of a basement membrane glycoprotein in embryonic kidney: possible role of laminin in morphogenesis, *Proc. Natl. Acad. Sci. USA,* 77:485.

Ekblom, P., Ekblom, M., Fecker, L., Klein, G., Zhang, H.-Y., Kadoya, Y., Chu, M.-L., Mayer, U., and Timpl, R., 1994, Role of mesenchymal nidogen for epithelial morphogenesis in vitro, *Development* 120:2003.

Engel, J., 1992, Laminins and other strange proteins, *Biochemistry* 31:10643.

Engel, J., Odermatt, E., Engel, A., Madri, J.A., Furthmayr, H., Rohde, H., and Timpl, R., 1981, Shapes, domain organization,and flexibility of laminin and fibronectin, two multifunctional proteins of the extracellular matrix, *J. Mol. Biol.* 150:97

Engvall, E., Earwicker, D., Haaparanta, T., Ruoslhti, E., and Sanes, J., 1990, Distribution and isolation of four laminin variants; tissue restricted distribution of heterotrimers assembled from different subunits, *Cell Regul.* 1:731.

Fox, J.W., Mayer, U., Nischt, R., Aumailley, M., Reinhardt, D., Wiedemann, H., Mann, K., Timpl, R., Krieg, T., Engel, J., Chu, M.-L., 1991, Recombinant nidogen consists of three globular domains and mediates binding of laminin to collagen type IV, *EMBO J.* 10:3137.

Funanage, V.L., Smith, S.M., and Minnich, M.A., 1992, Entactin promotes adhesion and long term maintenance of cultured regenerated skeletal myotubes, *J. Cell Physiol.* 150:251.

Garrison, K., MacKrell, A., and Fessler, J.H., 1991, *Drosophila* laminin A chain sequence, interspecies comparison, and domain structure of a major carboxyl portion, *J. Biol. Chem.* 266:22899.

Gehlsen, K.R., Sriramarao, P., Furcht, L.T., and Skubitz, A.P.N., 1992, A synthetic peptide from the carboxyl terminus of the laminin A chain represents a binding site for the alpha-3 beta-1 integrin, *J. Cell Biol.* 117:449.

George, E.L., Georges-Labouesse, E.N., Patel-King, R.S., Rayburn, H., and Hynes, R.O., 1993, Defects in mesoderm, neural tube and vascular development in mouse embryos lacking fibronectin, *Development* 119:1079.

Gerecke, D.R., Wagman, W.D., Champliaud, M.F., and Burgeson, R.E., 1994, The complete primary structure of a novel laminin chain, the laminin B1k chain, *J. Biol. Chem.* 269:11073.

Gresham, H.D., Chung, A.E., and Senior, R.M. (1993), Entactin stimulates neutrophil phagocytosis via alpha-3 beta-1, *Mol. Biol. Cell* 4:285a.

Hammarback, J.A., McCarthy, J.B., Palm, S.L., Furcht, L., and Letourneau, P.C., 1988, Growth cone guidance by substrate-bound laminin pathways is correlated with neuron-to-pathway adhesivity, *Dev. Biol.* 126:29.

Henchcliffe, C., Garcia-Alonso, L., Tang, J., and Goodman, C.S., 1993, Genetic analysis of laminin A reveals diverse functions during morphogenesis in *Drosophila*, *Development* 118:325.

Holm, K., Risteli, L., and Sariola, H., Differential expression of the laminin A and B chains in chimeric kidneys, *Cell Differ.* 24:331.

Hsieh, J.-C., Wu, C., and Chung, A.E., 1994, The binding of fibronectin to entactin is mediated through the 29 kDa amino terminal fragment of fibronectin and the G2 domain of entactin, *Biochem. Biophys. Res. Comm.* 199:1509.

Hunter, D.D., Shah, V., Merlie, J.P., and Sanes, J.R., 1989, A laminin-like adhesive protein concentrated in the synaptic cleft of the neuromuscular junction, *Nature (London)* 338:229.

Hynes, R.O., 1987, Integrins: a family of cell surface recpetors, *Cell* 48:549.

Hynes, R.O., 1992, Integrins: versatility, modulation, and signaling in cell adhesion, *Cell* 69:11.

Hynes, R.O., and Lander, A.D., 1992, Contact and adhesive specificities in the associations, migrations, and targeting of cells and axons, *Cell* 68:303.

Kallunki, P., Sainio, K., Eddy, R., Byers, M., Kallunki, T., Sariola, H., Beck, K., Hirvonen, H., Shows, T.B., and Tryggvason, K., 1992, A truncated laminin chain homologous to the B2 chain: structure, spatial expression, and chromosomal assignment. *J. Cell Biol.* 119:679.

Klein, G., Langegger, M., Timpl, R., and Ekblom, P., 1988, Role of laminin A chain in the development of epithelial cell polarity, *Cell* 55:331.

Kleinman, H.K., Cannon, F.B., Laurie, G.W., Hassell, J.R., Aumailley, M., Terranova, V.P., Martin, G.R., and Dubois-Dalc M., 1985, Biological activities of laminin, *J. Cell. Biochem.* 27:317.

Knibbs, R.N., Perini, F., and Goldstein, I.J., 1989, Structure of the major concanavalin A reactive oligosaccharides of the extracellular matrix component laminin, *Biochemistry* 28:6379.

Kusche-Gullberg, M., Garrison, K., MacKrell, A.J., Fessler, L.I., and Fessler, J.H., 1992, Laminin A chain: expression during *Drosophila* development and genomic sequence, *EMBO J.* 11:4519.

Laurie, G.W., Horikoshi, S., Killen, P.D., Segui-Real, B., and Yamada, Y., 1989, In situ hybridization reveals temporal and spatial changes in cellular expression of mRNA for a laminin receptor, laminin, and basement membrane (type IV) collagen in the developing kidney, *J. Cell Biol.* 109:1351.

Leivo, I., and Engvall, E., 1988, Merosin, a protein specific for basement membranes of Schwann cells, striated muscle and trophoblast, is expressed in nerve and muscle development, *Proc. Natl. Acad. Sci. USA* 85:1544.

Letourneau, P.C., Madsen, A.M., Palm, S.L., and Furcht, L.T., 1988, Immunoreactivity for laminin in the developing ventral longitudinal pathway of the brain, *Dev. Biol.* 125:135.

Lin, C.Q., and Bissell, M.J., 1993, Multi-faceted regulation of cell differentiation by extracellular matrix, *FASEB J.* 7:737.

Mann, K., Deutzmann, R., Aumailley, M., Timpl, R., Raimondi, L., Yamada, Y., Pan, T.-C., Conway, D., and Chiu, M.-L., 1989, Amino acid sequence of mouse nidogen, a multidomain basement membrane protein with binding activity for laminin, collagen IV, and cells, *EMBO J.* 8:65.

Matter, M.L., and Laurie, G.W., 1994, A novel laminin E8 cell adhesion site required for lung alveolar formation in vitro, *J. Cell Biol.* 124:1083.

Mayer, U., Nischt, R., Poschl, E., Mann, K., Fukuda, K., Gerl, M., Yamada, Y., and Timpl, R., 1993, A single EGF-like motif of laminin is responsible for high affinity nidogen binding, *EMBO J.* 12:1879.

Mecham, R.P., 1991, Receptors for laminin on mammalian cells, *FASEB J.* 5:2538.

Montell, D.J., and Goodman, C.S., 1988, Drosophila substrate adhesion molecule: sequence of laminin B1 chain reveals domains of homology with mouse, *Cell* 53:463.

Montell, D.J., and Goodman, C.S., Drosophila laminin:Sequence of B2 subunit and expression of all three subunits during embryogenesis, *J. Cell Biol.* 109:2441-2453.

Nagayoshi, T., Sanborn, D., Hickok, N.J., Olsen, D.R., Fazio, M.J., Chu, M.-L., Knowlton, R., Mann, K., Deutzmann, R., Timpl, R., and Uitto, J., 1989, Human nidogen: complete amino acid sequence and structural domains deduced from cDNAs, and evidence for polymorphism of the gene, *DNA* 8:581.

Nakae, H., Sugano, M., Ishimori, Y., Endo, T., and Obinata, T., 1993, Ascidian entactin/nidogen. Implication of evolution by shuffling two kinds of cysteine-rich motifs, *Eur. J. Biochem.* 213:11.

Nissinen, M., Vuolteenaho, R., Boot-Handford, R., Kallunki, P. and Tryggvason, K., 1991, Primary structure of the human laminin A chain, Limited expression in human tissues. *Biochem. J.* 276:369.

Ohno, M., Martinez-Hernandez, A., Ohno, N., and Kefalides, N.A., 1986, Laminin M is found in placental basement membranes but not in basement membranes of neoplastic origin, *Connective Tiss. Res.* 15:199.

Olsen, D.R., Nagayoshi, T., Fazio, M., Mattei, M.-G., Passage, E., Weil, D., Timpl, R., Chu, M.-L., and Uitto, J., 1989, Human nidogen: cDNA cloning, cellular expression, and mapping of the gene to chromosome 1q43, *Am. J. Hum. Genet.* 44:876.

Paulsson, M., Deutzmann, R., Dziadek, M., Nowack, H., Timpl, R., Weber, S.,.and Engel, J., Purification and structural characterization of intact and fragmented nidogen obtained from a tumor basement membrane, *Eur. J. Biochem.* 156:467.

Pikkarainen, T., Eddy, R., Fukushima, Y., Byers, M., Shows, T., Pihlajaniem, T., Saraste, M., and Tryggvason, K., 1987, Human laminin B1 chain. A multidomain protein with gene (LAMB1) locus in the q22 region of chromosome 7, *J. Biol. Chem.* 262:10454.

Pikkarainen, T., Kallunki, T., and Tryggvason, K., 1988, Human laminin B2 chain. Comparison of the complete amino sequence with the B1 chain reveals variability in sequence homology between different structural domains, *J. Biol. Chem.* 263:6751.

Pulkkinen, L., Christiano, A.M., Airenne, T., Haakana, H., Tryggvason, K., and Uitto, J., 1994, Mutations in the gamma-2 chain of kalinin/laminin 5 in the junctional forms of epidermolysis bullosa, *Nature Genet.* 6:293.

Reing, J, Durkin, M.E., and Chung, A.E., 1992, Laminin B1 expression is required for laminin deposition into the extracellular matrix of PC12 cells, *J. Biol. Chem.* 267:23143.

Reinhardt, D., Mann, K., Nischt, R., Fox, J.W., Chu, M.-L., Kreig, T., Timpl, R., 1993, Mapping of nidogen binding sites for collagen type IV, heparan sulfate proteoglycan, and zinc, *J. Biol. Chem.* 268:10881.

Ruoslahti, E., 1988, Fibronectin and its receptors, *Ann. Rev. Biochem.* 57:375.

Saga, Y., Yagi, T., Ikawa, Y., Sakakura, T., and Aizawa, S., 1992, Mice develop normally without tenascin, *Genes and Dev.* 6:1821.

Sanes, J.R., 1989, Extracellular matrix molecules that influence neural development, *Ann. Rev. Neurosci.* 12:491.

Sasaki, M., and Yamada, Y., 1987, The laminin B2 chain has a multidomain structure homologous to the B1 chain, *J. Biol. Chem.* 262:17111.

Sasaki, M., Kato, S., Kohno, K., Martin, G.R., and Yamada, Y., 1987, Sequence of the cDNA encoding the laminin B1 chain reveals a multidomain protein containing cysteine-rich repeats. *Proc. Natl. Acad. Sci. USA* 84:935.

Sasaki, M., Kleinman, H.K., Huber, H., Deutzmann, R. , and Yamada, Y., 1988, Laminin, a multidomain protein: the A chain has a unique globular domain and homology with the basement membrane proteoglycan and the laminin B chains, *J. Biol. Chem.* *263*:16536.

Schwoegler, S., Neubauer, K., Knittel, T., Chung, A.E., and Ramadori, G., 1994, Entactin gene expression in normal and fibrotic liver and in rat liver cells, *Lab. Inves.* 70:525.

Senior, R.M., Gresham, H.D., Griffin, G.L., Brown, E.J., Chung, A.E., 1992, Entactin stimulates neutrophil adhesion and chemotaxis through interactions between its Arg-Gly-Asp (RGD) domain and the leukocyte response integrin, *J. Clin. Inves.* 90:2251.

Sonenberg, A., Linders, C.J.T., Modderman, P.W., Damsky, C.H., Aumailley, M., and Timpl, R., 1990, Integrin recognition of different cell binding fragments of laminin (P1, E3, E8) and evidence that alpha-6 beta-1 but not alpha-6 beta-4 functions as a major receptor for E8, *J. Cell Biol.* 110:2145.

Sunada, Y., Bernier, S.M., Kozak, C.A., Yamada, Y., and Campbell, K.P., 1994, Deficiency of merosin in dystrophic *dy* mice and genetic linkage of laminin M chain gene to *dy* locus, *J. Biol. Chem.*, 269:13729.

Timpl, R., 1989, Structure and biological activity of basement membrane proteins, *Eur. J. Biochem.* 180:487.

Timpl, R., and Aumailley, M., 1993, Other basement membrane proteins and their calcium binding potential, in *Molecular and Cellular Aspects of Basement Membranes,* (D.H. Rohrbach and R. Timpl, eds.), pp. 211-235, Academic Press, San Diego.

Timpl, R., Rohde, H., Robey, P.G., Rennard, S.I., Foidart, J.-M., and Martin, G.R., 1979, Laminin-a glycoprotein from basement membranes, *J. Biol. Chem.* 254:9933

Tsao, T., Hsieh, J.-C., Durkin, M.E., Wu, C., Chakravarti, S., Dong, L.-J., Lewis, M., and Chung, A.E., 1990, Characterization of the basement membrane glycoprotein entactin synthesized in a baculovirus expression system, *J. Biol. Chem.* 265:5188.

Utani, A., Nomizu, M., Timpl, R., Roller, P.P., and Yamada, Y., 1994, Laminin chain assembly, *J. Biol. Chem.* 269:19167.

Wu, C., and Chung, A.E., 1991, Potential role of entactin in hemostasis. Specific interaction of entactin with fibrinogen A alpha and B beta chains, *J. Biol. Chem.* 266:18802.

Wu, T.-C., Wan, Y.-J., Chung, A.E., and Damjanov, I., 1983, Immunohistochemical localization of entactin and laminin in mouse embryos and fetuses, *Dev. Biol.* 100:496.

Xu, H., Christmas, P., Wu, X.-R., Wewer, U.M., and Engvall, E., 1994, Defective muscle basement membrane and lack of M-laminin in the dystrophic *dy/dy* mouse, *Proc. Natl. Acad. Sci. USA* 91:5572.

Yamada, Y., and Kleinman, H.K., 1992, Functional domains of cell adhesion molecules, *Curr. Opin. Cell Biol.* 4:819.

Yang, J.T., Rayburn, H., and Hynes, R.O., 1993, Embryonic mesodermal defects in alpha-5 integrin deficient mice, *Development* 119:1093.

Yelian, F.D., Edgeworth, N.A., Dong, L.-J., Chung, A.E., Armant, D.R., 1993, Recombinant entactin promotes mouse primary trophoblast cell adhesion and migration through the RGD recognition sequence, *J. Cell Biol.* 121:923.

Zagris, N., Stavridis, V., and Chung, A.E., 1993, Appearance and distribution of entactin in the early chick embryo, *Differentiation* 54:67.

THE EXPRESSION OF THE GENES FOR LAMININ
IN THE EARLY EMBRYO

Nikolas Zagris and Vassilis Stavridis

Department of Biology
University of Patras
Patras, 26500
Greece

INTRODUCTION

The extracellular matrix (ECM) is composed of glycoproteins, proteoglycans and glycosaminoglycans locally secreted by resident cells. It is assembled into an organized fibrillar network in the forming extracellular spaces before the organization of a basement membrane in the early embryo. The ECM interacts with cells ("dynamic reciprocity") and promotes and regulates cellular functions such as migration, adhesion and proliferation which result in induction, differentiation and morphogenesis (Grobstein, 1954; Bissel et al., 1982; Martin and Timpl, 1987; Engel, 1989; Hay, 1991; Sage and Bornstein, 1991; Adams and Watt, 1993; Chung, 1993; DeSimone, 1994; Horwitz and Thiery, 1994). Laminin is the first ECM glycoprotein to appear in the matrix in the extracellular spaces from which it mostly disappears during development and is later organized in basement membranes (Leivo et al, 1980; Cooper and MacQueen, 1983; Dziadek and Timpl, 1985; Rogers et al., 1986; Kucherer-Ehret et al., 1990; Zagris and Chung, 1990). In the adult, the ECM is confined to basement membranes. This paper focuses on the biological activity of laminin and its expression in the early embryo. The appearance of laminin early in embryogenesis may implicate it in regulating cell migration, adhesion and differentiation.

LAMININ STRUCTURE AND ACTIVITY

Laminin has the ability to self-associate and to interact with collagen IV, entactin/nidogen, perlecan and other proteins. This property is essential for the assembly of a specialized form of the ECM called the basement membrane (Martin, 1987; see Yurchenco and O'Rear, 1994). Basement membranes are thin layers which underly epithelial and endothelial cell layers and surround muscle fibers and peripheral nerves, serve as surfaces for cell adhesion and support and for migration of cells during wound healing and development. They also provide developmental signals and function as barriers to the passage of macromolecules and cells.

Organization of the Early Vertebrate Embryo
Edited by N. Zagris et al., Plenum Press, New York, 1995

The synthesis of laminin, along with other matrix molecules, is developmentally regulated and is involved in cell and tissue differentiation throughout embryogenesis. There is sequential assembly of components during early basement membrane formation (Dziadek and Timpl, 1985) rather than secretion of pre-formed complexes (Martin et al., 1984). In the mouse embryo: laminin and heparan sulfate proteoglycan are already present on the cell surface at the two-cell stage (Wu et al., 1983; Dziadek and Timpl, 1985); entactin/nidogen is first detected on compacted 8- to 16- cell stage morulae; and collagen IV and fibronectin are first detected in the blastocyst inner cell mass (Adamson and Ayers, 1979; Wartiovaara et al., 1979; Leivo et al., 1980). The appearance of collagen IV coincides with assembly of the first embryonal basement membranes. The structural organization of basement membranes is not homogeneous as, for example, is suggested by the variation in laminin-entactin ratio extracted from different tissues (Bernfield, 1981; Wan et al., 1984; Dziadek and Timpl, 1985; Sanes et al., 1990; Engel, 1992; Thomas and Dziadek, 1994). Collagen IV and laminin account for 70-80% of the total protein in the basement membrane. Collagen is thought to have a primary structural role whereas the multidomain nature of laminin also mediates interactions with a variety of basement membrane components and influences the behaviour of the cells with which it comes into contact. Important biological activities of laminin include directional cellular migration, enhancement of cell attachment, stimulation of growth and differentiation of a number of cell types, organization of the cytoskeleton, promotion of neurite outgrowth and survival of nerve cells (Baron-van Evercooren et al., 1982; Edgar et al., 1984; Martin, 1987; Engel, 1989, 1992; Lein and Higgins, 1989; Beck et al., 1990; Sage and Bornstein, 1991; Adams and Watt, 1993; Chung, 1993; Juliano and Haskill, 1993; Tryggvason, 1993; Yurchenco and O'Rear, 1994).

The fact that laminin is the first ECM glycoprotein to appear in the developing embryo and the highly conserved basic structure of laminin in diverse species such as human, (Wewer et al., 1983; Ohno et al., 1986), mouse (Engel et al., 1981) Drosophila (Fessler et al., 1987; Montell and Goodman, 1988) sea urchin (McCarthy et al., 1987), leech (Chiquet et al., 1988; Beck et al., 1989), the cnidarian hydra (Sarras et al., 1991; Sarras et al., 1994), and anthomedusa (Schmid et al., 1991) underscores its fundamental importance. The basic structure of the molecule is conserved but there are substantial variations in amino acid sequence among species the differences being more pronounced at the carboxyl ends of the chains.

The first models of the structure of laminin were based on laminin obtained from tumor cells. The prototype of laminin purified from the extracellular deposit of mouse Parietal Yolk Sac (PYS) carcinoma cells (Chung et al., 1979) and from the mouse Engelbreth-Holm-Swarm (EHS) tumor matrix (Timpl et al., 1979) consists of three distinct polypeptide chains designated as A or α1 (400 kDa), B1 or β1 (210 kDa), and B2 or γ1 (210 kDa) (Burgeson et al., 1994; Chung, this volume). The three chains are linked by disulfide bonds and associate through their COOH-terminal halves in a coiled-coil triple helix forming a cross-shaped molecule folded into a large number of structural and functionally autonomous domains (Chung et al., 1979; Timpl et al., 1979; Engel et al., 1981; Von der Mark and Kuhl, 1985; Engel, 1989, 1992; Beck et al., 1990; Mecham, 1991; Yamada, 1991; Yurchenco and O'Rear, 1994). The α1 chain has considerable sequence and structural homology with the smaller β1 and γ1 chains, but differs in that it has a large globular domain (G-domain) at the C-terminus that represents about 25% of its mass (Hartl et al., 1988). The γ1 chain is structurally related to the β1 chain except that the α domain is missing and the number of internal cysteine-rich repeats is lower (Sasaki and Yamada, 1987; see Chung, this volume).

The diverse activities of laminin are assigned to distinct functional domains which exist on the individual chains or are formed by the folding of all three chains of the molecule (Martin and Timpl, 1987; Engel, 1989, 1992; Beck et al., 1990; Mecham 1991; Yamada, 1991;

Yamada and Kleinman, 1992; Adams and Watt, 1993; Yurchenco and O'Rear, 1994). Laminin itself forms a particularly stable complex with entactin/nidogen at the Cys-rich EGF-like repeats (domain III) of the γ1 chain (Martin and Timpl, 1987; Paulsson et al., 1987; Engel, 1989; Beck et al., 1990; Mayer et al., 1993; Poschl et al., 1994; Yurchenco and O'Rear, 1994). Laminin self assembles to form a reversible polymer that contributes to the network structure of basement membrane (Yurchenco et al., 1985; see Yurchenco and O'Rear, 1994). The globular regions at the tips of the short arms (domain VI) of laminin mediate self association of the molecule (Charonis et al., 1985; Paulsson et al., 1987; Paulsson, 1988; Bruch et al., 1989; Beck et al., 1990; Schittny and Schittny, 1993; Yurchenco and O'Rear, 1994). The terminal globules of the short arms and the terminal globules (domain G1-G5) of the long arm bind to collagen IV. The long arm globule (G1-G5) contains clusters of basic amino acids and is believed to be the major binding site for heparin and may also be involved in the binding of perlecan and of other heparan sulfate proteoglycans anchored in cell membranes (Ott et al., 1982; Deutzmann et al., 1988; Sasaki et al., 1988; Frenette et al., 1989; Beck et al., 1990).

The laminin molecule has several potential cell binding sites on both its short and long arms (Martin and Timpl, 1987; Engel, 1989; Beck et al., 1990; Yamada, 1991; Engel, 1992; Adams and Watt, 1993; Chung, 1993; Juliano and Haskill, 1993; Tryggvason, 1993). The major cell binding site in the short arm central core region of the laminin cross has been shown to be latent or cryptic in the native trimeric protein and becomes exposed or activated upon disassembly of the trimeric stucture (Aumailley et al., 1987, 1990; Nurcombe et al., 1989). Laminin has a mitogenic effect on a number of cells in culture (Kleinman et al., 1985) which in dose response and time dependence is comparable to that of EGF (Panayotou et al., 1989). It has been suggested that the β1 and γ1 chains in the mesenchyme could act as autocrine mitogens for embryonic mesenchyme (Engel, 1989; Ekblom et al., 1990). It is the fragment which consists of Cys-rich repeats in domain III that stimulates this activity (Engel, 1989; Panayotou et al., 1989). The sequence YIGSR (single-letter code for amino acids) which represents part of the Cys-rich repeats with closest homology to EGF in domain III has been identified as a cell attachment determinant (Graf et al., 1987a,b; Iwamoto et al., 1987) but has no mitogenic activity (Panayotou et al., 1989). An Arg-Gly-Asp (RGD) sequence in the central region of the A chain stimulates angiogenesis (Grant et al., 1989). Many laminin - cell interactions involve sites either at or near to the end of the long arm (Edgar et al., 1984; Aumailley et al., 1987; Dillner et al., 1988; Gehlsen et al., 1989; Deutzmann et al., 1990). Fragment E8 comprising the end of the long arm including the G domain is a high affinity cell attachment site and promotes neurite outgrowth in tissue culture (Edgar et al., 1984; Deutzmann et al., 1990). The peptide IKVAV within E8 promotes cell attachment and neurite outgrowth (Tashiro et al., 1989). Sequences within fragment E8 are necessary for development of polarity in mouse embryonic mesenchymal cells during formation of a polarized, differentiated kidney epithelium (Klein et al., 1988).

The multiplicity of putative cell receptors for laminin reflects the presence of distinct cell binding sites on the laminin molecule (Edgar, 1989; Timpl, 1989; Albelda and Buck, 1990; Beck et al., 1990; Mecham, 1991; Sastry and Horwitz, 1993; DeSimone, 1994; Haas and Plow, 1994). The different families of receptors include: a) integrin integral membrane heterodimeric α/β molecules. For example, the α6β1 integrin appears to recognize the E8 major cell binding site of laminin exclusively, while α3β1 integrin binds to laminin, entactin and fibronectin (Albelda and Buck, 1990; Hynes, 1992; DeSimone, 1994; Haas and Plow, 1994); b) non-integrin proteins such as the 67-κDa group which are single chain polypeptides that appear to be peripheral rather than integral membrane proteins. The 67-κDa laminin receptor binds with high affinity to the region in the β1 chain of laminin that contains the amino acid sequence YIGSR (Liota et al., 1985; Graf et al., 1987b; Mecham et al., 1989; Mecham, 1991); c) galactoside-binding lectins. For example, the globular region of the A chain includes a high affinity binding site for the sulfated glycolipid, galactosylsulfatide (Taraboletti et al., 1990; Mecham, 1991).

The laminin assembly may occur via the formation of a disulfide-linked β-γ intermediate (Peters et al., 1985; Hunter et al., 1990; Tokida et al., 1990) but initial random association of laminin chains has been reported (Wu et al., 1988). It has been suggested that the α chain is necessary for secretion of laminin into the extracellular space (Peters et al., 1985; Frenette et al., 1989; Tokida et al., 1990). Laminin is synthesized by a variety of cultured cells including epithelial, endothelial and striated muscle cells as well as tumor cells (Chung et al., 1979; Hogan, 1980; Lander et al., 1985). The synthesis of the three chains of laminin has been shown to be coordinated in a few cell lines (Durkin et al., 1986; Kleinman et al., 1987; Klein et al., 1990) but may not be coordinated in all cells. For example, both Schwann cells and peripheral cells of the sciatic nerve have been shown to synthesize only the β1 and γ1 chains of laminin (Cornbrooks et al., 1983; Jaakkola et al., 1989). Different tissues synthesize unequal quantities of mRNA for the α1, β1 and γ1 chains of laminin (Cooper and MacQueen, 1983; Boot-Handford et al., 1987; Kleinman et al., 1987; Klein et al., 1988; Senior et al., 1988; Laurie et al., 1989; Kucherer-Ehret et al., 1990). Antibodies against the prototype (PYS/EHS) laminin could detect only β1 and γ1 chains in placenta, lens capsule, and glomeruli (Ohno et al., 1985; Mohan and Spiro, 1986), in the ganglion cell layer of retina (Dong and Chung, 1991), in undifferentiated mesenchyme of the developing kidney (Klein et al., 1988) and in most basement membranes of the developing lung (Thomas and Dziadek, 1994). Klein et al. (1990) report that laminin α1 chain polypeptide is mainly detected in basement membranes of epithelial cells but not in most endothelial basement membranes and not in all enbryonic mesenchyme matrices during organogenesis from embryonic day 11 to the newborn mouse. These findings have shown that many cell types produce variant forms of laminin and that laminin is not one distinct protein but belongs to a family of isoproteins.

Variant laminins that have substitutions for their α1 and/or β1 chains have been identified in basement membranes in the same organism or even by the same cells and show interesting spatial and temporal expression. Two variant laminin chains that have been identified are merosin, an α1 chain homologue (Ehrig et al. 1990) and s-laminin, a β1 chain homologue (Hunter et al., 1989). Merosin (α2β1γ1) appears to be restricted to trophoblast basement membrane of human placenta and in basement membranes of striated muscle and Schwann cells (Leivo and Engrall, 1988; Ehrig et al., 1990; Engel, 1992). In the mouse it is not present in the developing embryo but appears in the basement membranes of muscle and nerve after birth. s-laminin (α1β2γ1) is localized in synaptic clefts of the neuromuscular junction and is also found in kidney glomeruli (Hunter et al., 1989; Ehrig et al., 1990; Engvall et al. 1990). O'Rear (1992) reported the synthesis of a novel laminin β1 (β1-2) chain variant in avian eye, and Brubacher et al. (1991) described the isolation of a mixture of laminin variants from adult chick heart which exhibit different tissue distributions in the 6-day-old chick embryo. The role of the variant chains is not understood but it has been proposed that they are a means of providing diversity in function to the many different basement membranes.

EXPRESION OF LAMININ IN THE EARLY EMBRYO

Studies in the development of sea urchin show laminin in the thin fibrous layer of basement membrane on the basal surface of blastula cells (Spiegel et al., 1983; McCarthy et al., 1987). An early study (Wessel et al., 1984) reports presence of laminin even in the unfertilized egg. Reports in <u>Drosophila</u> <u>melanogaster</u> have suggested a coordinated synthesis of the three chains of laminin during embryogenesis (Montell and Goodman, 1988, 1989). In situ hybridization and antibody studies reveal no laminin transcripts or protein until the end of gastrulation and the beginning of germ band elongation (Montell and Goodman, 1989; Hortsch and Goodman, 1991). Fessler et al. (1987) show laminin in basement membranes of muscles and the nervous system in the embryo, and imaginal discs in larvae in Drosophila.

In Pleurodeles embryos, laminin-related polypeptides are present in the ECM underlying the blastocoel roof (Darribere et al., 1986). In situ hybridization, using cloned Xenopus cDNA fragment homologous to the mouse β1 chain (region coding domains I, II and III), showed that laminin transcripts accumulate from mid-gastrula/yolk plug stage (stage 11- Nieuwkoop and Faber, 1956) onwards in the Xenopus embryo. Immunological techniques detect laminin in dorsal mesoderm at stage 12 about two hours after the infered onset of transcription (Fey and Hausen, 1990). It could be that laminin is present in isoforms not recognized by the antibodies used.

Laminin appears to be developmentally regulated in the mouse. Labeling with ^{35}S-methionine and immunoprecipitation with antibodies to EHS tumor laminin showed that mouse oocytes, unfertilized and fertilized eggs synthesize only the β1 chains of laminin. The failure to detect synthesis of any laminin subunits at the two-cell stage may be due to the rapid degradation of maternal messages known to occur during this stage of embryonic gene activation (Flach et al., 1982; Clegg and Pico, 1983; 1985; Davidson, 1986; Monk, this volume). The synthesis of laminin polypeptides detected from the 4-cell stage onwards would then be from embryonic mRNA produced at the mid to late two-cell stage (Cooper and MacQueen, 1983). Only β1 and γ1 chains are made by 4- to 8-cell embryos before compaction (Cooper and MacQueen, 1983). Immunofluorescence with antibodies to laminin applied to unfixed embryos demonstrated a patchy distribution of laminin over the cell surface of blastomeres at the 2-, 4- and 8- cell non-compacted and compacted stages (Dziadek and Timpl, 1985). These immunofluorescence data imply secretion and deposition of these chains onto the cell surface. The 400 -KDa α1 chain of laminin appears as a synthetic product between the 8- and 16- cell stage (Cooper and MacQueen, 1983) which coincides with the first detection of entactin/nidogen on the cell surface (Dziadek and Timpl, 1985). Leivo et al. (1980) had reported that the appearance of newly synthesized laminin α1 chains in the cytoplasm of the 16-cell morula coincides with the appearance of extracellular laminin immunofluorescence. The presence of laminin may fascilitate cell adhesion during compaction at the morula stage, directionality of the morphogenetic movements, and may also be involved in the organization of the cytoskeleton and establishment of cell polarity. The synthesis of the three laminin polypeptides first detected at the 16-cell stage was continued in the blastocyst. During blastocyst formation and expansion, laminin and entactin/nidogen are located on the outer surface of trophoblast cells (Dziadek and Timpl, 1985) and are also deposited on the inner surface of the trophoblast layer as are collagen IV, fibronectin and entactin/nidogen (Adamson and Ayers, 1979; Wartiovaara et al., 1979; Wu et al., 1983). Ultrastructural studies show the appearance of distinct basement membrane at this time. Trophoblast basement membrane material may provide the substrate onto which parietal endoderm cells migrate and synthesize laminin in large amounts and begin secretion of Reinchert's membrane in vivo (Jollie, 1968; Hogan et al., 1980). Laminin is localized on the surface of the inner cell mass in blastocyst and in all basement membranes and fetal tissues in the mouse (Leivo et al., 1980; Timpl et al., 1983).

In the chick embryo, we (Zagris and Chung, 1990) studied the temporal and spatial profile of the β1/γ1 chain of laminin from the morula (stage 0, Vakaet, 1970 and Vakaet and Bortier, this volume; stage X, Eyal-Giladi and Kochav, 1976, and Eyal-Giladi, this volume) to the neurula (stage 6, Hamburger and Hamilton, 1951; Stage 9, Vakaet and Bortier, this volume) (Fig. 1). Antibodies were to the β1 / γ1 complex of mouse laminin (Chung et al., 1979). The antibodies reveal no laminin in the blastoderm which is a flat, compacted disc at

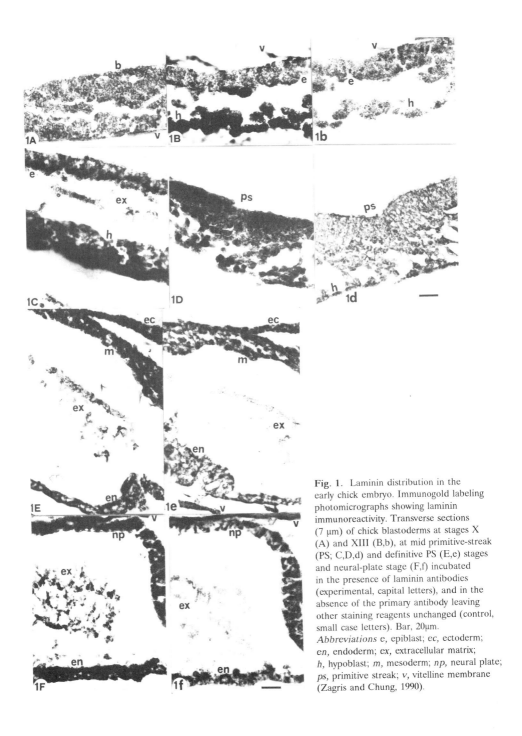

Fig. 1. Laminin distribution in the early chick embryo. Immunogold labeling photomicrographs showing laminin immunoreactivity. Transverse sections (7 μm) of chick blastoderms at stages X (A) and XIII (B,b), at mid primitive-streak (PS; C,D,d) and definitive PS (E,e) stages and neural-plate stage (F,f) incubated in the presence of laminin antibodies (experimental, capital letters), and in the absence of the primary antibody leaving other staining reagents unchanged (control, small case letters). Bar, 20μm.
Abbreviations e, epiblast; *ec,* ectoderm; *en,* endoderm; *ex,* extracellular matrix; *h,* hypoblast; *m,* mesoderm; *np,* neural plate; *ps,* primitive streak; *v,* vitelline membrane (Zagris and Chung, 1990).

stage X (homologus to the morula in amphibia - Eyal-Giladi, this volume) (Fig. 1A). In the embryo at stage XIII (blastula), laminin immunoreactivity is detected on the ventral surface of the epiblast and on the entire hypoblast (Fig. 1B). The intense labeling of the hypoblast indicates that these cells are active in laminin synthesis. As development proceeds, the migrating primitive streak cells in the epiblast that signal the onset of gastrula formation show intense laminin immunoreactivity (Fig. 1D), and the entire hypoblast and the ventral surface of the epiblast are heavily labeled (Fig. 1C). At the gastrula stage (definitive streak, stage 4-Hamburger and Hamilton, 1951; Stage 6 - Vakaet and Bortier, this volume), the ectoderm shows a positive reaction, the endoderm is heavily labeled in the area below and neighbouring the primitive streak and the mesoderm is intensely labeled (Fig. 1E). At the early neurula stage (stage 5-6- Hamburger and Hamilton, 1951; stage 7- Vakaet and Bortier, this volume), the endoderm below the neural plate is labeled intensely by the laminin antibodies and the labeling becomes weaker and disappears laterally (Fig. 1F). The neural plate and the somatic ectoderm show weak labeling. Laminin antibodies label the ECM intensely in the space (blastocoele) between the epiblast and the hypoblast and in the cavity below the neural plate (Fig. 1C, E, F). Laminin is detected as punctate deposits resembling strung beads as stained and unstained material alternate. In the early embryo, laminin immunoreactivity appears transiently in the developing tissues and defines loose meshworks rather than basement membranes (Fig. 1). Laminin due to the absence of masking by other basement membrane components may express a larger repertoire of its potential cellular interactions during development such as providing an adhesive surface for highly specific contact points guiding migrating cells and for the folding of epithelial sheets during the early phases of organization of the embryo.

In situ hybridization has been used to study the localization of laminin mRNAs during organogenesis in mice (Dong and Chung, 1991; Thomas and Dziadek, 1994). We used

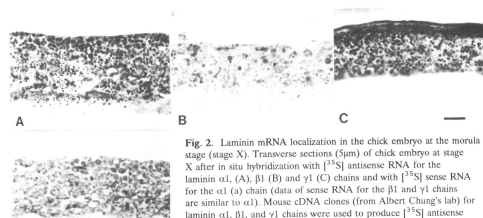

Fig. 2. Laminin mRNA localization in the chick embryo at the morula stage (stage X). Transverse sections (5µm) of chick embryo at stage X after in situ hybridization with [35S] antisense RNA for the laminin α1, (A), β1 (B) and γ1 (C) chains and with [35S] sense RNA for the α1 (a) chain (data of sense RNA for the β1 and γ1 chains are similar to α1). Mouse cDNA clones (from Albert Chung's lab) for laminin α1, β1, and γ1 chains were used to produce [35S] antisense RNA (experimental) and [35S] sense RNA (control). RNA probes were applied for hybridization to sections (5µm), washed, coated with nuclear emulsion and exposed for 10 days. The presence of silver grains indicates the location of hybridized [35S] RNA. Few or practically no silver grains are seen in the sense probe control (α). Bar, 20µm.

mouse cDNA for the three chains of laminin as templates to produce ^{35}S-UTP labeled cRNA probes for in situ hybridization in the early chick embryo. Laminin α1 and γ1 chains co-localize throughout early development. The β1 chain is non-detected or produced in very low levels. In Figure 2, the chick blastoderm at the morula stage (stage X) shows heavy expression of the α1 (Fig. 2A), and γ1 (Fig. 2C) chain mRNAs and low (practically background) expression of the β1 (Fig. 2B) transcripts. It would seem that the laminin isoform in the early chick either consist of α1/γ1 chains only or are complexed to polypeptides other than the β1 chain. This change in subunit composition can influence their interaction with cellular receptors and with other ECM molecules. Correlation between morphogenetic changes and alterations in the composition and organization of ECM poses an interesting problem in development.

Acknowledgments

This work has been supported by grants from the General Secretariat of Research and Technology of Greece, the Ministry of Health of Greece, and from NATO.

REFERENCES

Adams, J.C. and Watt, F.M. (1993) Regulation of development and differentiation by the extracellular matrix. Development 117: 1183-1198.

Adamson, E.D., and Ayers, S.E. (1979) The localization and synthesis of some collagen types in developing mouse embryos. Cell 16: 953-965.

Albelda, S.M. and Buck, C.A. (1990) Integrins and other cell adhesion molecules. FASEB J. 4: 2868-2880.

Aumailley, M., Gerl, M., Sonnenberg, A., Deutzmann, R., and Timpl, R. (1990) Identification of the Arg-Gly-Asp sequence in laminin A chain as a latent cell binding site being exposed in fragment P1. FEBS letters 262: 82-87.

Aumailley, M., Nurcombe, V., Edgar, D., Paulsson, M., and Timpl, R. (1987) The cellular interactions of laminin fragments. J. Biol. Chem. 262: 11532-11538.

Baron-van Evercooren, A., Kleinman, H.K., Ohno, S., Marangos, P., Schwartz, J.P., Dubois-Dalcq, M.E. (1982) Nerve growth factor, laminin and fibronectin promote neurite with in human fetal sensory ganglia cultures. J. Neurosci. Res. 8: 179-194.

Beck, K., Hunter, I., and Engel, J. (1990) Structure and function of laminin: anatomy of a multidomain glycoprotein. FASEB J. 4: 148-160.

Beck, K., McCarthy, R.A., Chiquet, M., Masuda-Nakagawa, L., and Schlage, W.K. (1989) Structure of basement membrane protein laminin: variations on a theme, in: "Cytoskeletal and Extracellular Proteins: Structure, Interactions and Assembly" (V. Aebi and J. Engel, eds) pp 102-105, Springer-Verlag, Berlin.

Bernfield, M.R. (1981) Organization and remodelling of the extracellular matrix in morphogenesis, in: "Morphogenesis and Pattern Formation" (T.G. Connelly, L.L. Brinkley, and B.M. Carlson, eds) pp 139-162, Raven Press, New York.

Bissel, M.J., Hall, G., and Parry, G. (1982) How does the extracellular matrix direct gene expression? J. Theor. Biol. 99: 31-68.

Boot-Handford, R.P., Kurkinen, M., and Prockop, D.J. (1987) Steady state levels of mRNAs coding for the type IV collagen and laminin polypeptide chains of basement membranes exhibit marked tissue-specific stoichiometric variations in the rat. J. Biol. Chem. 262: 12475-12478.

Brubacher, D., Wehrle-Haller, B., and Chiquet, M. (1991) Chick laminin: Isolation by monoclonal antibodies and differential distribution of variants in the embryo. Exp. Cell Res. 197: 290-299.

Bruch, M., Landwehr, R., and Engel, J. (1989) Dissection of laminin by cathepsin G into its long and short arm structures and localization of regions involved in calcium dependent stabilization and self-association. Eur. J. Biochem. 185: 272-279.

Burgeson, R.E., et al. (1994) A new nomenclature for the laminins. Matrix Biology 1: 209-211.

Charonis, A.S., Tsilibary, E.C., Yurchenco, P.D., and Furthmayr, H. (1985) Binding of laminin to type IV collagen: A morphological study. J.Cell Biol. 100: 1848-1853.

Chiquet, M., Masuda-Nakagawa, L., Beck, K. (1988) Attachment to an endogenous laminin-like protein initiates sprouting by leech neurons. J. Cell Biol. 107: 1189-1198.

Chung, A.E. (1993) Laminin, in: "Extracellular Matrix", (M.A. Zern and L.M. Reid, eds.), pp 25-48, Marcel Dekker, Inc., New York.

Chung, A.E., Jaffe, R., Freeman, I.L., Vergnes, J.-P. Braginski, J.E., and Carlin, B. (1979) Properties of a basement membrane-related glycoprotein synthesized in culture by a mouse embryonal carcinoma derived cell line. Cell 16: 277-287.

Clegg, K.B. and Piko, L. (1983) Poly (A) length, cytoplasmic adenylation, and synthesis of poly (A)$^+$ RNA in early mouse embryos. Devl. Biol. 95: 331-341.

Cooper, A.R. and MacQueen, H.A. (1983) Subunits of laminin are differentially synthesized in mouse eggs and embryos. Devl. Biol. 96: 467-471.

Cornbrooks, C.J., Carey, D.J., McDonald, J.A., Timpl, R., and Bunge, R.P. (1983) In vivo and in vitro observations on laminin production by Schwann cells. Proc. Natl. Acad. Sci. USA 80: 3850-3854.

Darribere, T., Riou, J.F., Shi, D.L., Delarue, M., Boucaut, J.C. (1986) Synthesis and distribution of laminin-related polypeptides in early amphibian embryos. Cell Tissue Res. 246: 45-51.

Davidson, E.H. (1986) "Gene Activity in Early Development", 3rd ed., Academic Press, New York.

DeSimone, D.W. (1994) Adhesion and matrix in vertebrate development. Cur. Opin. Cell Biol. 6: 747-751.

Deutzmann, R., Aumailley, M., Wiedemann, H., Pysny, W., Timpl, R. and Edgar, D. (1990) Cell adhesion spreading and neurite stimulation by laminin fragment E8 depend on maintenance of secondary and tertiary structure in its rod and globutar domain. Eur. J. Biochem. 191: 513-522.

Deutzmann, R., Huber, J., Schmetz, K.A., Oberbaumer, I., and Hartl, L. (1988) Structural study of long arm fragments of laminin. Eur. J. Biochem. 177: 35-45.

Dillner, L., Dickerson, K., Manthorpe, M., Ruoslahti, E., and Engvall, E. (1988) The neurite-promoting domain of human laminin promotes attachment and induces characteristic morphology in non-neuronal cells. Exp. Cell Res. 177: 186-198.

Dong, L.-J. and Chung, A.E. (1991) The expression of the genes for entactin, laminin A, laminin B1 and laminin B2 in murine lens morphogenesis and eye development. Differentiation 48: 157-172.

Durkin, M.E., Phillips, S.L., Chung A.E. (1986) Control of laminin synthesis during differentiation of F9 embryonal carcinoma cells. Differentiation 32: 260-266.

Dziadek, M. and Timpl, R. (1985) Expression of nidogen and laminin in basement membranes during mouse embryogenesis and in teratocarcinoma cells. Devl. Biol. 111: 372-382.

Edgar, D. (1989) Neuronal laminin receptors. Trends Neurosci. 72: 248-251.ʻ

Edgar, D., Timpl, R., and Thoenen, H. (1984) The heparin-binding domain of laminin is responsible for its effects on neurite outgrowth and neuronal survival. EMBO J. 3: 1463-1468.

Ehrig, K., Leivo, I., Argraves, W.S., Ruoslahti, E., and Engvall, E. (1990) Merosin, a tissue-specific basement membrane protein, is a laminin-like protein. Proc. Natl. Acad. Sci., USA 87: 3264-3268.

Ekblom, M., Klein, G., Mugrauer, G., Fecker, I., Deutzmann, R., Timpl. R., and Ekblom, P. (1990) Transient and locally restricted expression of laminin A chain mRNA by developing epithelial cells during kidney organogenesis. Cell 60: 337-346.

Engel, J. (1989) EGF-like domains in extracellular matrix proteins: localized signals for growth and differentiation? FEBS letters 251: 1-7.

Engel, J. (1992) Laminins and other strange proteins. Biochemistry 31: 10643-10651.

Engel, J., Odermatt, E., Engel, A., Madri, J.A., Furthmayr, H., Rohde, H., and Timpl, R. (1981) Shapes, domain organization, and flexibility of laminin and fibronectin, two multifunctional proteins of the extracellular matrix. J. Mol. Biol. 150: 97-120.

Engvall, E., Earwicker, D., Haaparanta, T., Ruoslahti, E., and Sanes, J. (1990) Distribution and isolation of four laminin variants; tissue restricted distribution of heterotrimers assembled from different subunits. Cell Regul. 1: 731-740.

Eyal-Giladi, H., Kochav, S. (1976) From cleavage to primitive streak formation: A complementary normal table and a new look at the first stages of the development of the chick. I. General morphology. Devl. Biol. 49: 321-337.

Fessler, L.I., Campbell, A.G., Duncan, K.G., and Fessler, J.H. (1987) Drosophila laminin: Characterization and localization. J. Cell. Biol. 105: 2383-2391.

Fey, J. and Hausen, P. (1990) Appearance and distribution of laminin during development of Xenopus laevis. Differentiation 42: 144-152.

Flach, G., Johnson, M.H., Braude, P.R., Taylor, R., and Bolton, V.N. (1982) Transition from maternal to embryonic control in the early mouse embryo. EMBO J. 1: 681-686.

Frenette, G.R., Ruddon, R.W., Krzesicki, R.F., Naser, J.A., and Peters, B.P. (1989) Biosynthesis and deposition of a non-covalent laminin-heparan sulfate proteoglycan complex and other basal lamina components by a human malignant cell line. J. Biol. Chem. 264: 3078-3088.

Gehlsen, K.R., Dickerson, K., Argraves, W.S., Engvall, E., and Ruoslahti, E. (1989) Subunit structure of a laminin-binding integrin and localization of its binding site on laminin. J. Biol. Chem. 264: 19034-19038.

Graf, J., Iwamoto, Y., Sasaki, M., Martin, G.R., Kleinman, H.K., Robey, F.A., Yamada, Y. (1987 a) Identification of an amino acid sequence in laminin mediating cell attachment, chemotaxis, and receptor binding. Cell 48: 989-996.

Graf, J., Ogle, R.C., Robey, F.A., Sasaki, M., Martin, G.R., et al. (1987 b) A pentapeptide from the laminin B1 chain mediates cell adhesion and binds the 67,000 laminin receptor. Biochemistry 26: 6896-6900.

Grant, D.S., Tashiro, K.-I., Segui-Real, B., Yamada, Y., Martin, G.R., and Kleinman, H.K. (1989) Two different laminin domains mediate the differentiation of human endothelial cells into capillary-like structures in vitro. Cell 58: 933-943.

Grobstein, C. (1954) Tissue interaction in the morphogenesis of mouse embryonic rudiments in vitro, in: "Aspects of Synthesis and Order in Growth", (D.Rudnik, ed.), pp 233-256, Princeton University Press, Princeton.

Haas, T.A., and Plow, E.F. (1994) Integrin-ligand interactions: a year in review. Cur. Opin. Cell Biol. 6: 656-662.

Hamburger, V., Hamilton, H.L. (1951) A series of normal stages in the development of the chick embryo. J. Morph. 88: 49-92.

Hartl, L., Oberbaumer, I., and Deutzmann, R. (1988) The N terminus of laminin A chain is homologous to the B chains. Eur. J. Biochem. 173: 629-635.

Hay, E.D., ed. (1991) "Cell Biology of Extracellular Matrix", 2nd ed., Plenum Press, New York.

Hogan, B.L.M. (1980) High molecular weight extracellular proteins synthesized by endoderm cells derived from mouse teratocarcinoma cells and normal extraembryonic membranes. Devl. Biol. 76: 275-285.

Hogan, B.L.M., Cooper, A.R., Kurkinen, M. (1980) Incorporation into Reichert's membrane of laminin-like extracellular proteins synthesized by parietal endoderm cells of the mouse embryo. Devl. Biol. 80: 289-300.

Hortsch, M. and Goodman, C.S. (1991) Cell and substrate adhesion molecules in Drosophila. Ann. Rev. Cell Biol. 7: 505-557.

Horwitz, A.F. and Thiery, J.P. (1994) Cell-to-cell contact and extracellular matrix. Cur. Opin. Cell Biol. 6: 645-647.

Hunter, D.D., Shah, V., Merlie, J.P. and Sanes, J.R. (1989) A laminin-like adhesive protein concentrated in the synaptic cleft of the neuromuscular junction. Nature 338: 229-234.

Hunter, I., Schulthess, T., Bruch, M., Beck, K. and Engel, J. (1990) Evidence for a specific mechanism of laminin assembly. Eur. J. Biochem 188: 205-211.

Hynes, R.O. (1992) Integrins: versatility, modulation and signalling in cell adhesion. Cell 69: 11-25.

Iwamoto, Y., Robey, F.A., Graf, J., Sasaki, M., Kleinman, H.K., Yamada, Y., Martin, G.R. (1987) YIGSR, a synthetic laminin pentapeptide, inhibits experimental metastasis formation. Science 238: 1132-1134.

Jaakkola, S., Peltonen, J., Uitto, J.J. (1989) Perineurial cells coexpress genes encoding interstitial collagens and basement membrane zone components. J. Cell Biol. 108: 1157-1163.

Jollie, W.P. (1968) changes in the fine structure of the parietal yolk sac of the rat placenta with increasing gestational age. Amer. J. Anat. 122: 513-532.

Juliano, R.L., Haskill, S. (1993) Signal transduction from the extracellular matrix. J. Cell Biol. 120: 577-585.

Klein, G., Ekblom, M., Fecker, L., Timpl, R., and Ekblom, P. (1990) Differential expression of laminin A and B chains during development of embryonic mouse organs. Development 110: 823-837.

Klein, G., Langegger, M., Timpl, R., and Ekblom, P. (1988) Role of laminin A chain in the development of epithelial cell polarity. Cell 55: 331-341.

Kleinman, H.K., Cannon, F.B., Laurie, G.W., Hassell, J.R., Aumailley, M., Terranova, V.P., Martin, G.R., and Dubois-Dalc, M. (1985) Biological activities of laminin. J. Cell Biochem. 27: 317-325.

Kleinman, H.K., Ebihara, I., Killen, P.D., Sasaki, M., Cannon, F.B., Yamada, Y., and Martin G.R. (1987) Genes for basement membrane proteins are coordinately expressed in differentiating F9 cells, but not in normal adult murine tissues. Devl. Biol. 122: 373-378.

Kucherer-Ehret, A., Pottgiesser, J., Kreutzberg, G.W., Thoenen, H., and Edgar, D. (1990) Developmental loss of laminin from the interstitial extracellular matrix correlates with decreased laminin gene expression. Development 110: 1285-1293.

Lander, A.D., Fujii, D.K., Reichardt, L.F. (1985) Purification of a factor that promotes neurite outgrowth: Isolation of laminin and associated molecules. J.Cell Biol. 101: 898-913.

Laurie, G.W., Horikoshi, S., Killen, P.D., Segui-Real, B., and Yamada, Y. (1989) In situ hybridization reveals temporal and spatial changes in cellular expression of mRNA for a laminin receptor, laminin and basement membrane (type IV) collagen in the developing kidney. J. Cell Biol. 109: 1351-1362.

Lein, P.J. and Higgins, D. (1989) Laminin and a basement membrane extract have different effects on axonal and dendritic outgrowth from embryonic rat sympathetic neurons in vitro. Devl. Biol. 136: 330-345.

Leivo, I. and Engvall, E. (1988) Merosin, a protein specific for basement membrane of Schwann cells, striated muscle, and trophoblasts, is expressed late in nerve and nuscle development. Proc. Natl. Acad. Sci. USA 85: 1544-1548.

Leivo, I., Vaheri, A., Timpl, R., and Wartiovaara, J. (1980) Appearance and distribution of collagens and laminin in the early mouse embryo. Devl. Biol. 76: 100-114.

Liotta, L.A., Wewer, U.M., Rao, C.N., and Bryant, G. (1985) Laminin receptor, in: "The Cell in Contact", (G.M. Edelman and J.P. Thiery, eds.), pp 333-344, John Wiley and Sons, New York.

Martin, G.R., Kleinman, H.K., Terranova, V.P., Ledbetter, S., Hassell, J.R. (1984) The regulation of basement membrane formation and cell-matrix interactions by defined supramolecular complexes. Ciba Symp. 108: 197-212.

Martin, G.R., Timpl, R. (1987). Laminin and other basement membrane components. Ann. Rev. Cell Biol. 3: 57-85.

Mayer, V., Nischt, R., Poschl, E., Mann, K., Fukuda, K., Gerl, M., Yamada, Y. and Timpl, R. (1993) A single EGF-like motif of laminin is responsible for high affinity nidogen binding. EMBO J. 12: 1879-1885.

McCarthy, R.A., Beck, K., and Burger, M.M. (1987) Laminin is structurally conserved in the sea urchin basal lamina. EMBO J. 6: 1587-1593.

Mecham, R.P. (1991) Laminin receptors. Ann. Rev. Cell Biol. 7: 71-91.

Mecham, R.P., Hinek, A., Griffin, G.L., Senior, R.M., and Liotta, L.A. (1989). The elastin receptor shows structural and functional similarities to the 67-kDa tumor cell laminin receptor. J. Biol. Chem. 264: 16652-16657.

Mohan, P.S. and Spiro, R.G. (1986) Molecular organization of basement membranes. J. Biol. Chem. 261: 4328-4336.

Montell, D.J. and Goodman, C.S. (1988) Drosophila substrate adhesion molecule: Sequence of laminin B1 chain reveals domains of homology with mouse. Cell 53: 463-473.

Montell, D. J. and Goodman, C.S. (1989) Drosophila laminin: Sequence of B2 subunit and expression of all three subunits during embryogenesis. J. Cell Biol. 109: 2441-2453.

Nieuwkoop, P.D., Faber, J. (1956) "Normal Table of Xenopus laevis" North-Holland Pub. Co., Amsterdam.

Nurcombe, V., Aumailley, M., Timpl, R., and Edgar D. (1989). The high affinity binding of laminin to cells. Assignment of a major cell-binding site to the long arm of laminin and of a latent cell-binding site to its short arms. Eur. J. Biochem. 180: 9-14.

Ohno, M., Martinez-Hernandez, A., Ohno, N., Kefalides, N.A. (1985) Comparative study of laminin found in normal placental membrane with laminin of neoplastic origin, in: "Basement Membranes" (S. Shibata, ed.), pp 3-11, Elsevier, Amsterdam.

Ohno, M., Martinez-Hernandez, A., Ohno, N., and Kefalides, N.A. (1986) laminin M is found in placental basement membranes but not in basement membranes of neoplastic origin. Connective Tiss. Res. 15: 199-

O'Rear, J.J. (1992) A novel laminin B1 chain variant in avian eye. J. Biol. Chem. 267: 20555-20557.

Ott, U., Odermatt, E., Engel, J., Furthmayr, H., and Timpl, R. (1982) Protease resistance and conformation of laminin. Eur. J. Biochem. 123: 63-72.

Panayotou, G., End, P., Aumailley, M., Timpl. R., Engel, J. (1989) Domains of laminin with growth factor activity. Cell 56: 93-101.

Paulsson, M. (1988) The role of Ca^{2+} binding in the self-aggregation of laminin-nidogen complexes. J. Biol. Chem. 263: 5425-5430.

Paulsson, M., Aumailley, M., Deutzmann, R., Timpl, R., Beck, K., and Engel, J. (1987) Laminin-nidogen complex: extraction with chelating agents and structural characterization. Eur. J. Biochem. 166: 11-16.

Peters, B.P., Hartle, R.J., Krzesick, R.F., Kroll, T.G., Perini, F., Balun, J.E., Goldstein, I.J., and Ruddon, R.W. (1985) The biosynthesis, processing and secretion of laminin by human choriocarcinoma cells. J.Biol. Chem. 260: 14732-14742.

Poschl, E., Fox, J.W., Block, D., Mayer, V., and Timpl, R. (1994) Two non-contiguous regions contribute to nidogen binding to a single EGF-like motif of the laminin γ1 chain. EMBO J. 13: 3741-3747.

Rogers, S.L., Edson, K.J., Letourneau, P.C., and McLoon, S.C. (1986) Distribution of laminin in the developing peripheral nervous system of the chick. 113: 429-435.

Sage, E.H. and Bornstein, P. (1991) Extracellular proteins that modulate cell-matrix interactions. J. Biol. Chem. 23: 14831-14834.

Sanes, J.R., Engvall, E., Butkowski, R., and Hunter, D.D. (1990) Molecular heterogeneity of basal laminae: Isoforms of laminin and collagen IV at the neuromuscular junction and elsewhere, J. Cell Biol. 111: 1685-1699.

Sarras, M.P., Jr., Madden, M.E., Zhang, X., Gunwar, S., Huff, J.K., and Hudson, B.G. (1991) Extracellular matrix (mesoglea) of Hydra vulgaris. I.Isolation and characterization. Devl. Biol. 148: 481-494.

Sarras, M.P., Jr., Yan, L., Grens, A., Zhang, X., Agbas, A., Huff, J.K., St. John, P.L., and Abrahamson, D.R. (1994) Cloning and biological function of laminin in Hydra vulgaris. Devl. Biol. 164: 312-324.

Sasaki, M., Kleinman, H.K., Huber, H., Deutzmann, R., and Yamada, Y. (1988) Laminin, a multidomain protein. J. Biol. Chem. 263: 16536-16544.

Sasaki, M. and Yamada, Y. (1987) The laminin B2 chain has a multidomain structure homologous to the B1 chain. J. Biol. Chem. 262: 17111-17117.

Sastry, S.K. and Horwitz, A.F. (1993) Integrin cytoplasmic domains: mediators of cytoskeletal linkages and extra-and intracellular initiated transmembrane signalling. Cur. Opin. Cell Biol. 5: 819-831.

Schittny, J.C. and Schittny, C.M. (1993) Role of the B1 short arm in laminin self-assembly. Eur. J. Biochem. 216: 437-441.

Schmid, V., Bally, A., Beck, K., Haller, M., Schlage, W.K., Weber, C. (1991) The extracellular matrix mesoglea of hydrozoan jellyfish and its ability to support cell adhesion and spreading. Hydrobiologia 216/217: 3-10.

Senior, R.V., Critchley, D.R., Beck, F., Walker, R.A., and Varley, J.M. (1988) The localization of laminin mRNA and protein in the postimplantation embryo and placenta of the mouse: an in situ hybridization and immunocytochemical study. Development 104: 431-436.

Spiegel, E., Burger, M.M., Speigel, M. (1983) Fibronectin and laminin in the extracellular matrix and basement membranes of sea urchin embryos. Exp. Cell Res. 144: 47-55.

Taraboletti, G., Rao, C.N., Krutasch, H.C., Liotta, L.A., Roberts, D.D. (1990) Sulfatide-binding domain of the laminin A chain. J. Biol. Chem. 265: 12253-12258.

Tashiro, K.-I., Sephel, G.C., Weeks, B., Sasaki, M., Martin, G.R., Kleinman, H.K., and Yamada, Y. (1989) A synthetic peptide containing the IKVAV sequence from the A chain of laminin mediates cell attachment, migration and neurite outgrowth. J. Biol. Chem. 264: 16174-16182.

Thomas, T. and Dziadek, M. (1994) Expression of collagen α1 (IV), laminin and nidogen genes in the embryonic mouse lung: implications for branching morphogenesis. Mech. Devl. 45: 193-201.

Timpl, R. (1989) Structure and biological activity of basement membrane proteins. Eur. J. Biochem. 180: 487-502.

Timpl, R., Johannson, S., van Delden, V., Oberbaumer, I., and Hook, M. (1983) Characterization of protease-resistant fragments of laminin mediating attachment and spreading of rat hepatocytes. J. Biol. Chem. 258: 8922-8927.

Timpl, R. and Rohde, H, Robey, P.G., Rennard, S.I., Foidart, J.-M. Martin, G.R. (1979) Laminin-A glycoprotein from basement membranes. J. Biol. Chem. 254: 9933-9937.

Tokida, Y., Aratani, Y., Morita, A., Kitagawa, Y. (1990) Production of two variant laminin forms by endothelial cells and shift of their relative levels by angiostatic steroids. J. Biol. Chem. 265: 18123-18129.

Tryggvason, K. (1993) The laminin family. Cur. Opin. Cell Biol. 5: 877-882.

Vakaet, L. (1970) Cinephotomicrographic investigations of gastrulation in the chick blastoderm. Arch. Biol. (Liege) 81: 387-426.

Von der Mark, K., Kuhl, U. 1985. Laminin and its receptor. Biochim. Biophys. Acta 823: 147-160.

Wan, Y.-J., Wu, T.-C., Chung, A.E., and Damjanov, I. (1984) Monoclonal antibodies to laminin reveal the heterogeneity of basement membranes in the developing and adult mouse tissues. J. Cell Biol. 98: 971-979.

Wartiovaara, J., Leivo, I., Vaheri, A. (1979) Expression of the cell-surface-associated glycoprotein, fibronectin, in the early mouse embryo. Devl. Biol. 69: 247-257.

Wessel, G.M., Marchase, R.B., and McClay, D.R. (1984) Ontogeny of the basal lamina in the sea urchin embryo. Devl. Biol. 103: 235-245.

Wewer, U., Alberchtsen, R., Manthorpe, M., Varon, S., Engvall, E., and Ruoslahti, E. (1983) Human laminin isolated in a nearly intact, biological active form from placenta by limited proteolysis. J. Biol. Chem. 258: 12654-12660.

Wu, C., Friedman, R., and Chung A.E. (1988) Analysis of the assembly of laminin and the laminin-entactin complex with laminin chain specific monoclonal and polyclonal antibodies. Biochemistry 27: 8780-8787.

Wu, T.-C., Wan, Y.-J., Chung, A.E., and Damjanov, I. (1983) Immunohistochemical localization of entactin and laminin in mouse embryos and fetuses. Devl. Biol. 100: 496-505.

Yamada, K.M. (1991) Adhesive recognition sequences. J. Biol. Chem. 266: 12809-12812.

Yamada, Y., and Kleinman, H.K. (1992) Functional domains of cell adhesion molecules. Cur. Opin. Cell Biol. 4: 819-823.

Yurchenco, P. D., and O'Rear, J.J. (1994) Basal lamina assembly. Cur. Opin. Cell Biol. 6: 674-681.

Yurchenco, P.D., Tsilibary, E.C., Charonis, A.S., and Furthmayr, H. (1985) Laminin polymerization in vitro: evidence for a two-step assembly with domain specificity. J. Biol. Chem. 260: 7636-7644.

Zagris, N. and Chung, A.E. (1990) Distribution and functional role of laminin during induction of the embryonic axis in the chick embryo. Differentiation 43: 81-86.

STRUCTURE AND FUNCTION OF CADHERINS

Florence Broders and Jean Paul Thiery

Laboratoire de Physiopathologie du Développement
Ecole Normale Supérieure CNRS URA 1337
46, rue d'Ulm 75230 Paris cedex 05, France

INTRODUCTION

Cadherins form a superfamily of transmembrane glycoproteins involved in Ca^{2+}-dependent cell-cell adhesion (Takeichi, 1991; Kemler et al., 1989). Cadherins mediate homophilic interactions and are thought to be responsible for the selective cell-cell adhesion or cell sorting that is necessary to position different cell types during development (Takeichi, 1988). Cadherins also play a fundamental role in maintaining the integrity of multicellular structures. Most of the cadherins identified so far are from the vertebrates, but several cadherins have recently been characterized in Drosophila. During embryogenesis, the expression of multiple members of the cadherin superfamily is spatio-temporally regulated and correlates with morphogenetic events, particularly with the remodeling of epithelia and the segregation of cell collectives. Several cadherins are expressed in each cell type simultaneously, although in a defined repertoire and with variable levels of expression. These different combinations of cadherins are likely to be responsible for distinct adhesive properties of cells.

Members of the cadherin superfamily have been classified in several categories based on sequence alignment, including type I cadherins, T-cadherins (Sacristàn et al., 1993), type II cadherins (Suzuki et al., 1991), new members such as OB-cadherin (Okasaki et al., 1994), K-cadherin (Xiang et al., 1994) and LI-cadherin (Berndorff et al., 1994), protocadherins (Sano et al., 1993), cadherins from desmosomes (Arnemann et al., 1993), and cadherin-related molecules such as fat (Mahoney et al., 1991) and c-ret (Iwamoto et al., 1993).

Two major categories of calcium-sensitive junctions associated with cells in contact can be distinguished. The first category consists of junctions whose plaques anchor actin microfilaments and are called "zonula adherens", "adherens junctions" or "intermediate junctions"; these junctions contribute to the establishment and maintainance of epithelial cell polarity. Type I cadherins (E-, N, P-) are concentrated at the adherens junctions generally at the apex on the lateral

Organization of the Early Vertebrate Embryo
Edited by N. Zagris *et al.*, Plenum Press, New York, 1995

portion of polar epithelial and endothelial cells. Cadherins are linked to the actin filaments through specific proteins called catenins (Kemler, 1993; Gumbiner, 1993) and cytoplasmic plaque proteins such as plakoglobin (Franke et al., 1987), vinculin (Geiger, 1979), alpha actinin, tenuin, plectin (Wiche et al., 1991) and radixin (Tsukita et al., 1989). Cadherins are also found more diffusely distributed outside these differentiated junctions.

The second category of junctions consists of desmosomes or "maculae adhaerentes", which are disc-shaped cell adhesion structures linked to the intermediate filaments. A family of cadherins is localized in the desmosome networks that occur in epithelial but also in myocardial cells and Purkinje fibers cells. The desmosomal cadherins interact with plaque components including desmoplakins (Green et al., 1988), plakoglobin (Franke et al., 1987), desmocalmin (Tsukita and Tsukita, 1985), lamin B-like protein, plectin, band 6 protein which is related to plakoglobin (Hatzfeld et al., 1994), and "IFAP 300 protein" (Skalli et al., 1994).

CLASSICAL (TYPE I) CADHERINS

In mammals, classical cadherins (type I) include neural (N)-cadherin, epithelial (E)-cadherin (uvomorulin), placental (P)-cadherin, retinal (R) or cadherin-4, and muscle (M)-cadherin; each shows a unique pattern of distribution during development and in the adult (Hatta et al., 1987; Nagafuchi et al., 1987; Nose et al., 1987; Inuzuka et al., 1991; Tanihara et al., 1994a; Donalies et al., 1991). Most of these cadherins are also present in birds: L-CAM is considered to be the equivalent of E-cadherin in mammals (Gallin et al., 1987); chicken K-CAM (Gally and Edelman, 1992), which is most likely the chicken B-cadherin (Napolitano et al., 1991), may be related to P-cadherin in the mouse (Redies et Müller, 1994). In Xenopus, homologous cadherins have been characterized and sequenced, including N-cadherin (Detrick et al,. 1990; Ginsberg et al., 1991; Simonneau et al, 1992), E-cadherin (Broders et al., 1993; Levine et al., 1994), and cadherin-4 (Broders et al., unpublished observations); EP-/C- cadherin (Ginsberg et al., 1991; Choi et al., 1990) and XB-/U-cadherin (Müller et al., 1994) have not been found so far in higher vertebrates. Classical cadherins share a common 723-748 amino acid sequence and have an apparent molecular mass of 120-140kD. They are synthesized as precursor polypeptides which require a series of post-translational modifications (glycosylation, phosphorylation and proteolytic cleavage) to form a mature protein. Classical cadherins exhibit an extracellular domain subdivided in 5 subdomains (EC1-EC5) of 110 amino acids, a transmembrane hydrophobic region and a highly conserved cytoplasmic tail (Kemler, 1992).

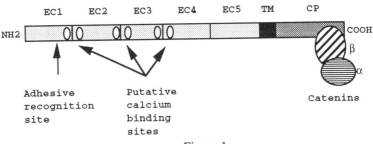

Figure 1

Amino acid sequence similarities

Classical cadherins exhibit on average 50% amino acid identity within a single species. Most of the aromatic amino acids are shared by all cadherins. The percentage of identity is higher among one subclass of cadherins in different species; for instance, Xenopus, chicken and mouse N-cadherin share about 85% identity (Detrick et al., 1990). On the other hand, only 63% amino acid identity was found between Xenopus E-cadherin and the E-cadherins of other species; Xenopus E-cadherin shares a similar level of identity with Xenopus EP-cadherin (Broders et al., 1993). Comparison of amino acid identities of the extracellular subdomains EC1 to EC5 of cadherin subclasses revealed that: EC1 is the most conserved subdomain showing an average identity of 70%, the subdomains EC2 to EC4 show 50% to 60% identity; the EC5 subdomain is the least conserved, exhibiting an average of 40% identity; the transmembrane regions show an identity of 50% to 80%, and the cytoplasmic regions show 80% to 90% identity. T-cadherin, which lacks the cytoplasmic domain, exhibits sequences characteristic of classical cadherins (Tanihara et al., 1994b).

Functional sites of the extracellular domain of type I cadherins

Regions determining homophilic interactions. The HAV sequence (His-Ala-Val) located in the EC1 subdomain around position 80 is considered to be a recognition sequence (Blaschuk et al., 1990). Site-directed mutagenesis of E-cadherin at only two positions on either side of the HAV sequence (position 78 and 83) alters the specificity of adhesion of classical cadherins (Nose et al., 1990). Homophilic binding can be observed in cells expressing chimeric molecules of P- and E-cadherin having at least the same 113 amino terminal sequence, strongly suggesting that this region is essential for the binding of identical cadherins. This region is well conserved within a given type of cadherin from various species. However, T-cadherin was reported to have homophilic cell adhesion activity despite the absence of a HAV sequence (Ranscht and Dours-Zimmermann, 1991); moreover, the amino acids at position 78 and 83 do not show any identity with those previously identified for type I cadherin.

The maturation of the precursor forms of cadherins can be prevented by mutations in the putative recognition sequence for endoproteolytic enzymes RXKR, where X can be any residue. These mutated precursor forms are expressed at the cell surface but lack adhesive function: this result stresses the importance of the mature amino terminal region (Ozawa and Kemler, 1990). Classical cadherins show homophilic cell adhesion, but heterophilic adhesion is sometimes observed; as described in Inuzuka et al. (1991), R-cadherin can interact transiently with N-cadherin but cells expressing either N- or R-cadherin forming a mosaic aggregate eventually sort out. Binding of monoclonal antibodies to the amino-terminal region results in inhibition of cadherin adhesion. However, one inhibitory monoclonal antibody (DECMA-1) has been shown to bind to the cysteine-rich EC5 juxtamembrane subdomain (Ozawa et al., 1991). Much work remains to be done to fully characterize the sites participating in the homophilic interactions.

Calcium binding sites. Cadherins require calcium ions for their adhesive function (Takeichi, 1990); calcium can also confer stability on the cadherin molecule against degradation by proteases. The extracellular region revealed 6 short specific sequences possibly involved in the interactions with calcium ions.

The putative calcium-binding motifs (DXNDN and DXD) are located in hydrophilic loops preceded by a beta sheet and alpha helix structures (Kemler et al., 1989). The DXNDNXP sequence appears with minor variations in the carboxy-terminal region of the extracellular subdomains EC1 to EC3 and the DXD sequence is located near the amino-terminal region of EC2, EC3, EC4. However, recently, a new sequence alignment of the extracellular domains revealed additional DXND/ENXP and DXD sequences located near the carboxy-terminal region of EC4 and the amino-terminal region of EC5, respectively (Tanihara et al., 1994b), but their role is still unknown. A single amino acid substitution (aspartic acid in position 134 to alanine or lysine) abolishes the adhesive function of E-cadherin (Ozawa et al., 1990), and provides evidence that one of the possible Ca2+ binding sites in EC2 contains the binding activity and is necessary for cell adhesion. These conserved sequences are absent from the amino-terminal region of EC1 and from the carboxy-terminal region of EC5.

Other sequences. Other stretches of very well conserved amino acids (DXE) are found in the middle region of EC1, EC2 and EC4 (DRE) and in the middle region of EC3 (D(F/Y)E), but their function is unknown. The new alignment between cadherin sequences revealed additional structural features such as the AXDXGXP motif near the C-terminus of all the 5 subdomains (EC1-EC5), which had not been recognized before; this better alignment of cadherin extracellular domains shows that EC5 contains, despite its high divergence, the DXD motif located in its amino-terminal region and some conserved amino acids (Tanihara et al., 1994b). Since cadherins are glycoproteins, an average of five consensus sequences for N-linked glycosylation are usually found (NXS or NXT sequences).

Functional sites of the cytoplasmic region of type I cadherins

The sequences of the cytoplasmic domain are highly conserved among cadherin subclasses and exhibit long stretches of conserved amino acids; some of which, located in the 40 amino acid region described in Suzuki et al. (1991) are particularly specific for one type of cadherin. Cadherins co-localize with cortical actin bundles at cell-cell boundaries, particularly in adherens junctions (Hirano et al., 1987; Geiger and Ginsberg, 1991), and the catenins, through their association to the cytoplasmic domains of cadherins, are thought to mediate the association between cadherins and the cytoskeleton (Ozawa et al., 1989; McCrea and Gumbiner, 1991). Catenins link cadherins to other transmembrane proteins such as the Na+/K+ ATPase and cytoplasmic proteins such as fodrin (McNeill et al., 1990).

Another post-translational modification which affects cadherins is phosphorylation of serine or threonine residues: consensus sequences for cAMP-dependent protein kinase phosphorylation of serine and threonine has been defined as KRXXST and RRXST (Krebs and Beavo, 1979); a corresponding sequence is located in the juxtamembrane domain of the mouse E-cadherin.

Recently, a short highly phosphorylated core region has been identified within the 72-amino acid residues of the carboxy-terminal region of E-cadherin, which corresponds to the catenin binding domain (Stappert and Kemler, 1994). This 30 amino acid region contains 8 serine residues which are well-conserved among cadherins; the substitution of the whole serine cluster results in unphosphorylated E-cadherin and is correlated with the loss of catenin association and cadherin-mediated cell-cell adhesion. 4 serine residues may be targets for casein kinase II (S/TXXA/EXA/E sequences), as described in Marin et al. (1986). Truncated cadherin lacking this region cannot mediate cell adhesion. Three proteins with an apparent molecular mass of 102, 88 and 80 kD, corresponding to alpha-, beta- and gamma-catenins respectively, co-precipitate with antibodies against E-cadherin (Kemler, 1993). However, these proteins are also present in cells devoid of cadherins.

Sequence analysis of alpha-catenin revealed two isoforms, alphaE- and alphaN-catenin. Alpha catenin displays three distinct regions which share partial identity with those of vinculin, a protein also involved in the cytoplasmic anchoring of integrins in focal contact (Geiger et al., 1980). Alpha-catenin has been cloned in the mouse, Drosophila, chicken and human (Herrenknecht et al., 1991, Oda et al., 1994, Hirano et al., 1992, Claverie et al., 1993). Alpha-catenin binds to the carboxy terminal region of beta-catenin.

Beta-catenin exhibits strong identity to plakoglobin, a component of the desmosomal plaque and adherens junctions (Franke et al., 1989), and to the product of the Drosophila armadillo gene, a member of the wingless class of segment polarity genes (McCrea et al., 1991). The central region of beta-catenin contains 13 repeats of approximately 40 amino acids (Hülsken et al., 1994); these repeats were originally described in the armadillo protein. Beta-catenin binds tightly and directly to the cytoplasmic domain of cadherins through its central domain; deletions of the amino and carboxy terminal regions including at least some repeats abolish the binding of beta-catenin to cadherins (Hülsken et al., 1994).

Data from pulse-chase experiments demonstrated that beta-catenin associates with the precursor cadherin in the endoplasmic reticulum, while alpha-catenin is added later to the complex during proteolytic processing of cadherin (Hinck et al., 1994).

Gamma-catenin is identical to plakoglobin (Knudsen and Wheelock, 1992). The relative amount of gamma-catenin in the cadherin-catenin complex varies depending on cell type, and was not always found in the complex; beta- and gamma-catenin are mutually exclusive in a cadherin-catenin complex (Näthke et al., 1994).

TYPE II CADHERINS

Recent studies have allowed the identification of type II cadherins (Suzuki et al., 1991), which share higher homologies between themselves than with classical cadherins. Eight members have been isolated and characterized using the polymerase chain reaction (PCR) technique from human and rat nervous tissues. They have been numbered from cadherin-5 to cadherin-12. Comparison of the extracellular domains of type II cadherins revealed more than 40% identity; this percentage is higher than that obtained for comparison between the extracellular domains of type I and type II cadherins (less than 35% identity). Cadherin-5 is the endothelial cell cadherin and shows the lowest homology with the other type II cadherins.

Type I and type II cadherins have distinctive features: most of the aromatic amino acids are conserved but some of them are found only in one of the two groups; each group has characteristic deletions or additions of short amino acid sequences in the extracellular domain (Tanihara et al., 1994a). Type II cadherins share the characteristic motifs representing calcium-binding sites (DXNDN and

DXD) and the DXE sequence located in the middle of the subdomains EC1 to EC4. Amino acid sequence comparisons between the extracellular subdomains revealed that EC2 was better conserved than EC1; comparisons between cadherin-8 and cadherin-11 revealed a high identity in both the cytoplasmic domain (75%) and the extracellular domain (66%) (Tanihara et al., 1994a).

Type II cadherins contain no HAV sequence but a XAV sequence, and their adhesion properties may be different from those of type I cadherins. They do not show strong cell adhesion activity, as shown by aggregation experiments on transfected cells (Tanihara et al., 1994b; Kido et al., in preparation). Interestingly, in situ hybridization of cadherin-11 during rat embryonic development revealed a high expression in mesenchymal cells, including the migratory neural crest cells and sclerotomal cells (Simonneau et al., 1985); cadherin-11 might have a role in signal transduction or in regulation of cell adhesion. The biological properties of cadherin-5 were analyzed by using a cDNA transfection approach. This type II cadherin exhibits the following characteristic features: 1) cadherin-5 was digested easily with trypsin, even in the presence of high calcium concentration; 2) transfectants, as well as parental L-cells, did not show cell aggregation activity even after prolonged incubation; 3) immunoprecipitation revealed a weak interaction with alpha-catenin which might explain the lack of strong cell aggregation, since it has been reported that an association of the cytoplasmic domain of classical cadherin with alpha-catenin, was essential for cell adhesion activity (Hirano et al., 1992). Similarly, the transfectants of cadherin-8 showed similar properties to those of cadherin-5 as regards Ca2+ protection against trypsin digestion, cell adhesion activity and catenin association. Transfection of chimeric cadherin-4, containing the cytoplasmic region of cadherin-5, showed cell aggregation activity comparable to that of wild-type cadherin-4 transfectants, whereas transfectants of cadherin-5 containing the cadherin-4 cytoplasmic domain, did not show appreciable cell aggregation. This suggests that the extracellular domain in conjunction with the cytoplasmic domain play an important role in cell aggregation activity (Tanihara et al., 1994b).

NEW MEMBERS OF THE CADHERIN FAMILY

Newly identified cadherins include OB-cadherin, found in the human and mouse osteoblast lineage (Okasaki et al., 1994), K-cadherin expressed in mouse fetal kidney and kidney carcinoma (Xiang et al., 1994) and LI-cadherin found in rat liver and intestine (Berndorf et al., 1994).

OB-cadherin expressed in mouse osteoblast cell lines contains most of the amino acid sequence motifs shared in the cadherin superfamily such as the DXNDN, DXD and the DXE sequences found in the extracellular domain and the well conserved 70 amino acids found in the carboxyterminal region involved in the binding of catenins. OB-cadherin exhibits a QAV sequence in place of the HAV sequence motif in the EC1 subdomain at position 80. Comparison of the amino acid sequences revealed 33 to 39% sequence identity to the type I cadherin, and a strong similarity of 97% with a partial sequence of cadherin-11 (Suzuki et al., 1991). However, it is still questionable whether the rat cadherin-11 is the counterpart of the mouse OB-cadherin. Transfection experiments and aggregation assays of OB-cadherin revealed calcium-dependent cell adhesive properties. In humans, two different forms of OB-cadherin were found: one is a counter part of the mouse OB-cadherin in the osteoblast lineage and the other is a truncated form lacking the cytoplasmic domain, in which the sequence coding for the transmembrane domain has a short deletion of 30 bp replaced by an insertion of an additional sequence of 179bp. This truncated form is detected only in tumorigenic cells (and absent from normal osteoblasts). The high sequence conservation between human and mouse OB-cadherin and the up-regulation of OB-cadherin mRNA during differentiation of the osteoblastic cell line suggest a specific function of this cadherin in bone cell differentiation and bone formation.

K-cadherin is a calcium-dependent adhesive cadherin showing 32 to 38% homology with type I cadherins and high homology with the partially identified human cadherin-6 (Suzuki et al., 1991). K-cadherin exhibits a QAI sequence instead of HAV in the extracellular subdomain EC1 at position 80. The level of K-cadherin expression is high in fetal rat kidney and brain; it is also found in kidney carcinoma in which an alternatively spliced isoform was detected. As has been previously mentioned, truncated OB-cadherin and K-cadherin in adults are only expressed in carcinoma cells and may play a significant role in tumorigenesis; no soluble forms of OB- and K-cadherin have been reported.

LI-cadherin is a calcium-dependent cell adhesion molecules containing the DXNDN, DXD and DXE motifs; it is expressed in the rat liver and intestine. Major characteristics make this protein unique among the known cadherins. LI-cadherin has two extracellular subdomains termed EC1a, which has no significant homology with the repeated subdomains, and EC2a, which contains the characteristic motifs DXNDN and DXD representing calcium-binding sites and the DXE sequence. The HAV motif of the extracellular subdomain EC1 is replaced by an AAL sequence similar to the FAL sequence of M-cadherin (Donalies et al., 1991); there is no AAL sequence in EC1a. LI-cadherin has the signal peptide but lacks the precursor segment and the endogenous proteolytic cleavage site RXKR at the carboxyterminal region of this fragment (Ozawa and Kemler, 1990). Finally, LI-cadherin has a very short cytoplasmic tail of only 18 amino acids potentially constituting a catenin binding site. Despite these structural differences, LI-cadherin mediates intercellular adhesion in a calcium- dependent manner and may have a role in the development of the liver and the intestine during embryogenesis.

PROTOCADHERINS

A novel family of proteins related to cadherins have been named protocadherins; they retain many features of the primordial cadherin motif in the extracellular domain (Sano et al., 1993) found in type I cadherins and in the Drosophila fat EC18 subdomain which represents a typical fat extracellular subdomain. These proteins initially identified in the rat and human, have also been characterized in the mouse, Xenopus laevis, Drosophila and Caenorhabditis elegans. They are characterized by 6 to 7 extracellular subdomains homologous to those of type I cadherins and a specific cytoplasmic tail which does not share any identity with previously identified cadherins. Preliminary results showed that protocadherins do not co-precipitate with catenins.

Two clones of human protocadherins PC42 and PC43 have been fully sequenced, they contain 7 and 6 extracellular subdomains respectively. The putative calcium binding sites DXNDNXPXF, DXD and DXE are the major conserved motifs among protocadherins and type I cadherins. There are other conserved sequence motifs, such as DXDXGXN and AXDXGXP located near the

N-terminal and C-terminal of the extracellular subdomains respectively. The lengths between the well-conserved sequences and amino acid residues are identical among the repeats except in the C-terminal region of EC1, in the middle of EC2 and in the C-terminal region of the last repeat (EC6 or EC7), which does not exhibit any identity with the corresponding regions of other repeats. In type I cadherin, only the EC2 subdomain has some features of the protocadherins. The similarity percentage between the extracellular subdomains of protocadherins is below 40% which is the average value between the subdomains of type I and type II cadherins and between type II cadherins (Tanihara et al.,1994b).

Transfectants do not show any significant cell aggregation during short periods of incubation, but prolonged incubations resulted in gradual formation of small aggregates. Many protocadherins are highly expressed in specific areas of the brain and in neuroglial cell lines. Their cell adhesion mechanism may be different from that of classical cadherins.

DESMOSOMAL CADHERINS

In desmosomes, two types of transmembrane glycoproteins have been found, called desmogleins (Dsg) and desmocollins (Dsc), which form a special subclass of cadherins connected to the intermediate filament network of the cytoskeleton. Additionally, two major constitutive plaque proteins have been identified and named plakoglobin and desmoplakin. Two isoforms of desmoplakin generated by differential splicing are found in desmosome forming cells. In contrast to the desmosomal cadherins, plakoglobin is not exclusively found in the desmosomes, but is also found at the zonulae adherens of polar epithelial cells within which are associated the classical cadherins such as E-cadherin.

Some important differences between desmogleins and desmocollins are found particularly in the cytoplasmic domain and in the content of cyteine residues. There is 40% protein sequence identity between the extracellular domains of desmocollin Dsc and bovine N-cadherin (Legan et al., 1994) and 28% between the extracellular domains of desmoglein Dsg1 and bovine N-cadherin (Collins et al., 1991). The cytoplasmic domains of desmocollins have only 24 and 23% identity with the cytoplasmic domain of bovine N-cadherin. To date, three different desmogleins (Dsg1, Dsg2, Dsg3) and desmocollins (Dsc1, Dsc2, Dsc3) have been identified (Koch et al., 1992; Schäfer et al., 1994).

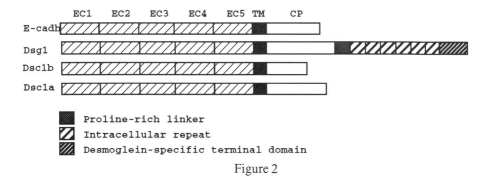

Figure 2

Classical and desmosomal cadherins (Koch and Franke, 1994) show similar structural characteristics, such as in their extracellular domain which is organized into five subdomains. Subdomains EC1 to EC4 of desmosomal cadherins contain internally repetitive sequences which are believed to be essential for homophilic interaction between classical cadherins.

Desmocollins

Desmocollins (Dsc1, Dsc2, Dsc3) are synthesized as two protein isoforms generated by differential splicing of a pre-mRNA derived from a single gene (spliced variants a and b). Spliced variant b has a shortened carboxyterminal domain including a unique sequence of eleven amino acids; in the variant a, whose size is similar to that of classical cadherins, the carboxy-terminal region is involved in the assembly of the desmosomal plaque (Troyanovski et al., 1993). Recently, Dsc2 has been found to be present ubiquitously in bovine epithelia, but is absent in humans. In certain stratified squamous epithelia Dsc1 and Dsc3 may become predominant.

Desmogleins

Desmogleins are characterized by a large cytoplasmic domain that contains five, six or two copies in Dsg1, Dsg2, and Dsg3 respectively of a repetitive sequence of 29 amino acids of unknown function (Schäfer et al., 1994). Only Dsg2 has been found to be expressed in all human and bovine desmosome-bearing tissues and cultured cell lines and serves as a general marker for desmosomes. Dsg1 and Dsg3 are restricted to stratified squamous epithelia and certain tumors and are recognized by autoantibodies in patients suffering from pemphigus vulgaris and foliaceus respectively.

Function

Little is known about the interactions between desmocollins and desmogleins; Intra- and intercellular homophilic and heterophilic interactions could all be envisaged. A chimeric construct of the extracellular domain of Dsg3 and the cytoplasmic domain of E-cadherin led unexpectedly to a weak adhesion (Amagai et al., 1994). Analysis of the cytoplasmic interactions by chimeric construction of the transmembrane of the gap junction protein Connexin 32 and the cytoplasmic region of Dsc1a, produced a functional connexon that bound intermediate filaments (Troyanovsky et al., 1993). The fusion protein Connexin 32-Dsc1b did not bind plakoglobin and did not anchor intermediate filaments. The chimera connexin 32-Dsg1 did not lead to the formation of junctions but acted as a dominant-negative mutant which disassembled the endogenous desmosomes. Proteins that may link the desmosomal cadherins to the intermediate filament network include the band 6 protein, plakoglobin and desmoplakin.

Important questions are now addressed concerning the respective function of desmogleins and desmocollins in their homophilic or heterophilic binding and in the regulation of assembly and disassembly during cell differentiation and during invasive growth and metastasis of cancer cells.

CADHERINS IN DROSOPHILA

The fat locus of Drosophila encodes a member of the cadherin superfamily with 34 tandem repeats of cadherin subdomains, four epidermal growth factor-like repeats, a transmembrane domain, and a novel cytoplasmic domain (Mahoney et al., 1991). Fat is required for correct morphogenesis. Fat is a tumor suppressor gene involved in the control of cell proliferation suggesting that this cadherin directly or indirectly regulate cell behavior. The dachsous gene also codes for multiple cadherin repeats in its extracellular domain (Cowin and Brown, 1993).

C-ret

The human and murine ret-proto-oncogene have extracellular cadherin-like domains, that are known to be important for Ca2+-dependent homophilic binding of the cadherins (DXND and DXD) and a cytoplasmic tyrosine kinase domain (Iwamoto et al., 1993). Moreover, the other highly conserved motif of the cadherin (LDRE) is also found in an exact corresponding position in c-ret. One crucial issue is to identify the ligand of c-ret and see whether it is involved in an adhesive process, which in this special case could be directly transduced through the tyrosine kinase domains typical of growth factor receptors.

CADHERIN GENES

Mapping

The chromosomal locations of most of the type I cadherins have been established. The murine E-, P- and M-cadherin genes have been mapped to mouse chromosome 8 and the human E- and M-cadherin genes to human chromosome 16q; in both species, the M-cadherin locus was near the E-cadherin locus. The mouse P-cadherin locus is localized in the central region of mouse chromosome 8 and highly linked to the E-cadherin locus (Hatta et al., 1991). It has been shown that the L-CAM and K-CAM (B-cadherin) were arranged in tandem in the chicken genome and are separated by less than 700bp (Sorkin et al., 1991), but their chromosomal mapping is unknown. The mouse and human N-cadherin genes have been mapped on chromosome 18. The gene loci of desmosomal cadherins desmocollins and desmogleins have been mapped in human to chromosome 18 (King et al., 1993).

Structure

The mouse E-cadherin gene is encoded by 16 exons distributed over a region of more than 40 kb of genomic DNA; the mouse P-cadherin gene spans over 45 kb and consists of 16 exons (Hatta et al., 1991; Faraldo and Cano, 1993). A marked feature of both genes is the length of the second intron -23 kb long in P-cadherin and more than 40 kb in E-cadherin. The mouse N-cadherin gene consists of 16 exons dispersed over 200 kb; its large size compared to its cDNA (4.3 kb) is ascribed to the fact that the first and second introns are 34.2 kb and greater than 100 kb long, respectively (Miyatani et al., 1992). E-, P- and N-cadherin genes exhibit a similar genomic organization with conserved localization of the intron-exon boundaries.

In human, E- and P-cadherin are arranged in tandem and separated by 32 kb (Bussemakers et al., 1994a).

The chicken L-CAM is encoded by a single gene of 10kb long containing 16 exons. The chicken K-CAM gene (B-cadherin) and L-CAM gene have 11 exons of identical sizes but the sizes of the introns are dissimilar; the last two exons (15 and 16) of K-CAM and L-CAM are almost identical (Sorkin et al., 1991).

The organization of the bovine dsg1 gene consists of 15 exons distributed over more than 37.5 kb of genomic DNA (Puttagunta et al., 1994). The comparison with classical cadherins reveals a striking conservation of exon boundaries in regions encoding the extracellular domains.

REGULATION OF CADHERIN EXPRESSION

Promoters

The promoter region of the mouse P- and E-cadherin gene has been recently analyzed to characterize their regulatory elements (Faraldo and Cano, 1993; Behrens et al., 1991). The results show that both promoters have several potential binding sites for transcription factors such as SP1, and have a CAAT box but no TATA box. The P-cadherin promoter region has no palindromic E-pal sequences, which are found specifically in the E-cadherin promoter, and may have a role in the regulation of the epithelium specific expression of the E-cadherin gene.

It has been shown recently that the human E-cadherin promoter is inactive in a human prostate cancer cell line which does not express E-cadherin (Bussemakers et al., 1994b); this might be due to the binding of a repressor protein to the promoter.

Factors involved

Recently, for the first time, evidence has been provided that the E-cadherin gene transcription is controlled by a tyrosine kinase receptor (D'souza and Taylor-Papadimitriou, 1994); the overexpression of the ERBB2 receptor (homologous to the EGF receptor) in a non-tumorigenic human mammary epithelial cell line, MTSV1-7, is associated with a reduced ability to undergo morphogenesis in vitro and with a decreased level of expression of the E-cadherin and alpha2 integrin genes. The ERBB2 receptor is expressed at low levels by most normal tissues but is dramatically increased in a significant proportion of breast cancers.

Estradiol has been shown to regulate E-cadherin mRNA levels in the surface epithelium of the mouse ovary (Mac Calman et al., 1994): immature mice injected with 17-β estradiol display a rapid and significant increase of E-cadherin mRNA in the ovary.

During chicken retinal development, insulin has been shown to enhance the down-regulation of N-cadherin expression (Roark et al., 1992).

REGULATION OF CADHERIN FUNCTION

Different factors involved

Cadherin-mediated adhesion is dependent upon the Ca2+ concentration, and may be modulated by factors such as kinases controling the connexion between the cadherin-catenin complex and the cytoskeleton. Signalling molecules and intracellular regulators including the APC protein can modulate the function of cadherins. Recently, the mesoderm inducing factor activin has been shown to reduce the EP-cadherin adhesive function in animal cap explants, by an unknown mechanism (Brieher and Gumbiner, 1994); this may occur by disassembly of clusters of cadherins, which would lower the binding affinity of EP-cadherin and decrease cell-cell adhesion, or alternatively by inducing a conformational change in EP-cadherin.

In tumor cells, E-cadherin function can be modulated by different factors. An increased E-cadherin-mediated cell adhesion can be induced by activation of G protein-coupled receptors in small cell lung carcinoma (Williams et al., 1993). A decrease of E-cadherin adhesive function has been reported by proteoglycans, most likely by a steric hindrance mechanism (Vleminckx et al., 1994). The MCF7 cells are invasive human breast cancer cells which have a functionally inactive E-cadherin. Insulin-like growth factor I (IGF-1) activates the invasion suppressor function of E-cadherin present at the surface of MCF-7 human mammary carcinoma cells, and leads to the aggregation of the carcinoma cells and their inability to invade in an in vitro assay (Bracke et al., 1993). Tamoxifen, an antioestrogen used in adjuvant therapy of breast carcinoma, restores E-cadherin function in the adenocarcinoma MCF-7/6-treated cells, and suppresses their invasive phenotype (Bracke et al., 1994).

Factors such as IL-6 and acidic Fibroblast growth factor (FGF-1) can induce cell-cell separation, which is not correlated with a down-regulation of E-cadherin expression. Interleukin 6 (IL-6) treatment of ductal breast carcinoma does not shut off E-cadherin expression but down-regulates this molecule at cell borders; this localized decrease may play a role in cell-cell separation of IL-6 treated cells (Tamm et al., 1994). Acidic fibroblast growth factor causes NBT-II rat bladder carcinoma to undergo an epithelial-mesenchymal transformation; the level of E-cadherin remains unchanged but this cadherin is redistributed over the entire cell surface (Boyer et al., 1992).

Ca2+ concentration

The depletion of Ca2+ ions from culture medium promptly induces the proteolysis of cadherins and the disassembly of the adherens junctions followed by the endocytosis of cadherins and the undercoat of the adherens junctions and desmosomes (Kartenbeck et al., 1991); calcium ions do not act as simple bridges but maintain a particular conformation of the cadherin molecules.

Binding of catenins

Beta-catenin is thought to be a central regulator of cadherin-mediated adhesion. Recently, experiments have been carried out with chimeric proteins encoding nonfunctional E-cadherin partially deleted in its cytoplasmic domain, linked to a functional alpha-catenin or in just an amino-terminal or carboxy-terminal region of alpha-catenin (Nagafuchi et al., 1994); the analysis of transfected L cells expressing these different fusion proteins revealed that the carboxy-terminal half-region of alpha catenin fused to the cytoplasmic domain of E-cadherin was sufficient to restore the adhesive function of truncated E-cadherin for mediating cell adhesion activity. Most interestingly, transfected L cells expressing the chimeric E-cadherin/alpha-catenin molecule grow as stable clusters whereas L cells expressing the wild type E-cadherin can exchange neighbours. These experiments show that removing beta-catenin from the complex reinforces adhesion, suggesting that beta-catenin might work as a negative regulator of cell adhesion.

Phosphorylation

The epithelial adherens junctions of normal cells are major targets for tyrosine phosphorylation which leads to the disassembly of these intercellular junctions (Volberg et al., 1992); phosphorylation may control the ratio of free and actin-bound cadherin-catenin complexes, thereby regulating the adhesive forces of cadherins.

Recently, Stappert and Kemler (1994) have reported that E-cadherin-catenin interactions may be regulated by phosphorylation of the whole serine cluster (8 residues) within the catenin binding domain, as mentioned in a previous subchapter; the inability to bind catenin is only observed when the entire cluster is mutated. E-cadherin-catenin interactions seem to be regulated by phosphorylation of the catenin-binding domain and represent a molecular mechanism regulating cadherin-mediated cell adhesion.

It has been shown that cadherins are co-localized with protein kinase C and tyrosine kinases of the src-family (c-src, c-lyn and c-yes) in adherens junctions (Tsukita et al., 1991). Multiple proteins, including cadherins and catenins may be phosphorylated: beta-catenin has been shown to be a good substrate for tyrosine kinase as ezrin and radixin.

Orthovanadate, an inhibitor of protein-tyrosine phosphatase increases the level of tyrosine phosphorylation at the adherens junction of rat liver cells (Tsukita et al., 1991), suggesting the existence of a protein-tyrosine phosphatase working in adherens junctions. In endothelial cell lines derived from the human umbilical vein (HUVEC) or bovine aortic endothelial cells (BAEC), pervanadate leads immediately to a dramatic increase in adherens junction-associated phosphotyrosine, but with longer incubations (30 minutes), the intercellular adhesion sites disintegrate and the tyrosine phosphorylation of vinculin and actin in focal contacts increases (Ayalon et al., 1994).

In transformed cells, the level of beta-catenin phosphorylation has been correlated with a decrease of cadherin cell adhesion (Matsuyoshi et al., 1992). Enhanced tyrosine phosphorylation of cadherins and alpha- and beta- catenins correlates with decreased adhesion and increased migration and invasiveness in cells transformed with a temperature-sensitive v-src gene (Behrens et al., 1993). Elevation of tyrosine phosphorylation at the adherens junction of RSV-transformed chick lens cells induces its disassembly (Volberg et al., 1991). Therefore, tyrosine kinases associated with zonulae adherens junctions, may regulate the assembly of the cadherin-cytoskeleton complex.

Growth factors appear to be good candidates to modulate phosphorylation during development and cell differentiation. Beta-catenin is tyrosine phosphorylated in cells treated with hepatocyte growth factor (HGF) or epidermal growth factor (EGF) (Shibamoto et al., 1994), and a decrease in E-cadherin concentration at cell-cell boundaries and the scattering of these treated cells were observed. Some protein kinase surface receptors have been found to be associated with cadherin-catenin complexes and could regulate the adhesive properties of different cell types. The EGF receptor co-localize with cadherins on the baso-lateral membrane of epithelial cells (Fukuyama and Shimizu, 1991). Recently, a direct binding of beta-catenin to the EGF receptor has been reported (Hoschutesky et al., 1994), suggesting an important role of catenins between the EGF-induced signal transduction and cadherin function.

Cadherins and Wnt signalling

The Wnt gene family encodes a group of secreted signalling factors involved in mammary tumorigenesis and in patterning events during development. Wnt-1 is the best studied member, whose locus is activated in response to proviral insertion of the mouse mammary tumor virus (MMTV), and contributes to mammary tumorigenesis. In addition, Wnt-1 plays an important role in neural development. Recent studies in Drosophila showed that armadillo, a homologue of beta-catenin, was involved in the signalling pathway of wingless, a member of the Wnt family. Wnt-1 was found to modulate cell-cell adhesion in mammalian cells by stabilizing beta-catenin binding to the cadherin, resulting in an increase in the strength of the calcium-dependent cell adhesion. Beta-catenin has been implicated in axis formation during Xenopus development (McCrea et al., 1993), a process that may also require Wnt signalling.

Recently, it has been shown that the inactivation of the Wnt-1 gene caused an up-regulation of E-cadherin and a down-regulation of alphaN-catenin expression at both protein and mRNA levels in distinct areas of the embryonic brain (Shimamura et al., 1994). These results suggest that the Wnt-1 signal is involved in the regulation of expression and function of both cadherin and catenins and acts upstream in the signalling pathway.

Cell surface turn-over

Usually, cadherins appear to be regulated at the transcriptional level but modulation of cadherin expression can occur by cell surface turnover or endocytic mechanism. The down-regulation of N-cadherin during retinal development is due to the release of a soluble proteolytic form NCAD90, corresponding to the N-terminal fragment of N-cadherin at the cell surface which retains adhesive and neurite-promoting functions (Paradies and Grunwald, 1993). Differential cadherin expression could contribute to the cell polarity of epithelial cells. E-cadherin and desmoglein are initially addressed uniformly to the plasma membrane, but they are rapidly internalized at the apical surface, while at the lateral surface these molecules have a much longer half-life (Wollner et al., 1992). Once localized, E-cadherin seems to serve as an important organizer of other cell surface and cytoskeletal molecules during epidermal morphogenesis. The activity of cadherins influences the formation of tight, gap junctions and desmosomes (Jongen et al., 1991); the inhibition of gap and adherens junction assembly by connexin and N-cadherin antibodies has been reported (Meyer et al., 1992).

CADHERINS DURING MORPHOGENESIS

The function of cadherins in morphogenetic events has been addressed by the use of different perturbation assays. One of the earliest experiments was the inhibition of compaction of the 8 cell-stage mouse embryo by anti E-cadherin antibodies (Kemler et al., 1990).

In the chick, antibodies to N-cadherin (A-CAM) were very effective in disrupting epithelial somites (Duband et al., 1987). Also, these antibodies microinjected in vivo into the cranial mesenchyme adjacent to the midbrain disrupted cranial development in the chick, leading to a distorted, folded and overgrown neural tube with aggregates of neural crest cells outside the neural tube (Bronner-Fraser et al., 1992).

In vivo, these antibodies inhibit the formation of mesenchymal condensates. N-cadherin is necessary in mediating cell-cell interactions between mesenchymal cells involved in chondrogenesis, in micromass culture in vitro, and in the intact limb bud in vivo (Oberlender and Tuan, 1994).

E-cadherin gene knock-out experiments have been successfully carried out using the homologous recombination technique in embryonic stem cells (Larue et al., 1994). Nullizygous E-cadherin mouse embryos underwent compaction at the 8 cell-stage due to the presence of residual maternal E-cadherin, but died around the time of implantation because they failed to form a trophectodermal epithelium. The inactivation of several other cadherin genes is currently being performed in different laboratories.

Many recent studies have been based on the use of Xenopus embryos. In Xenopus, the two maternal cadherins, EP-/C-cadherin and its pseudo-allele XB-/U-cadherin, are expressed in oocytes, fertilized eggs, and later during development. E-cadherin is first expressed at the gastrula stage in the presumptive ectoderm which will give rise to epidermis, whereas N-cadherin is first expressed at the neurula stage in neural tissues. The ectopic expression of chicken N-cadherin in Xenopus embryos resulted in abnormal histogenesis such as thickening, clumping or fusion of the cell layers (Fujimori et al., 1990). Over-expression of Xenopus N-cadherin led to severe morphological defects including a dramatic alteration of the ectodermal cell layers and defects in the neural tube closure (Detrick et al., 1990); the morphological defects obtained suggest that quantitative differences in cadherin expression can control morphogenesis of cell layers and the sorting out in distinct compartments. In contrast, the over-expression of N-CAM in Xenopus embryos did not perturb the neural tube (Kintner, 1988), suggesting that different cell adhesion molecules have distinct roles during morphogenesis. The depletion of EP-cadherin mRNA, using antisense deoxynucleotides injected into full-grown oocytes, reduced adhesion between blastomeres, leading to a disaggregation phenotype; this dose-dependent phenotype could be rescued by injection of mRNA coding for E-cadherin (Heasman et al., 1994). The role of cadherins as morphoregulatory molecules has also been demonstrated in vivo by using truncated forms of cadherins which act as dominant-negative mutants. The importance of cytoskeletal interactions was illustrated by experiments where N-cadherin lacking the extracellular domain was expressed in embryos (Kintner, 1992) and in cultured cells (Fujimori and Takeichi, 1993), resulting in abnormal development and disruption of cell-cell adhesion respectively. The competition for catenins was thought to be responsible for the observed perturbations.

Recently, we showed in our laboratory that truncated N-cadherin, expressed in anterior dorsal blastomeres of Xenopus laevis embryos produced a moderate perturbation in neural morphogenesis, whereas truncated XB-cadherin provoked larger malformations in the brain and in the eye. A synergistic effect was observed when the two truncated cadherins were expressed together (Dufour et al., 1994). From these studies, it appeared that the cytoplasmic domain of one cadherin type can perturb the function of several different endogenous cadherins since at least N-cadherin is not expressed until the neurula stage. A differential perturbation effect was observed with two cytoplasmic domains sharing 60% identity. These two closely related cytoplasmic domains of cadherins may exert their differential effects by interacting with different affinities with catenins or other components of the cytoskeleton (Dufour et al., 1994).

We have also perturbed the function of cadherins expressed during early development by over-expressing truncated E- and EP- cadherins in which the extracellular domain has been deleted (Broders et al., submitted). Truncated cadherin mRNA was injected into different blastomeres of stage 6 embryos, chosen for their known contribution to defined tissues. Injection of similar amounts of mRNA coding for truncated E- or EP-cadherin, into different blastomeres gave rise to different levels of perturbation of embryonic development. Defects resulting from injection of the truncated cadherins were already seen at the late blastula stage when the descendants of the blastomeres lose cohesion, migrate and mingle extensively within clones derived from non-injected blastomeres. Truncated EP-cadherin produced more severe morphogenetic defects than truncated E-cadherin in tissues derived from the injected blastomeres. Injections of mRNA encoding truncated E- and EP-cadherins, into blastomeres involved in neural induction (B1B2 and C1) led to the duplication of the antero-posterior axis as well as to malformations in the anterior neural structures. The over-expression of truncated E-cadherin mRNA led to a hyperplasia of several structures. From these results and those obtained with truncated N- and XB-cadherins (Dufour et al., 1994), we can conclude that even though type I cadherin cytoplasmic domains share a strong sequence identity, they induce distinct perturbations in the patterning of axial structures. These different levels of perturbation of embryonic development may reflect the different affinity of the cadherin cytoplasmic domain for catenins or for other cytoskeletal structures. These results emphasize the importance of specific cadherin-mediated cell adhesion during histogenesis and during cell migration in the early stages of Xenopus development.

Other types of experiments have been carried out with injections of E-cadherin dominant negative mutants, with their cytoplasmic regions deleted. In this case, the defects can be targeted to a tissue which expresses the same endogenous cadherins (Levine et al., 1994); overexpression of EP-cadherin was unable to compensate the disruption of E-cadherin function, indicating that E-cadherin was specifically required for maintaining the integrity of the ectoderm after gastrulation.

OTHER FUNCTIONS FOR CADHERINS

Recently, additional functions for cadherins have been reported. Cadherins play a role in the down-regulation of integrin expression that occurs during the terminal differentiation of human keratinocytes (Hodivala and Watt, 1994): antibodies against P- and E-cadherin prevent the selective loss of integrins from terminally differentiating cells; thus cadherins can negatively regulate integrin expression in keratinocytes.

Adhesion of epidermal Langerhans cells and melanocytes to keratinocytes mediated by E-cadherin has been reported (Tang et al., 1993; 1994).

During differentiation of human colon adenocarcinoma Caco-2 cells, an intestinal peptide transporter HPT-1 was found to be related to the cadherin superfamily: HPT-1 contains the DXNDN and LDRE sequences but lacks a cytoplasmic domain corresponding to the highly conserved region of cadherins (Dantzig et al., 1994). The presence of a single transmembrane domain in HPT-1, and the cysteine residues in the extracellular region, may indicate that the protein self-aggregates to form a fully functional intestinal peptide transporter or may associate with another transporter.

PATHOGENESIS

Cancer

The role of cadherins in cancer received considerable attention when many clinical specimens and tumor cell lines revealed a correlation between reduction of E-cadherin expression and progression of various cancers and increased cellular invasiveness (Umbas et al., 1992).

Tumor progression

Human carcinomas can be subdivided into two types, depending of the level of expression of cadherins (Takeichi, 1993; Birchmeier and Behrens, 1994). Differentiated tumors (type 1), which retain an epithelial morphology maintain high levels of cadherin expression, whereas dedifferentiated tumors (type 2) show reduced levels; a similar trend is observed in tumor cell lines. Type 2 tumors such as breast and prostate are more infiltrative than those of the type 1 and show a higher frequency of lymph node metastasis (Umbas et al., 1992; Oka et al., 1993); however, in gastric cancers, type 1 carcinoma cells can metastasize to the liver (Oka et al., 1992).

One likely explanation is that E-cadherin expressed at the surface of tumor cells is not functional, perhaps as a result of a transient down-regulation of cadherin or a phosphorylation of catenins (Matsuyoshi et al., 1992; Hamaguchi et al., 1993), weakening cell-cell adhesion and facilitating the detachment of cells (Matsuura et al., 1992).

Interestingly, several cancers are associated with deletions in the human chromosome 16 where the E-cadherin gene is localized. The analysis of a significant number of human diffuse type gastric carcinomas (Oda et al., 1994) revealed that loss of cell-cell adhesion was correlated in 50% of the cases with several types of mutation in the E-cadherin gene. In a majority of cases, the mutation caused an in frame skipping of exon 8 or 9, both of which encode calcium binding domains. In one case, the loss of heterozygosity with mutations in the remaining allele was found, supporting the hypothesis that E-cadherin is a tumor suppressor gene (Becker et al., 1994).

Other mechanisms, including proteolysis of cadherins could be involved in cancer pathogenesis; indeed soluble E-cadherin fragments were found to be increased in the serum of cancer patients (Katayama et al., 1994) and can be considered as a tumor marker. It has been reported that soluble forms of OB- and K-cadherin are specifically expressed in carcinoma cells.

Cell lines. A decreased expression of E-cadherin and an increased invasive capacity was observed in Epstein-Barr virally transfected human epithelial and murine adenocarcinoma cells (Fahraeus et al., 1992). It was shown that transformation of MDCK cells with the Harvey-ras oncogene resulted in the loss of epithelial morphology in culture and the acquisition of malignant properties (Mareel et al., 1991). Analysis of these cells revealed a reduction in E-cadherin expression; transfection of these cells with E-cadherin cDNA restored an epithelial morphology and a non-malignant growth pattern. Human retinoblastoma cell lines vary in their adhesive properties and in their N-cadherin level expression (Schiffman and Grunwald, 1992).

Dysfunction of cell-cell adhesion can be mediated by abnormalities of associated proteins such as catenins: the human lung cancer cell line PC9, which grows as isolated cells in spite of beta-catenin expression and a strong E-cadherin expression, showed a deletion of the alpha catenin gene (Hirano et al., 1992); transfection of PC9 cells with alphaN-catenin cDNA led to a transition to a cell-cell adhesive phenotype (Hirano et al., 1992). Alpha-catenin transfection is correlated with the acquisition of an apical-basal polarity including the formation of junctional complexes and the polarized distribution of cell surface proteins (Watabe et al., 1994). The authors also found a retardation of cell growth due to the activation of E-cadherin.

The APC tumor suppressor protein. Germinal mutations in the APC protein are responsible for the familial adenomatous polyposis (FAP). Most interestingly, beta catenin, linked or not to the alpha catenin, was found to bind to the APC protein. However, this complex did not bind to cadherins (Su et al., 1993); it is is conceivable that APC modulates the interactions between catenins and cadherins. Rubinfeld et al. (1993) postulated that mutant APC could have a reduced affinity for beta-catenin, which itself is linked to alpha-catenin, leading to the APC phenotype; in this case, the formation of colorectal cancer might result from the loss of intercellular contacts dependent upon an interaction between APC and beta-catenin. These results suggest that APC is involved in cell-cell adhesion (Hülsken et al., 1994).

Auto-immune disease

Two human auto-immune disease leading to the blistering of the skin and mucous membrane involve circulating anti-cadherin autoantibodies. Two of the desmosomal cadherins are antigens in this severe human auto-immune skin disease: Dsg1 "PF-antigen" and Dsg3 "PV-antigen" have been identified by auto-antibodies in pemphigus foliaceus and pemphigus vulgaris respectively (Allen et al., 1993; Amagai et al., 1992).

Bacterial infection

Recently, it has been shown that cadherins were required by pathogens for the spread of infection (Sansonetti et al., 1994). The mouse fibroblastic sarcoma cell line S180, devoid of endogenous cadherins, was transfected with the L-CAM or N-cadherin cDNA then infected with the gram negative Shigella flexneri which causes bacillary dysentry in humans by invading epithelial cells of the colon. It was shown that the formation of protrusions that bring bacteria inside adjacent cells is localized in the area of intermediate junctions and requires actin and cadherins.

CONCLUDING REMARKS

During the last few years, molecular biology has allowed the identification of an increasing number of cadherins and cadherin-related molecules. Considerable progress has been made in our understanding of the function of type I cadherins during early embryonic development, histogenesis and disease. Recently, studies have focused on the regulation of the function of type I cadherins, through their dynamic association with catenins and the cytoskeleton and through phosphorylation. Type II cadherins and the protocadherins do not exhibit strong adhesive properties and their precise roles need to be determined. The critical issue is now to elucidate the transduction events mediated by type I and type II cadherins. Much work remains to be done to understand the control mechanisms of regulation of their expression and function, in particular to better define the adhesion signaling complex machinery and the function of cadherins in the hierarchy of genes implicated in morphogenesis.

ACKNOWLEDGEMENTS

We thank Dr. M. L. Faraldo and Dr. B. Barbour for reading the manuscript and providing helpful comments.

REFERENCES

Allen, E. M., Giudice , G. J. and Diaz, L. A. (1993) Subclass reactivity of pemphigus foliaceus autoantibodies with recombinant human desmoglein. J. Invest. Dermatol. 100: 685-691.

Amagai, M., Karpati, S., Klaus-Kovtun, V., Udey, M. C. and Stanley, J. R. (1994) Extracellular domain of pemphigus vulgaris antigen (desmoglein) mediate weak homophilic adhesion. J. Invest. Dermatol. 102: 402-408.

Amagai, M., Karpati, S., Prussick, R., Klaus-Kovtun, V. and Stanley, J. R., (1992) Autoantibodies against the amino-terminal cadherin-like binding domain of pemphigus vulgaris antigen are pathogenic. J. Clin. Invest. 90: 919-926.

Arnemann, J., Sullivan, K. H., Magee, A. I., King, I. A. and Buxton, R.S. (1993) Stratification-related expression of isoforms of the desmosomal cadherins in human epidermis. J. Cell Sci. 104: 741-750.

Ayalon, O., Yarden, Y. and Geiger, B. (1994) Components of the signal transduction cascade are concentrated in adherens type junction. in: "Molecular Mechanisms of Transcellular Signalling: from the Membrane to the Gene". NATO/FEBS ed. Publisher .

Becker, K. F., Atkinson, M. J., Reich, U., Nekarda, H., Siewert, J. R. and Höfler, H. (1994) E-cadherin gene mutations provide clues to diffuse type gastric carcinomas. Cancer Res. 54: 3845-3852.

Behrens, J., Löwrick, O., Klein-Hitpass, L; and birchmeier, W. (1991) The E-cadherin promoter: functional analysis of a GC-rich region and an epithelial cell-specific palindromic regulatory element. Proc. Natl. Acad. Sci. USA. 88: 11495-11499.

Behrens, J., Vakaet, L., Friis, R., Winterhager, E., Ven Roy, F., Mareel, M. M. and Birchmeier, W. (1993) Loss of epithelial differentiation and gain of invasiveness correlates with tyrosine phosphorylation of the E-cadherin: beta-catenin complex in cells transformed with a temperature-sensitive v-SRC gene. J. Cell Biol. 120: 757-766.

Berndorff, D., Gessner, R., Kreft, B., Schnoy, N., Lajous-Petter, A. M., Loch, N., Reutter, W., Hortsch, M. and Tauber, R. (1994) Liver-intestine cadherin: molecular cloning and characterization of a novel Ca^{2+}-dependent cell adhesion molecule expressed in liver and intestine. J. Cell Biol. 125: 1353-1369.

Birchmeier, W. and Behrens, J. (1994) Cadherin expression in carcinomas: role in the formation of cell junctions and the prevention of invasiveness. B. B. A. 1198: 11-26.

Blaschuk, O. W., Sullivan, R., David, S. and Pouliot, Y. (1990) Identification of a cadherin cell adhesion recognition sequence. Dev. Biol. 139:227-229.

Boyer , B., Dufour, S. and Thiery, J. P. (1992) E-cadherin expression during the aFGF-induced dispersion of a rat bladder carcinoma cell line. Exp. Cell Res. 201: 347-357.

Bracke , M. E., Vyncke, B. M., Bruyneel, E. A., Vermeulen, S. J., De Bruyne, G. K., Van Larebeke, N. A., Vleminckx, K., Van Roy, F. M. and Mareel, M. M. (1993) Insulin-like growth factor I activates the invasion suppressor function of E-cadherin in MCF-7 human mammary carcinoma cells in vitro. Br. J. Cancer. 68: 282-289.

Bracke, M. E., Charlier, C., Bruyneel, E. A., Labit, C., Mareel, M. M and Castronovo, V. (1994) Tamoxifen restores the E-cadherin function in human breast cancer MCF-7/6 cells and suppresses their invasive phenotype. Cancer Res. 54: 4607-4609.

Brieher, M. W. and B. M. Gumbiner. (1994) Regulation of C-cadherin function during activin induced morphogenesis of Xenopus animal caps. J. Cell Biol. 126: 519-527.

Broders, F., Girault, J. M., Simmoneau, L., Suzuki , S. and Thiery, J. P. (1993) Sequence and distribution of Xenopus laevis E-cadherin transcripts. Cell. Adhes. Comm. 1: 265-277.

Broders, F., Girault, J. M. and Thiery, J. P. (1994) Contribution of cadherins to directional cell migration and histogenesis in Xenopus embryos. submitted.

Bronner-Fraser, M., Wolf, J. J. and Murray, B. A. (1992) Effects of antibodies against N-cadherin and N-CAM on the cranial neural crest and neural tube. Dev. Biol. 153: 291-301.

Bussemakers, M. J. G., van Bokhoven, A., Völler, M., Smit, F. P. and Schalken, J. A. (1994a).The genes for the calcium-dependent cell adhesion molecules P- and E-cadherin are tandemly arranged in the human genome. B. B. R. C. 203: 1291-1294.

Bussemakers, M. J. G., Giroldi, L. A., van Bokhoven, A. and Schalken, J. A. (1994b) Transcriptional regulation of the human E-cadherin gene in human prostate cancer cell lines. B. B. R. C. 203: 1284-1290.

Choi, Y, S., Sehgal, R., McCrea, P. and Gumbiner, B. (1990) A cadherin-like protein in eggs and cleaving embryos of Xenopus laevis is expressed in oocytes in response to progesterone. J. Cell Biol. 110: 1575-1582.

Claverie, J. M., Hardelin, J. P., Legouis, R., Levilliers, J., Bougueleret, L., Mattei, M. G. and Petit, C. (1993) Characterization and chromosomal assignment of a human cDNA encoding a protein related to the murine 102-kDa cadherin-associated protein (alpha-catenin). Genomics. 15: 13-20.

Collins, J. E., Legan, P. K., Kenny, T. P., Garcie, J. M., Holton, J. L. and Garrod, D. R. (1991) Cloning and sequence analysis of desmosomal glycoproteins 2 and 3 . Desmocollins: cadherin-like desmosomal adhesion molecules with heterogenous cytoplamsic domains. J. Cell Biol. 113: 381-391.

Cowin, P. and Brown, A. M. C. (1993) in: "Molecular Basis of Morphogenesis", Bernfield, M., eds, Wiley-Liss, Inc., New York.

Dantzig, A. H., Hoskins, J., Tabas, L. B., Bright, S., Shepard, R. L., Jenkins, I. L., Duckworth, D. C., Sportsman, J. R., Mackensen, D., Rosteck Jr., P. R. and Skatrud, P. L. (1994). Association of intestinal peptide transport with a protein related to the cadherin superfamily. Science. 264: 430-443.

Detrick, R. J., Dickey, D. and Kintner, C. R. (1990) The effects of N-cadherin misexpression on morphogenesis in Xenopus embryos. Neuron 4: 493-506.

D'souza, B. and Taylor-Papadimitriou, J. (1994). Overexpression of ERBB2 in human mammary epithelial cells signals inhibition of transcription of the E-cadherin gene. Proc. Natl. Acad. Sci. USA. 91: 7202-7206.

Duband, J. L., Dufour, S., Hatta, M., Takeichi, M., Edelman, G. M. and Thiery, J. P. (1987). Adhesion molecules during somitogenesis in the avian embryo. J. Cell Biol. 104: 1361-1374.

Dufour, S., Saint-Jeannet, J. P., Broders, F., Wedlich, D. and Thiery, J. P. (1994). Differential perturbations in the morphogenesis of anterior structures induced by overexpression of truncated XB- and N-cadherins in Xenopus embryos. J. Cell Biol. 127: 521-535.

Donalies, M., Cramer, M., Ringwald, M. and Starzinski-Powitz, A. (1991). Expression of M-cadherin, a member of the cadherin multigene family, correlates with differentiation of skeletal muscle cells. Proc. Natl. Acad. Sci. USA. 88: 8024-8028.

Fahraeus, R., Chen, W., Trivedi, P., Klein, G. and Obrink, B. (1992). Decreased expression of E-cadherin and increased invasive capacity in EBV-LMP-transfected human epithelial and murine adenocarcinoma cells. Int. J. Cancer. 52: 834-838.

Faraldo, M. L. and Cano, A. (1993) The 5' flanking sequences of the mouse P-cadherin gene. Homologies to 5' sequences of the E-cadherin gene and identification of a first 215 base-pair intron. J. Mol. Biol. 231: 935-941.

Franke, W. W., Kapprell, H. P. and Cowin, P. (1987). Immunolocalization of plakoglobin in endothelial junctions: identification as a special type of zonulae adhaerentes. Biol. Cell. 59: 205-218.

Franke, W. W., Goldschmidt, M. D., Zimbelmann, R., Mueller, H. M., Schiller, D. L. and Cowin, P. (1989). Molecular cloning and amino acid sequence of human plakoglobin, the common junctional plaque protein. Proc. Natl. Acad. Sci. USA. 86: 4027-4031.

Fujimori, T., Miyatani, S. and Takeichi, M. (1990). Ectopic expression of N-cadherin perturbs histogenesis in Xenopus. Development. 110: 97-104.

Fujimori, T. and Takeichi, M. (1993). Disruption of epithelial cell-cell adhesion by exogenous expression of a mutated nonfunctional N-cadherin. Mol. Cell Biol. 4: 37-47.

Fukuyama, R. and Shimizu, N. (1991) Detection of epidermal growth factor receptors and E-cadherins in the basolateral membrane of A431 cells by laser scanning fluorescence microscopy. Jap. J. Cancer Research. 82, 8-11.

Gallin, W. J., Sorkin, B. C., Edelman, G. M. and Cunningham, B. A. (1987). Sequence analysis of a cDNA clone encoding the liver cell adhesion molecule. Proc. Natl. Acad. Sci. USA. 84: 2808-2812.

Gally, J. A. and Edelman, G. M. (1992) Evidence for gene conversion in genes for cell-adhesion molecules. Proc. Natl. Acad. Sci. USA. 89, 3276-3279.

Geiger, B. (1979) A 130 K protein from chicken gizzards: its localization at the termini of microfilament bundles in cultured chicken cells. Cell 18, 193-205.

Geiger, B. and Ginsberg, D. (1991). The cytoplasmic domain of adherens-type junctions. Cell Motil. Cytoskeleton. 20: 1-6.

Geiger, B., Tokuyasu, K. T., Dutton, A. H. and Singer, S. J. (1980). Vinculin, an intercellular protein localized at specialized sites where microfilament bundles terminate at cell membranes. Proc. Natl. Acad. Sci. USA. 77: 4127-4131.

Ginsberg, D., DeSimone, D. and Geiger, B. (1991). Expression of a novel cadherin (EP-cadherin) in unfertilized eggs and early Xenopus embryos. Development 111: 315-325.

Green, K. J., Goldman, R. D. an d Chisholm, R. L. (1988). Isolation of cDNAs encoding desmosomal plaque proteins: evidence that bovine desmoplakins I and II are derived from two mRNAs and a single gene. Proc. Natl. Acad. Sci. USA. 85: 2613-2617.

Gumbiner, B. M. (1993). Proteins associated with the cytoplasmic surface of adhesion molecules. Neuron. 11: 551-564.

Hamaguchi, M., Matsuyoshi, N., Ohnishi, Y., Gotoh, B., Takeichi, M. and Nagai, Y. (1993). p60 v-src causes tyrosine phosphorylation and inactivation of the N-cadherin-catenin cell adhesion system. EMBO. J. 12: 307-314.

Hatta, K., Takagi, S., Fujisawa, H. and Takeichi, M. (1987). Spatial and temporal expression pattern of N-cadherin cell adhesion molecules correlates with morphogenetic processes of chicken embryos. Dev. Biol. 120: 215-227.

Hatta, M., Seiji, M., Copeland, N. G., Gilbert, D. J., Jenkins, N. A. and Takeichi, M. (1991). Genomic organization and chromosomal mapping of the mouse P-cadherin gene. Nucl. Acids. Res. 19: 4437-4441.

Hatzfeld, M., Kristjansson, G. I., Plessmann, U. and Weber, K. (1994). Band 6 protein, a major constituent of desmosomes from stratified epithelia, is a novel member of the armadillo multigene family. J. Cell Science. 107: 2259-2270.

Heasman, J., Ginsberg, D., Geiger, B., Goldstone, K., Pratt, T., Yoshida-Noro, C. and Wylie, C. (1994). A functional test for maternally inherited cadherin in Xenopus shows its importance in cell adhesion at the blastula stage. Development. 120: 49-57.

Herrenknecht, K., Ozawa, M., Eckerskorn, C., Lottspeich, F., Lenter, M. and Kemler, R. (1991). The uvomorulin-anchorage protein alpha catenin is a vinculin homologue. Proc. Natl. Acad. Sci. USA. 88: 9156-9160.

Hinck, L., Näthke, I. S., Papkoff, J. and Nelson, W. J. (1994) Dynamics of cadherin/catenin complex formation: novel protein interactions and pathways of complex assembly. J. Cell Biol. 125: 1327-1340.

Hirano, S., Nose, A., Hatta, K., Kawakami, A. and Takeichi, M. (1987). Calcium-dependent cell-cell adhesion molecules (cadherins): subclass specificities and possible involvement of actin bundles. J. Cell Biol. 105: 2501-2510.

Hirano, S., Kimoto, N., Shimoyama, Y., Hirohashi, S. and Takeichi, M. (1992). Identification of a neural alphaN-catenin as a key regulator of cadherin function and multicellular organization. Cell. 70: 293-301.

Hodivala, J. K. and Watt, F. (1994). Evidence that cadherins play a role in the downregulation of integrin expression that occurs during keratinocyte terminal differentiation. J. Cell. Biol. 124: 589-600.

Hoschuetzky, H., Aberle, H. and Kemler, R. (1994). Beta-catenin mediates the interaction of the cadherin-catenin complex with epidermal growth factor receptor. J. Cell Biol. In press.

Hülsken, J., J. Behrens and W. Birchmeier. (1994). Tumor suppressor gene products in cell contacts: the cadherin-APC-armadillo connection. "Current Opinion in Cell Biology" F. Horwitz and J. P. Thiery eds. 6: 711-716.

Inuzuka, H., Miyatani, S. and Takeichi, M. (1991). R-cadherin: a novel Ca2+-dependent cell-cell adhesion molecule expressed in the retina. Neuron. 7: 69-79.

Iwamoto, T., Taniguchi, M., Asai, N., Ohkusu, K., Nakashima, I. and M. Takahashi. (1993). cDNA cloning of mouse ret proto-oncogene and its sequence similarity to the cadherin superfamily. Oncogene. 8, 1087-1091.

Jongen, W. M. F., Fitzgerald, D. J., Asamoto, M., Piccoli, C., Slaga, T. J., Gros, D., Takeichi, M. and Yamasaki, H. (1991). Regulation of connexin43-mediated gap junctional intercellular communication by Ca2+ in mouse epidermal cells is controlled by E-cadherin. J. Cell. Biol. 114: 545-555.

Kartenbeck, J., Schmelz, M., Franke, W. W. and Geiger, B. (1991). Endocytosis of junctional cadherins in bovine kidney epithelial (MDBK) cells cultured in low Ca $^{2+}$ on medium. J. Cell Biol. 113: 881-892.

Katayama, M., Hirai, S., Kamihagi, K., Nakagawa, K., Yasumoto, M. and Kato, I. (1994). Soluble E-cadherin fragments increased in circulation of cancer patients. British J. Cancer. 69: 580-585.

Kemler, R., Ozawa, M. and Ringwald, M. (1989). Calcium-dependent cell adhesion molecules. Current Opin. Cell Biol. 1: 892-897.

Kemler, R., Gossler, A., Mansouri, A. and Vestweber, D. (1990) The Cell Adhesion Molecule Uvomorulin, in: "Morphoregulatory Molecules", G. M. Edelman, B. A. Cunningham, J. P. Thiery, eds., J. Wiley and Sons, New York.

Kemler, R. (1992). Classical cadherins. Seminar. Cell Biol. 3: 149-155.

Kemler, R. (1993). From cadherins to catenins: cytoplasmic protein interactions and regulation of cell adhesion. T. I. G. 9: 317-321.

Kemler, R., Ozawa, M. and Ringwald, M. (1989). Calcium-dependent cell adhesion molecules. Curr. Opin. Cell Biol. 1: 892-897.

King, I. A., Arnemann, J., Spurr, N. K. and Buxton, R. S. (1993). Cloning of the cDNA (DSC1) coding for human type 1 desmocollin and its assignment to chromosome 18. Genomics. 18: 185-194.

Kintner, C. (1988). Effects of altered expression of the neural cell adhesion molecule, N-CAM, on early neural development in Xenopus embryos. Neuron. 1: 545-555.

Kintner, C. (1992). Regulation of embryonic cell adhesion by the cadherin cytoplasmic domain. Cell 69: 225-236.

Knudsen, K. A. and Wheelock, M. J. (1992). Plakoglobin, or an 83 kD homologue distinct from beta-catenin, interacts with E-cadherin and N- cadherin. J. Cell Biol. 118: 671-679.

Koch, P. J., Goldschmidt, M. D., Zimbelmann, R., Troyanovsky, R. and Franke, W. W. (1992). Complexity and expression patterns of desmosomal cadherins. Proc. Natl. Acad. Sci. USA. 89: 353-357.

Koch, P. J. and Franke, W.W. (1994). Desmosomal cadherins: another growing multigene family of adhesion molecules. Current Opin. Cell Biol. 6, 682-687.

Krebs, E. G. and Beavo, J. A. (1979) Phosphorylation-dephosphorylation of enzymes. Annu. Rev. Biochem. 48: 923-959

Larue, L., Ohsugi, M., Hirchenhain, J. and Kemler, R. (1994). E-cadherin null mutant embryos fail to form a trophectoderm epithelium. Proc. Natl. Acad. Sci. USA. 91: 8263-8267.

Legan, P. K., Yue, K. K. M., Chidgey, M. A. J., Holton, J. L., Wilkinson, R. W. and Garrod, D. R. (1994). The bovine desmocollin family: a new gene and expression patterns reflecting epithelial cell proliferation and differentiation. J. Cell Biol. 126: 507-518.

Levine, E., Hyun Lee, C., Kintner, C. and Gumbiner, B. (1994). Selective disruption of E-cadherin function in early Xenopus embryos by a dominant negative mutant. Development. 120: 901-909.

Mac Calman, C. D., Farookhi, R. and Blaschuk, O. W. (1994) estradiol regulates E-cadherin mRNA levels in the surface epithelium of the mouse ovary. Clin. Exp. Metastasis. 12: 276-282.

Mahoney, P. A., Weber, U., Onofrechuk, P., Biessmann, H., Bryant, P. J. and Goodman, C. S. (1991). The fat tumor suppressor gene in Drosophila encodes a novel member of the cadherin gene superfamily. Cell. 67: 853-868.

Mareel, M. M., Behrens, J., Birchmeier, W., De Bruyne, G. K., Vleminckx, K., Hoogewus, A., Fiers, W. C. and Van Roy, F. M. (1991). Down-regulation of E-cadherin expression in Madin-Darby canine kidney (MDCK) cells inside tumors of nude mice. Int. J. Cancer. 47: 922-928.

Marin, O., Meggio, F., Marchiori, F., Borin, G. and Pinna, L. A. (1986) Site-specificity of casein kinase-2 (TS) from rat liver cytosol. A study with model peptide substrates. Eur. J. Biochem. 160: 239-244.

Matsuura, K., Kawanishi, J., Fujii, S., Imamura, M., Hirano, S., Takeichi, M. and Niitsu, Y. (1992). Altered expression of E-cadherin in gastric cancer tissues and carcinomatous fluid. Br. J. Cancer. 66: 1122-1130.

Matsuyoshi, N., Hamaguchi, M., Taniguchi, S., Nagafuchi, A., Tsukita, S. and Takeichi, M. (1992) Cadherin-mediated cell-cell adhesion is pertubed by v-src tyrosine phosphorylation in metastatic fibroblasts. J. Cell Biol. 118: 703-714.

McCrea, P. D. and Gumbiner, B. M. (1991). Purification of a 92-kDa cytoplasmic protein tightly associated with the cell-cell adhesion molecule E-cadherin (Uvomorulin). J. Biol. Chem. 266: 4514-4520.

McCrea, P. D., Turck, C. W. and Gumbiner, B. (1991). A homolog of the armadillo protein in Drosophila (plakoglobulin) associated with E- cadherin. Science. 254: 1359-1361.

McCrea, P. D., Brieher, W. M. and Gumbiner, B. M. (1993). Induction of a second body axis in Xenopus by antibodies to beta-catenin. J. Cell Biol. 123: 477-484.

McNeill, H., Ozawa, M., Kemler, R. and Nelson, W. J. (1990). Novel function of the cell adhesion molecule uvomorulin as an inducer of cell surface polarity. Cell. 62: 309-316.

Meyer, R. A., Larid, D. W., Revel, J. P. and Johnson, R. G. (1992) Inhibition of gap junction and adherens junction assembly by connexin and A-CAM antibodies. J. Cell Biol. 119: 179-189.

Miyatani, S., Copeland, N. G., Gilbert, D. J., Jenkins, N. A. and Takeichi, M. (1992) Genomic structure and chromosomal mapping of the mouse N-cadherin gene. Proc. Natl. Sci. USA. 89: 8443-8447.

Müller, H. A. J., Kühl, M., Finnemann, S., Schneider, S., Van Der Poel, S. Z., Hausen, P. and Wedlich, D. (1994). Xenopus cadherins: The maternal pool comprizes distinguishable members of the family. Mech. Dev. in press.

Nagafuchi, A., Shirayoshi, Y., Okasaki, K., Yasuda, K. and Takeichi, M. (1987). Transformation of cell adhesion properties by exogenously introduced E-cadherin cDNA. Nature. 329: 341-343.

Nagafuchi, A., Ishihara, S. and Tsukita, S. (1994). The roles of catenins in the cadherin-mediated cell adhesion: functional analysis of E-cadherin-alpha catenin fusion molecules. J. Cell Biol. 127: 235-245.

Napolitano, E. W., Venstrom, K., Wheeler, E. F. and Reichardt, L. F. (1991). Molecular cloning and characterization of B-cadherin, a novel chick cadherin. J. Cell Biol. 113: 893-905.

Näthke, I. S., Hinck, L., Swedlow, J. R., Papkoff, J., and Nelson, W. J. (1994) Defining interactions and distributions of cadherin and catenin complexes in polarized epithelial cells. J. Cell Biol. 125: 1341-1352.

Nose, A., Nagafuchi, A. and Takeichi, M. (1987). Isolation of placental cadherin cDNA: identification of a novel gene family of cell-cell adhesion molecules. EMBO. J. 6: 3655-3661.

Nose, A., Tsuji, K. and Takeichi, M. (1990). Localization of specificity determining sites in cadherin cell adhesion molecules. Cell. 61: 147-155.

Oberlender, S. A. and Tuan, R. S. (1994). Expression and functional involvement of N-cadherin in embryonic limb chondrogenesis. Development. 120: 177-187.

Oda, T., Kanai, Y., Oyama, T., Yoshiura, K., Shimoyama, Y., Birchmeier. W., Sugimura, T. and Hirohashi, S. (1994). E-cadherin gene mutation in human gastric carcinoma cell lines. Proc. Natl. Acad. Sci. USA. 91: 1858-1862.

Oka, H., Shiozaki, H., Kobayashi, K., Tahara, H., Tamura, S., Miyata, M., Doki, Y., Iihara, K., Matsuyoshi, N. and Hirano, S. (1992). Immunohistochemical evaluation of E-cadherin adhesion molecule expression in human gastric cancer. Virchows Arch. A. Pathol. Anat. Histopathol. 421: 149-156.

Oka, H., Shiozaki, H., Kobayashi, K., Inoue, M., Tahara, H., Kobayashi, T., Takatsura, Y., Matsuyoshi, N., Hirano, S., Takeichi, M. and Mori, T. (1993). Expression of E-cadherin cell adhesion molecules in human breast cancer tissues and its relationship to metastasis. Cancer Res. 53: 1696-1701.

Okazaki, M., Takeshita, S., Kawai, S., Kikuno, R., Tsujimura, A., Kudo, A. and Amann, E. (1994). Molecular cloning and characterization of OB-cadherin, a new member of cadherin family expressed in osteoblasts. J. Biol. Chem. 269: 12092-12098.

Ozawa, M., Baribault, H. and Kemler, R. (1989). The cytoplasmic domain of the cell adhesion molecule uvomorulin associates with three independent proteins structurally related in different species. EMBO. J. 8: 1711-1717.

Ozawa, M., Engel, J. and Kemler, R. (1990). Single amino acid substitutions in one Ca2+ binding site of uvomorulin abolish the adhesive function. Cell. 63: 1033-1038.

Ozawa, M. and Kemler, R. (1990). Correct proteolytic cleavage is required for the cell adhesive function of uvomorulin. J. Cell Biol. 111: 1645-1650.

Ozawa, M., Hoschützky, H., Herrenknecht, K. and Kemler, R. (1991). A possible new adhesive site in the cell-adhesion molecule uvomorulin. Mech. Dev. 33: 49-56.

Paradies, N. E. and Grunwald, G. B. (1993). Purification and characterization of NCAD90, a soluble endogenous form of N-cadherin, which is generated by proteolysis during retinal development and retains adhesive and neurite-promoting function. J. Neurosci. Res. 36: 33-45.

Puttagunta, S., Mathur, M. and Cowin, P. (1994) Structure of DSG1, the bovine desmosomal cadherin gene encoding the pemphigus foliaceus antigen. Evidence of polymorphism. J. Biol. Chem. 269: 1949-1955.

Ranscht, B. and Dours-Zimmermann, M. T. (1991). T-cadherin, a novel cadherin cell adhesion molecule in the nervous system lacks the conserved cytoplasmic region. Neuron. 7: 391-402.

Redies, C. and Müller, H. A. J. (1994). Similarities in structure and expression between mouse P-cadherin, chicken B-cadherin and frog XB/U-cadherin. Cell. Adhes. Comm. in press.

Roark, E. F., Paradies, N. E., Lagunowich, L. A. and Grundwald, G. B. (1992) Evidence for endogenous proteases, mRNA level and insulin as multiple mechanisms of N-cadherin down-regulation during retinal development. Development. 114: 973-984

Rubinfeld, B., Souza, B., Albert, I., Müller, O., Chamberlain, S. H., Masiarz, F. R., Munemitsu, S. and Polakis, P. (1993). Association of the APC gene product with beta-catenin. Science. 262: 1731-1734.

Sacristàn, M. P., Vestal, D.J., Dours-Zimmermann, M. T. and Ranscht, B. (1993). T-cadherin 2: molecular characterization, function in cell adhesion, and coexpression with T-cadherin and N-cadherin. J. Neurosci. Res. 34: 664-680.

Sano, K., Tanihara, H., Heimark, R. L., Obata, S., Davidson, M., Saint.John, T., Taketani, S. and Suzuki, S. (1993). Protocadherins: a large family of cadherin-related molecules in central nervous system. EMBO. J. 12: 2249-2256.

Sansonetti, P. J., Mounier, J., Prévost, M. C. and Mège, R. M. (1994). Cadherin expression is required for the spread of shigella flexneri between epithelial cells. Cell. 76: 829-839.

Schäfer, S., Koch, P. J. and Franke,W. W. (1994). Identification of the ubiquitous human desmoglein, Dsg2, and the expression cataloque of the desmoglein subfamily of desmosmal cadherins. Exp. Cell Res. 211: 391-399.

Schiffman, J. S. and Grunwald, G. B. (1992). Differential cell adhesion and expression of N-cadherin among retinoblastoma cell lines. Invest. Ophthalmol. 33: 1568-1574.

Shibamoto, S., Hayakawa, M., Takeuchi, K., Hori, T., Oku, N., Miyazawa, K., Kitamura, N., Takeichi, M. and Ito, F. (1994) Tyrosine phosphorylation of beta-catenin and plakoglobin enhanced by hepatocyte growth factor and epidermal growth factor in human carcinoma cells. Cell Adhesion Communication. 1, 295-305.

Shimamura, K., Hirano, S., McMahon, A. P. and Takeichi, M. (1994). Wnt-1-dependent regulation of local E-cadherin and alphaN-catenin expression in the embryonic mouse brain. Development. 120: 2225-2234.

Simonneau, L., Broders, F. and Thiery, J. P. (1992). N-cadherin transcripts in Xenopus laevis from early tailbud to tadpole. Dev. Dynamics. 194: 247-260.

Simonneau, L., Kitagawa, M., Suzuki, S. and Thiery, J. P. (1995) Cadherin-11 expression marks the mesenchymal phenotype: towards new functions for cadherins? Cell Adhes. Comm., in press.

Skalli, O., Jones, J. C. R., Gagescu , R. and Goldman, R. D. (1994). IFAP 300 is common to desmosomes and hemidesmosomes and is a possible linker of intermediate filaments to these junctions. J. Cell Biol. 125: 159-170.

Sorkin, B. C., Gallin, W. J., Edelman, G. M. and Cunningham, B. A. (1991). Genes for two calcium-dependent cell adhesion molecules have similar structures and are arranged in tandem in the chicken genome. Proc. Natl. Acad. Sci. USA. 88: 11545-11549.

Stappert, J. and Kemler, R. (1994) A short core region of E-cadherin is essential for catenin binding and is highly phosphorylated. Cell Adhesion Comm. 2: 319-327.

Su, L. K., Vogelstein, B. and Kinzler, K. W. (1993). Association of the APC tumor suppressor protein with catenins. Science. 262: 1734-1737.

Suzuki, S., Sano, K. and Tanihara, H. (1991). Diversity of the cadherin family: evidence for eight new cadherins in nervous tissue. Cell Regulation 2: 261-270.

Takeichi, M. (1988). The cadherins: cell-cell adhesion molecules controlling animal morphogenesis. Development. 102: 639-655.

Takeichi, M. (1990). Cadherins: a molecular family important in selective cell-cell adhesion. Annu. Rev. Biochem. 59: 237-252.

Takeichi, M. (1991). Cadherin cell adhesion receptors as a morphogenetic regulator. Science. 251: 1451-1455.

Takeichi, M. (1993) Cadherins in cancer: implications for invasion and metastasis. Current Opin. Cell Biol. 5: 806-811.

Tamm, I., Cardinale, I., Kikuchi, T. and Krueger, J. (1994) E-cadherin distribution in interleukin 6-induced cell-cell separation of ductal breast carcinoma. Proc. Natl. Acad. Sci. USA. 91: 4338-4342.

Tang, A., Amagai, M., Granger, L. G., Stanley, J. R. and Udey, M. C. (1993) Adhesion of epidermal langerhans cells to keratinocytes mediated by E-cadherin. Nature. 361: 82-85.

Tang., A., Eller, M. S., Hara, M., Yaar, M., Hirohashi, S. and Gilchrest, B. A. (1994) E-cadherin is the major mediator of human melanocyte adhesion to keratinocytes in vitro. J. Cell Science. 107: 983-992.

Tanihara, H., M. Kido, S., Obata, S., Heimark, R. L., Davidson, M., St. John, T., and Suzuki, S. (1994a). Characterization of cadherin-4 and cadherin-5 reveals new aspects of cadherins. J. Cell Science. 107: 1697-1704.

Tanihara, H., Sano, K., Heimark, R. L., St. John, T. and S. Suzuki. (1994b). Cloning of five human cadherins clarifies characteristic features of cadherin extracellular domain and provides further evidence for two structurally different types of cadherin. Cell Adhes. Comm. 2: 15-26.

Troyanovsky, S. M., Eshkind, L. G., Troyanovsky, R. B., Leube, R. E. and Franke, W.W. (1993). Contribution of cytoplasmic domains of desmosomal cadherins to desmosome assembly and intermediate filament anchorage. Cell. 72: 561-574.

Tsukita, S. A. and Tsukita, S. A. (1985) Desmocalmin: a calmodulin-binding high molecular weight protein isolated from desmosomes. J. Cell Biol. 101, 2070-2080.

Tsukita, S. A., Hieda, Y. and Tsukita, S. H. (1989). A new 82-kD barbed end-capping protein (Radixin) localized in the cell-to-cell adherens junction: purification and characterization. J. Cell Biol. 108: 2369-2382.

Tsukita, S. A., Oishi, K., Akiyama,T., Yamanashi, Y., Yamamoto, T. and Tsukita, S. H.   (1991). Specific proto-oncogenic tyrosine kinases of src family are enriched in cell-to-cell adherens junctions where the level of tyrosine phosphorylation is elevated. J. Cell Biol. 113: 867-879.

Umbas, R., Schalken, J. A., Aalders, T. W., Carter, B. S., Kathaus, H. F. Schaafsma, H.E., Debruyne, F. M. and Isaacs, W. B. (1992). Expression of the cellular adhesion molecule E-cadherin is reduced or absent in high-grade prostate cancer. Cancer Res. 52: 5104-5109.

Vleminckx, K. L., Deman, J. J., Bruyneel, E. A., Vandenbossche, G. M. R., Keirsebilck, A. A., Mareel, M. M. and Van Roy, F. M. (1994). Enlarged cell-associated proteoglycans abolish E-cadherin functionality in invasive tumor cells. Cancer Res. 54: 873-877.

Volberg, T., Geiger, B., Dror, R. and Zyck, Y. (1991). Modulation of intercellular adherens-type junctions and tyrosine phosphorylation of their components in RSV-transformed cultured chick lens cells. Cell Regulation. 2: 105-120.

Volberg, T., Zick, Y., Dror, R., Sabanay, I., Gilon, C., Levitski, Z. A. and Geiger, B. (1992). The effect of tyrosine-specific protein phosphorylation on the assembly of adherens-type junctions. EMBO. J. 11: 1733-1742.

Watabe, M., Nagafuchi, A., Tsukita, S. and takeichi, M. (1994). Induction of polarized cell-cell association and retardation of growth by activation of the E-cadherin-catenin adhesion system in a dispersed carcinoma line. J. Cell Biol. 127: 247-256.

Wiche, G., Becker, B., Luber, K., Weitzer, G., Castanon, M. J., Hauptmann, R., Stratowa, C. and Stewart, M. (1991). Cloning and sequencing of rat plectin indicates a 466 kD polypeptide chain with a three-domain structure based on a central-alpha-helical coiled coil. J. Cell Biol. 114: 83-99.

Williams, C. L., Hayes, V. Y., Hummel, A. M., Tarara, J. E. and Halsey, T. J. (1993). Regulation of E-cadherin-mediated cell adhesion by muscarinic acetylcholine receptors in small lung carcinoma. J. Cell Biol. 12: 643-654.

Wollner, D. A., Krzeminski, K. A. and Nelson,W. J. (1992). Remodelling the cell surface distribution of membrane proteins during the development of epithelial cell polarity. J. Cell Biol. 116: 889-899.

Xiang, Y. Y., Tanaka, M., Suzuki, M., Igarashi, H., Kiyokawa, E., Naito, Y.,Ohtawara, Y., Shen, Q., Sugimura, H. and Kino, I.,1994, Isolation of complementary DNA encoding K-cadherin, a novel rat cadherin preferentially expressed in fetal kidney and kidney carcinoma, Cancer Res. 54: 3034-3041.

AN ELEVATION OF INTERNAL CALCIUM OCCURRING VIA L-TYPE CHANNELS MEDIATES NEURAL INDUCTION IN THE AMPHIBIAN EMBRYO

Catherine Leclerc, Marc Moreau, Lydie Gualandris-Parisot
Géraldine Dréan, Solange Canaux, and Anne-Marie Duprat
Centre de Biologie du Développement
UMR 9925, CNRS/Université Paul Sabatier
118, route de Narbonne
F-31062 Toulouse Cedex, France

INTRODUCTION

In amphibians, neural induction takes place during gastrulation, as a consequence of an interaction between the chordamesoderm (inductive tissue) and the ectoderm (target tissue). The mechanism of neural induction has been the subject of many investigations more than 60 years (Saxén, 1989; Duprat et al. 1990; Westenbroeck et al. 1990). Although the natural inducer still remains unidentified, numerous, apparently unrelated, substances have been found to act as inducers (Tiedemann and Born, 1978; Saxén, 1989). Recently an endogeneous soluble protein, noggin, has been shown to have neural inducing activity in *Xenopus* (Lamb et al. 1993). It has been also suggested that the inhibition of the signal transduced by the activin type II -receptor leads to neuralization (Hemmati-Brivanlou and Melton, 1994). Follistatin, that blocks activin activity by direct binding of activin protein, can induce neural tissue *in vivo* (Hemmati-Brivanlou et al. 1994). Furthermore, it has also been demonstrated that the inducing signal from the chordamesoderm is recognized at the level of the plasma membrane of the target tissue (Tiedemann and Born, 1978; Takata et al.

Organization of the Early Vertebrate Embryo
Edited by N. Zagris *et al.*, Plenum Press, New York, 1995

1981; Gualandris et al. 1985). Therefore it has become important to define the mechanism of transduction of the neuralizing signal. Using *Xenopus* embryos, it has been suggested that activation of protein kinase C (PKC) by phorbol esters, leads to neural induction in a limited part of ectodermal explants (Davids et al. 1987; Otte et al. 1988, 1989). An increase in cAMP dependent protein kinase (PKA) activity during neural induction, has also been observed (Otte et al. 1989), suggesting a cross-talk between PKA and PKC, during this process. These data suggest that PKC pathway is activated by neural induction to initiate neural-specific gene expression. Additional support for a role of a PKC pathway in neural induction is provided by the phorbol ester (TPA) induction of the neural-specific src^+ mRNA in dorsal competent ectoderm of *Xenopus* embryo (Collett and Steele, 1992, 1993).

On the other hand, induction of the target tissue toward the neural pathway can be provoked simply by modifying the extracellular concentration of divalent cations (Barth and Barth, 1964; Saint-Jeannet et al. 1990).

Recently, a direct examination of ionic signalling during neural induction (Sater et al. 1994), has revealed that dorsal ectoderm cells of *Xenopus* undergo an increase in internal pH (pH_i), in response to the neural inducing signal, suggesting that intracellular alkalinization may participate in gene expression associated with neural induction. Furthermore, in *Pleurodeles waltl* embryos it has been demonstrated that L-type calcium channels are directly implicated in the transduction of the neuralizing signal brought about by concanavalinA (Moreau et al. 1994). ConA is an effective inducer for both anurans and urodeles, acting at the plasma membrane level. In addition, specific binding sites for glycoconjugates have been identified in the plasma membrane. So far, ConA is the only lectin known to have inducing activity (Takata et al. 1981; Gualandris et al. 1985).

EVOLUTION OF $[Ca^{2+}]_i$ DURING NEURAL INDUCTION TRIGGERED *IN VITRO*

All stages for *Pleurodeles waltl* refer to those of Gallien and Durocher (1957). In *Pleurodeles waltl* the neural competence of the ectoderm starts at stage 7 and the target tissue can no longer be induced after stage 12 (Leikola, 1963; Gualandris et al. 1985). In response to 300 μg/ml ConA added to the external medium, $[Ca^{2+}]_i$ rose in competent (stage 8a) ectoderm explants of *Pleurodeles waltl*, as revealed by the fluorescent Ca^{2+} indicator fluo-3 (Figure 1A). This rise could be detected less than 20 minutes after addition of ConA to the medium and its maximum level corresponded to 10 to 25 % of the resting $[Ca^{2+}]_i$ (see Kao et al. 1989 for details of the measurements); the maximum amplitude was reached 25 minutes to 1 hour after adding ConA. In parallel control experiments, ConA treatment of ectoderm excised from the same batches of embryos clearly elicited neuralization as judged from the presence of numerous neurons and glial cells, differentiating after 2 to 3 days of culture at 20°C (Figure 2 A,B). No $[Ca^{2+}]_i$ increase was observed, in control experiments, when ConA was omitted from the medium. A strict correlation between an early increase in $[Ca^{2+}]_i$ and neural induction was subsequently observed in the following experimental situations :

First, no variation in $[Ca^{2+}]_i$ was detected following treatment with 300 µg/ml of the non-inducing lectin succinyl-ConA (Figure 1 A, inset)

Second, non competent ectoderm (stage 6), treated with ConA 300 µg/ml, under the same conditions as competent ectoderm, failed to show any increase in $[Ca^{2+}]_i$ above 2 % of the initial value (see Figure 1B).

Third, neural induction in response to ConA failed to occur when competent ectoderm was loaded with the Ca^{2+} chelator BAPTA (0.4 µM BAPTA-AM). It should be noticed that *in vitro* induction resulting from association of the blastoporal lip with dorsal ectoderm was also inhibited when ectodermal cells were loaded with BAPTA.

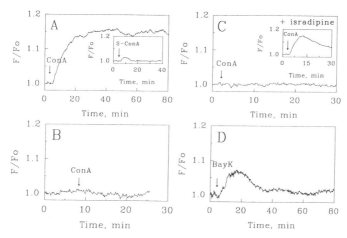

Figure 1. Neural induction is correlated to the activation of L-type calcium channels.
(A) Recording of fluorescence changes of the Ca^{2+} indicator fluo-3, revealing elevation of $[Ca^{2+}]_i$ in *Pleurodeles* target competent tissue (stage 8a) in response to the inducing lectin concanavalin A (ConA). 300 µg/ml.Inset : Fluorescence measurement of fluo-3AM loaded competent ectoderm of *Pleurodeles waltl* (stage 8a), treated with succinyl-ConA 300 µg/ml. (B) Fluorescence measurement of fluo-3 AM loaded uncompetent ectoderm (stage 6) stimulated by ConA 300 µg/ml. No $[Ca^{2+}]_i$ increase is detectable. (C) Effect of pretreatment for 30 minutes with 10 µM isradipine on the variation of $[Ca^{2+}]_i$ triggered by 600µM ConA. Under these conditions, the increase in Ca^{2+} is totally inhibited.(D) Action of 20 µM Bay K 8644 on $[Ca^{2+}]_i$ measured in competent ectoderm loaded with fluo-3 under the same conditions as in (A). Inset : control experiment showing the Ca^{2+} increase triggered by ConA 600 µg/ml. The experiment was performed separatly, but on embryos from the same batch.

The next question we addressed was the origin of the increased $[Ca^{2+}]_i$. The major route for Ca^{2+} entry into cells involves calcium voltage-operated channels (CaVOC). Using gold-ConA conjugates, ConA binding sites have been localized in the plasma membrane of the responsive ectoderm (Grunz, 1985a). Furthermore, it has been shown that ConA binds to the α_2 subunit of the L-type calcium channel and in fact lectin affinity is exploited in the

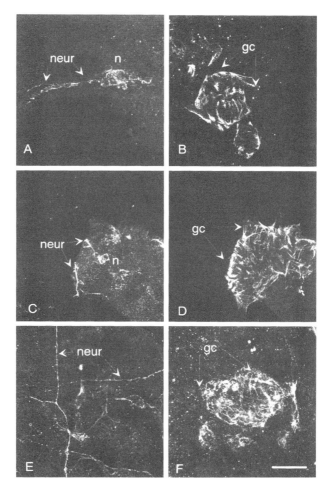

Figure 2. Development of cultured competent ectodermal cells of *Pleurodeles waltl* after various treatments. The assays for neuralization were performed on embryos from the same batch as those used for the calcium measurements. Left side : preparations visualised under epifluorescence in confocal microscopy by immunocytochemical staining using a monoclonal antibody (NC1) directed against gangliosides specific for the neuronal membrane at this stage. Right side : glial cells identified using GFAP antibodies (Soula et al. 1990) visualised under epifluorescence confocal microscopy. (A-B) Neural induction triggered by ConA 300 μg/ml on dorsal explants. Observations were performed after 5 days of culture. Note the presence of numerous neurons (n) with neurites (neur) and areas of glial cells (g.c.) (C-D) Action of 20 μM Bay K 8644. The ectoderm was treated for 3 hours. and maintained in culture for 5 days. Neurons extending neurites can be identified using the NC1 marker. (E-F) Effect of treatment for 30 minutes with 10 mM caffeine. In this experiment neurons (n) have differentiated after 5 days. Numerous neurons (n) and large areas of glial cells (g.c) can be visualized using antibodies NC$_1$ and anti-GFAP respectively.
bar : 50 μm

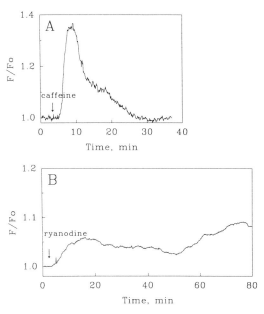

Figure 3. Evidence for the presence of internal calcium stores in competent ectoderm cells
(A) Action of the methylxanthine caffeine (20 mM) on $[Ca^{2+}]_i$ measured after loading competent *Pleurodeles* ectoderm with fluo-3. It should be noted that caffeine was inductive (see also Figure 2). (B) Action of 0.2 μM ryanodine on ectoderm loaded with fluo-3. An increase in $[Ca^{2+}]_i$ was observed. Ryanodine was also inductive.

purification of L-type calcium channels (Borsotto et al. 1984; Curtis and Catterall, 1984). In investigating the possible action of ConA on these channels, the use of a calcium-free medium was excluded since this triggers a complete dissociation of ectoderm cells and causes neural induction indirectly (Saint-Jeannet et al. 1990). However pharmacological agents that specifically block L-type calcium channels are available.

We tested the effect of 10 μM isradipine (a specific inhibitor of L-type calcium channels) on $[Ca^{2+}]_i$ in ectoderm after treatment with ConA. Isradipine totally inhibited the stimulation by ConA of increased $[Ca^{2+}]_i$ (Figure 1C), as well as neural induction, since only epidermal cells subsequently differentiated in culture. It should be noticed that nifedipine 100 μM, another specific inhibitor of L-type calcium channel also produced the same effects. Conversely, a transitory increase in $[Ca^{2+}]_i$ of 10 to 30 minutes was observed after addition of 10 or 20 μM BayK 8644 to the external medium (Figure 1D). This substance, a dihydropyridine, is a well-known agonist of CaVOC of the L-type. Furthermore, as shown in Figure 2 C,D, this transitory increase in $[Ca^{2+}]_i$ was sufficient to trigger neural induction in all cases, as revealed by the subsequent development in cultured explants of neurones and glial cells, identified immunocytochemically with antibodies specific for these cell types (NC$_1$ and anti-GFAP, respectively).

The above results demonstrate that ConA acts *via* CaVOC of the L-type (dihydropyridine sensitive). At this point, we sought to determine whether neural induction could also be triggered by calcium freed from intracellular stores. In order to answer this question we used drugs that liberate sequestred-calcium from intracellular organelles. One of the currently documented forms of calcium release involves the activation of ryanodine or/and methylxanthines sensitive stores (McPherson and Campbell, 1993). The actions of caffeine and ryanodine on dorsal ectoderm from *Pleurodeles waltl* early gastrulae were thus examined.

In competent ectoderm (stage 8a), caffeine at a concentration of 20 mM, in the presence of 1 or 10 mM external Ca^{2+}, had a pronounced effect on internal Ca^{2+} release, triggering a 30% increase of $[Ca^{2+}]_i$ (Figure 3A). The peak of $[Ca^{2+}]_i$ increase was reached in 1 minute and then declined gradually during 20 minutes. Ectoderm cells cultured for 30 minutes to 3 hours in the presence of caffeine differentiated into immunocytochemically identifiable neurones and glial cells after 5 to 6 days of culture, confirming the neuralizing effect of rising $[Ca^{2+}]_i$ (Figure 2 E,F). In order to test directly for the presence of receptors activated by ryanodine, this compound was added to the external medium. A continuous $[Ca^{2+}]_i$ increase took place (Figure 3B), although its amplitude was less than that provoked by caffeine. As in the case of caffeine, ryanodine triggered neural induction in the *in vitro* system despite the low amplitude of the Ca^{2+} release. However, neural induction triggered by ryanodine yielded less neurones. These data demonstrate that a rise in $[Ca^{2+}]_i$, above a threshold value, irrespective of the mechanism by which Ca^{2+} is increased within the cell, is sufficient to neuralize competent ectoderm cells. The preceeding results are summarized in table 1 .

Previous work has demonstrated that activation of protein kinase C (PKC) by phorbol esters triggers neural induction (Davids et al. 1987; Otte et al. 1988, 1989). Protein phosphorylation by the Ca^{2+}/phospholipid dependent protein kinase (PKC) is one of the possible mechanisms of Ca^{2+} channels regulation. Indeed, PKC has been reported to exert inhibitory or stimulatory effects on L-type calcium channels (Shearman et al. 1989; Yang and Tsien, 1993). This led us to the hypothesis that the inductive effect of phorbol esters might occur *via* the activation of calcium channels, either directly or indirectly.

Table 1. Effect of Ca^{2+} agonists or antagonists on $[Ca^{2+}]i$ and neural induction in competent presumptive ectoderm of the amphibian *Pleurodeles waltl*. The increase in $[Ca^{2+}]i$ was considered positive if it corresponded to at least 10 % of the basic level. Neural induction was scored by direct microscopic observation of neurones under the phase contrast microscope (objective X40) and by immunocytochemistry using NC1 and GFAP antibodies (see text and figure 2 for more details).

	calcium increase	neural induction
ConA	+	+
Succinyl ConA	-	-
Bay K	+	+
ConA+Nifedipine	-	-
Ryanodine	+	+
Caffeine	+	+

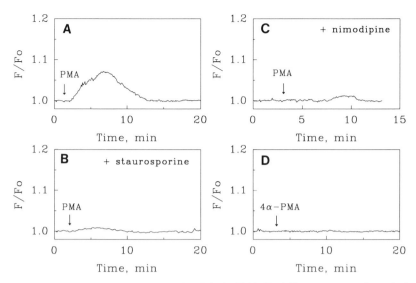

Figure 4. Modulation of L-type calcium channel activity by PMA. Each figure represent the evolution of $[Ca^{2+}]_i$ measured as fluorescence variation of competent ectoderm (stage 8a), loaded with fluo-3AM (A) Effect of PMA (500 nM) showing an increase in $[Ca^{2+}]_i$. (B) Inhibition of PMA 500 nM effect by staurosporine 500 nM. (C) Effect of nimodipine 10μM on the $[Ca^{2+}]_i$ increase provoked by PMA (500 nM) action. The explant was incubated 10 minutes in nimodipine before PMA addition. Nimodipine was continuously present during stimulation by PMA. (D) Fluorescence measurement of the competent explant, treated with 4-α PMA 500 nM, a phorbol ester analog inactive with respect to PKC.

To test this hypothesis, $[Ca^{2+}]_i$ was monitored during application of PMA, a potent stimulator of various isoforms of PKC. Continuous bath application of 50 to 500 nM of PMA transiently increased $[Ca^{2+}]_i$ in a dose dependent manner, with a variable time course. The fluorescence usually increased gradually, reaching a maximum in 5 to 10 minutes (Figure 4A) and returned to the resting level in 10 to 20 minutes. Control experiments designed to validate the direct action of PMA on PKC activation in our system involved the use of staurosporine, a protein kinase inhibitor which exhibits higher affinity for PKC than for other kinases, and 4α-phorbol 12-myristate, 13-acetate (4α-PMA), a phorbol ester that is inactive with respect to PKC.

In dorsal ectoderm incubated for 40 minutes in medium containing 500 nM staurosporine the effect of 500 nM PMA was dramatically decreased as shown on Figure 4B. In addition, 4α-PMA (500 nM) produced no significant effect on the fluorescence level recorded in dorsal explants (less than 1% increase) (Figure 4D).

In order to define the role of PKC on L-type calcium channels, we tested the effect of an antagonist of L-type calcium channels (nimodipine) on the $[Ca^{2+}]_i$ increase triggered by PMA. Nimodipine (10 μM) blocked the increase in $[Ca^{2+}]_i$ induced by PMA 500 nM, indicating that the calcium increase was mostly due to L-type channels (Figure 4C). This

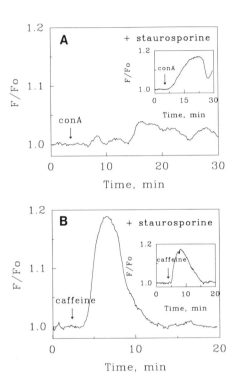

Figure 5. PKC mediates the activation of L-type calcium channels, but not calcium release from internal stores. (A) Effect of staurosporine 500 nM on competent ectoderm stimulated by ConA 300 µg/ml. Explants were preincubated 30 minutes in staurosporine before ConA addition. inset : Control, showing a fluorescence increase of the competent ectoderm loaded with fluo-3 AM, triggered by ConA 300 µg/ml (embryos of the same batch as in A). (B) Effect of caffeine 10 mM on competent ectoderm preincubated 30 minutes in staurosporine 500 nM. inset : Control of the effect of caffeine 10 mM on competent ectoderm.

result raised the question whether or not the ConA effect on $[Ca^{2+}]_i$ was also dependent on the activation of L-type calcium channels by PKC. In fact, the $[Ca^{2+}]_i$ increase triggered by ConA 300 µg/ml was abolished when dorsal explants were incubated 30 minutes in 500 nM ConA is indeed controlled by PKC-dependent phosphorylation processes. Furthermore neural induction was totally inhibited when ConA-stimulated presumptive ectoderm was preincubated 30 minutes in 500 mM staurosporine since in all cases only epidermal differentiation was observed.

A further confirmation that PKC acts on plasma membrane Ca^{2+} channels, was provided by the result that show that dorsal explants incubated 40 minutes in presence of staurosporine 500 nM were still able to release calcium upon 10 mM caffeine treatement (Figure 5B and inset). No significant difference was observed in fluorescence intensity between caffeine-stimulated dorsal explants incubated with or without staurosporine, although the duration of the transient was slightly reduced (30%) in the presence of staurosporine. This demonstrates that PKC plays an essential role by acting on the plasma membrane calcium channels and not on cytosolic Ca^{2+} stores.

We thus establish that neural induction, triggered *in vitro* by ConA, involved the activation of L-type calcium channels. In addition, an increase in internal calcium concentration ($[Ca^{2+}]_i$) obtained by direct stimulation of internal stores by caffeine or ryanodine is sufficient to trigger neural induction.

EXPRESSION OF L-TYPE Ca^{2+} CHANNEL DURING EARLY EMBRYOGENESIS IN *XENOPUS LAEVIS*

In order to substantiate the role of the L-type calcium channel in early development, we have detailed the kinetics of the expression and the localization of its α_1 subunit in early stages of *Xenopus laevis* embryogenesis using immunological techniques. Doing this, we hypothesised that as in *Pleurodeles waltl* embryos, the transduction of the neuralizing signal involved calcium influx through L-type calcium channels (Moreau et al. 1994).

We used a mouse anti-dihydropyridine binding complex monoclonal immunoglobulin directed against the α_1 subunit of L-type Ca^{2+} channel extracted from transverse tubules of rabbit skeletal muscle

The L-type channel is a multimeric protein composed of 5 subunits α_1, α_2, β, γ and δ (Catterall, 1991). The α_1 subunit carries the ionic pore and binding sites for agonists and antagonists of the channel.

Temporal expression of α_1

Embryos were staged according to Nieuwkoop and Faber (1967). Temporal expression of this subunit was first investigated in isolated cells from presumptive ectoderm dissected from early *Xenopus laevis* embryos.

When embryos where dissected at stage 7 (early blastula, 4 hours after fertilization), the cytoplasm was heterogeneously labelled : we observed different positive domains deep in the cytoplasm. A faint immunoreactivity was also found at the plasma membrane level.

When dissection was carried out at stage 8 (onset of midblastula transition, Gerhart, 1980; Newport and Kirschner, 1982a), the labelling was again seen to be both intracellularly and localized in patches at the plasma membrane.

At stage $10^{1/4}$ (early gastrula), the ectoderm is composed of two cell layers which possess some differences in neural competence (Grunz, 1984, 1985b). We therefore decided to study them independently. The two sheets were separated in Ca^{2+}/Mg^{2+} free medium. We did not observed any difference in the location nor the intensity of the labelling of these two cell layers. The immunoreactivity was seen at the plasma membrane and in the cytoplasm just beneath. We note here again that only a part of the cell was labelled.

The results show that the α_1 subunit of the L-type Ca^{2+} channel begins to be expressed during cleavage stages and is localized in the cytoplasm and at the plasma membrane of the ectoderm cells before neural induction.

In order to determine the location of the L-type Ca^{2+} channel in the three germ layers of early embryos we went on to study its distribution by immunocytochemistry on sections on the whole embryo.

Spatial expression of α_1

At stage 5 (16-cell embryo), no labelling was found, other than that due to non-specific trapping of antibodies by the vitellin membrane (data not shown).

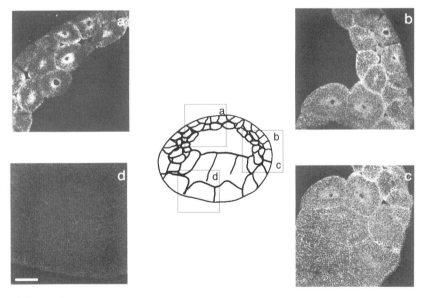

Figure 6. Expression of the α_1 subunit of L-type calcium channel, using monoclonal immunoglobulin directed against this α_1 subunit. Immunohistochemistry was carried out on 10 μm serial sections and observations were performed using an epifluorescence confocal microscope. In stage 7 embryos, marginal zone cells are positive whereas endoderm is not labelled. (a) ectoderm, (b) and (c) marginal zone, (d) endoderm.
bar = 50μm.

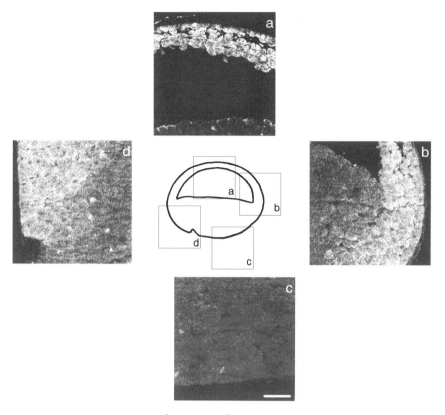

Figure 7. Immunolabelling at stage 10 $^1/4$. Stage 10 $^1/4$ embryos were treated and observed as in figure 6. Ectoderm and mesoderm are labelled whereas endoderm is negative. There is no difference between the dorsal and ventral sides of the embryo. (a) ectoderm, (b) ventral marginal zone, (c) endoderm, (d) dorsal marginal zone. bar = 50 µm.

At stage 7 the embryo is composed of the presumptive ectoderm consisting of small cells and the presumptive endoderm constitued by large rounded cells. At the site of contact between ectoderm and endoderm, i.e. the marginal zone, mesodermal induction is taking place (for reviews see Gurdon, 1987; Woodland, 1989; Slack, 1994). The α_1 labelling was observed in the whole presumptive ectoderm (Figure 6a) and in small cells of the marginal zone (Figure 6b, c) whereas endoderm was not labelled (Figure 6d). In the area between ectoderm and endoderm, we noted some bigger positively stained cells, probably induced mesodermal cells (Figure 6c). Strikingly, the whole marginal zone was labelled, indicating that α_1 was expressed in both the dorsal and the ventral side of the embryo (data not shown). At the cellular level the protein was expressed at the plasma membrane and around the nucleus, the nucleus itself being negative.

At stage $10^1/4$ the morphogenetic movements of gastrulation begin and appearance of the blastopore lip allows the dorsal and ventral sides of the embryo to be distinguished. The ectoderm is composed of two cell layers. Whereas the mesoderm is localized at the marginal zone and begins its invagination, the endoderm still occupied the vegetal half of the embryo. Immunoreactivity was observed throughout the ectoderm (Figure 7a) and the marginal zone

(Figure 7b, d) whereas endoderm remained unlabelled (Figure 7c). In the ectoderm, the internal layer was more positive than the external layer which was only faintly labelled. At the marginal zone we observed a clear boundary between positive cells and negative endodermal cells. At the dorsal marginal zone positive cells were localized above the blastopore lip, the region corresponding to the invaginating marginal zone described in the fate map (Dale and Slack, 1987). It should be underlined that there was no difference in labelling between dorsal and ventral sides both of ectoderm and mesoderm (Figure 7b and d).

In positive tissues we found that the labelling was stronger in the plasma membrane than in the cytoplasm.

These data indicate that the α_1 subunit of L-type Ca^{2+} channel is first expressed from stage 7 in presumptive ectodermal cells and mesodermal territories both in the cytoplasm and in the plasma membrane.

Colocalisation with G_o protein

In a previous study, the spatial and temporal expression of the α subunit of G_o ($G\alpha_o$) was shown to be related to the acquisition of neural competence in the presumptive ectoderm of *Pleurodeles waltl* embryos (Pituello et al. 1991). Consequently, it was important to study the respective localization of the α_1 subunit of the L-type Ca^{2+} channel and $G\alpha_o$ protein. Double labelling was thus carried out on isolated ectodermal cells. At every stage tested (8 to $10^{1/4}$) we found that most of α_1 positive cells were also $G\alpha_o$ positive. Moreover, both antibodies labelled the cells in the same way (data not shown). This strong colocalization suggests the possibility of some sort of interaction between these proteins.

DISCUSSION

Many studies have shown that the ionic environment plays important roles in early development. During meiotic maturation of amphibian oocytes changes in intracellular Na^+ and K^+ activities have been reported (Moreau et al. 1984). A down regulation of the Na/K-ATPase by endocytosis appears to be implicated in these changes in *Xenopus* oocytes (Schmalzing et al., 1990). In *Pleurodeles waltl* oocytes we have demonstrated that the same process occurs. In addition, reinsertion of the Na/K-ATPase in the plasma membrane can be provoked by increasing internal calcium concentration by ionomycin (Canaux, unpublished results).

The Na/K-ATPase has also been implicated during early embryogenesis. Slack et al. (1973) suggest that this pump may be involved in blastocoel formation. In *Xenopus* and *Ambystoma* embryos, the Na/K-ATPase has been reported to control neural differentiation (Messenger and Warner, 1979; Breckenridge and Warner, 1982).

Modulation of Na/K-ATPase activity has been shown, in oocytes, to be correlated with variations of intracellular calcium concentration (Moreau et al., 1984; Schmalzing and Kröner, 1991). A calcium-dependent control of Na/K-ATPase activity may also occurs in early steps of neural development. Indeed, two experimental arguments strengthen this hypothesis: i) The experiments reported here demonstrate that competent cells of dorsal ectoderm of *Pleurodeles waltl* undergo an increase in intracellular calcium in response to artificial neuralizing signals such as ConA. ii) The Na/K-ATPase only starts to be expressed at the mid-neural fold stages in *Xenopus* and is therefore a consequence of neural induction (Blackshaw and Warner, 1976).

The role played by the ionic environnement in neural induction was indicated early on, mainly by the work of Barth and Barth (1964), but the direct effect of ionic movements, in this process, had never been documented. Sater *et al.* (1994) have shown that planar neural induction in *Xenopus* is accompagnied by an increase in internal pH. Therefore, intracellular alkalinization and an internal calcium increase are among the earliest known responses to neural induction (Moreau et al., 1994). The idea that an intracellular calcium increase is in fact a direct cause in triggering neural induction is supported by the following lines of evidence : firstly, in our experimental paradigms, neural induction was always accompagnied by an increase in intracellular calcium; secondly, in the absence of neural induction (non competent ectoderms stimulated by ConA), $[Ca^{2+}]_i$ increase never developed; thirdly, when $[Ca^{2+}]_i$ increase was inhibited by the calcium chelator BAPTA, neural induction was abolished. Ectoderm cells treated by the BAPTA survived well and developed into epidermis. Although for technical reasons, we were unable to follow Ca^{2+} changes directly in ectoderm associated with the blastoporal lip, neuralization with this natural inducer was totally blocked by BAPTA. Therefore this direct relationship between $[Ca^{2+}]_i$ increase and neural induction is apparently not restricted to artificial neuralizing factors.

Our results suggest that the increase in $[Ca^{2+}]_i$, triggered by ConA is mainly due to the activation of L-type calcium channels (dihydropyridine sensitive calcium channels).

Our data, together with the fact that ConA binds, in PC_{12} cells, to the α_2 subunit of the L-type calcium channel (Greenberg et al. 1987), strongly suggest that this channel is indeed the primary molecular target of the lectin in neural induction. However, the relationship between a putative receptor of the neuralizing signal and L-type calcium channels, remains unknown at this stage of the work.

ConA treatment did not elicit a $[Ca^{2+}]_i$ increase in non competent ectoderm (*Pleurodeles waltl*, stage 6) suggesting that L-type calcium channels are absent or not functional at this stage. This in turn, raises the interesting possibility that the incorporation in the plasma membrane of dorsal ectoderm and/or acquisition of functionality of L-type calcium channels may play an important role in the appearance of competence. Indeed, in *Xenopus laevis* embryos, immunolabelling experiments indicate that the α_1 subunit of the L-type calcium channels are first expressed just before the beginning of neural competence (for neural competence see the recent review of Fukui and Asashima, 1994). The α_1 subunit is then clearly detected in the plasma membrane at the onset of the midblastula transition (MBT), the critical period which corresponds to the beginning of zygotic gene expression (Gerhart, 1980; Newport and Kirschner, 1982a,b). While before the MBT labelling for α_1 is mostly observed in the cytoplasm of presumptive ectoderm cells, at the early gastrula stage the α_1 subunits of the L-type calcium channel is present both in the cytoplasm and in the plasma membrane of ectodermal and mesodermal cells. These observations suggest that the pool of α_1 subunit detected is probably of maternal origin and that the L-type calcium channels incorporate in cell membranes in a progressive manner from the early blastula through to the beginning of gastrulation. Our data also show that the α_o subunit of the G_o protein strictly colocalized with the α_1 subunit of the L-type calcium channel. In *Pleurodeles waltl* the α_o subunit of the G_o protein is first expressed at the onset of neural competence (Pituello et al. 1990). Despite these results, little information is avaible about the molecular basis of neural competence. The immuno-detection of the α_1 subunit of the L-type calcium channels in *Xenopus laevis* ectoderm cells prior to neural induction, together with the demonstration that direct stimulation of the L-type calcium channels is able to induce early gastrula ectoderm cells of *Pleurodeles waltl* to differentiate into neural tissue suggest that the molecular basis of competence probably involves some kind of receptor system able to recognize the inducing signal. For neural competence we propose that the L-type calcium channel can play such a role. This hypothesis requires further analysis to determine the initial expression of L-type calcium channels at peculiar developmental stages, associated with patch clamp studies to determine the onset of function of these channels.

The sustained increase in $[Ca^{2+}]_i$ triggered by ConA was somewhat intriguing, when considering the rapid inactivation of L-type calcium channels. Three main hypothesis can be considered to explain this phenomenon : i) 3 hours of presence of ConA in the external medium are required to trigger neural induction (Takata et al. 1981; Grunz, 1985b; Gualandris et al. 1985). This rather long time is justified by the fact that ConA is internalized in a few minutes by the ectoderm cells (Grunz, 1985b), although it has been demonstrated that internalization is not a prerequisite for neural induction. The continuous presence of this lectin may thus exert an iterative effect on calcium channels, resulting in the long lasting $[Ca^{2+}]_i$ increase observed. ii) Another possibility would be to assume that this lectin exerts its effect on Ca^{2+} influx by increasing channel numbers as this has been suggested for PC_{12} cells (Greenberg et al. 1987). This possibility is currently being tested in our laboratory. iii) Finally, the observed time course of increase is rather suggestive of the intervention of an internal relay, which might be due to a process referred to as a *calcium induced calcium release* (CICR). In our system, internal calcium stores can be activated by ryanodine or methylxanthines. Caffeine is a powerful agonist of intracellular calcium release. Ryanodine is less efficient but also less permeant that caffeine in *Pleurodeles* ectoderm cells, which might explain the difference in amplitude of calcium release between both drugs. These experiments demonstrate that such stores are present in the ectodermal cells at competent stages. CICR may also involve inositol trisphosphate sensitive stores (InsP3) (Berridge,

1993). However we have no evidence for the existence of such stores in ectoderm cells, since no permeant agonist of InsP3 receptors are currently available.

With the two first hypotheses, which do not take into account the inactivation of L-channels, the sole effect of ConA on calcium channels should result in a more transitory $[Ca^{2+}]_i$ increase than that observed in our experiments. The third hypothesis is more likely, since we have demonstrated the presence of ryanodine-sensitive calcium stores. These releasable calcium stores can constitute an internal relay for calcium influx and the long lasting kinetics observed in presence of ConA (Figure 1A), could be due to a continuous refilling of internal stores from the external medium. However, experiments in calcium-free medium could not be performed in our system since it has been demonstrated that such conditions dissociate the ectoderm tissue and indirectly trigger neural induction (Saint-Jeannet et al. 1990). Therefore, no definitive conclusions can be drawn regarding an eventual CICR acting as an internal relay of the Ca^{2+} influx upon neural induction triggered *in vitro* by ConA.

The long lasting effect of caffeine could suggest that caffeine acts on its other target, *i.e.* inhibition of phosphodiesterase, in turn triggering an increase in the cAMP level. In addition, a direct control of L-type calcium channel by cAMP-dependant protein kinase (PKA) has been described (Armstrong and Eckert, 1987), suggesting that the observed effect of caffeine could be due to the activation of L-channels. However, this mechanism seems unlikely since the calcium release triggered by caffeine was not modified in the presence of staurosporine, which, as demonstrated in this study, totally inhibits the activation of L-channels by PMA.

The activation of PKC, particularly by phorbol esters (TPA or PMA) has been described to be involved in neural induction (Otte et al. 1988, 1989). Here we suggest that one role of PKC is to regulate L-type calcium channel activity. Such a role has been described already in numerous systems (for review see Shearman et al. 1989). However the effective role of phorbol esters remains unclear. They may either stimulate or inhibit the activation of calcium channels. Some non specific effects have been also described, but in these cases phorbol esters always appeared as inhibitors of calcium permeability (Shearman et al. 1989). A role for PKC in up modulation of L-type calcium channels has been recently demonstrated in frog sympathetic neurons (Yang and Tsien, 1993). In our system, the effect of PMA may also be explained by an up modulation of calcium channels. In fact, our results showing that the $[Ca^{2+}]_i$ increase triggered by ConA is abolished by a PKC inhibitor (staurosporine) and that $[Ca^{2+}]_i$ increase triggered by PMA is inhibited by nimodipine, confirm the control of this calcium permeability by PKC. Consequently, neural induction mediated by phorbol esters (Davids et al. 1987; Otte et al. 1988, 1989) may be partially interpreted as a direct action of PKC on calcium channels. Phosphorylation by PKC could increase channel activity by different mechanisms : PKC could act by direct phosphorylation of a component of the L-type calcium channel (Yang and Tsien, 1993). It has also been suggested in *Aplysia* neurons that stimulation of PKC recruits a covert type of Ca^{2+} channels

(Strong et al. 1987). On the other hand, it has been shown in many cell types, that PKC down modulates the activity of other voltage-activated channels, particularly K^+ and Na^+ channels. Therefore the induced depolarisation may indirectly activate voltage-dependent calcium channels (Shearman et al. 1989).

The total inhibition by staurosporine of ConA-triggered neural induction can be explained by at least two mechanisms : an inactivation of L-channels or a direct effect of ConA on PKC. The latter possibility is unlikely since it has been demonstrated that PKC is a regulatory element which is located downstream (Yang and Tsien, 1993). Nevertheless, patch clamp experiments are required to provide a definitive answer to this question.

It is well established that calcium changes constitute a common response to external signals (for review see Berridge, 1993). On the other hand, neural induction takes place as a result of the activation of specific sets of genes. How then might an increase of calcium activate genes ? Protein products of immediate early genes (IEG) appear to play a critical role in long term changes in cell function. IEG encode regulatory proteins that direct the cell genome in response to environmental stimuli (Sheng and Greenberg, 1990). During activation of PC_{12} cells, occurring when stimulated by depolarizing agents such as KCl, an increase in $[Ca^{2+}]_i$ resulting from the activation of L-type calcium channels is observed. This calcium increase leads to the activation of IEG, such as c-fos or Jun-B, in less than 20 minutes after the opening of calcium channels. The mechanism by which Ca^{2+} influx activates transcription of c-*fos* or *Jun*-B has been characterized (Sheng et al. 1990). It involves the phosphorylation of transcription factor CREB (cAMP response element binding protein) which is directly mediated by a calmodulin-dependent protein kinase. The function of CREB, then could be to integrate various signalling pathways resulting in gene expression. Among these, one can include follistatin, which can cause neural induction by blocking the activin activity (Hemmati-Brivanlou and Melton, 1994; Hemmati-Brivanlou et al. 1994). On the other hand, the characterization and cloning of the noggin receptor would provide clues about the possible ionic control of the noggin signal transduction pathway. In our system, a similar mechanism of CREB phosphorylation following a Ca^{2+} signal might be responsible for specific gene activation involved in neural induction. In this respect, it is interesting to mention that a transient variation of $[Ca^{2+}]_i$ (whatever its provenance, either from external medium or from internal stores) of less than 30 minutes (*i.e.* caffeine or BayK effects) is sufficient to drive the competent ectoderm in the way of neural determination.

Acknowledgements

We are grateful to Pierre Guerrier and Philippe Cochard for criticisms, suggestions and to Françoise Foulquier for technical assistance.This work was supported by CNRS (Centre National de la Recherche Scientifique), by a research grant from AFM (Association Française contre les Myopathies) n°4350432.

REFERENCES

Armstrong, D., and Eckert, R .(1987) Voltage activated calcium channels that must be phosphorylated to respond to membrane depolarization, *Proc.Natl.Acad.Sci.(USA)* **84**: 2518-2522.

Barth, L.G., and Barth, L.J. (1964) Sequential induction of the presumptive epidermis of the *Rana pipiens* gastrula, *Biol.Bull.* **127**: 413-427.

Berridge, M.J. (1993) Inositol trisphophate and calcium signalling, *Nature* **361**: 315-325.

Blackshaw, S.E., and Warner, A.E. (1976) Alterations in resting membrane properties at neural plate stages of development of the nervous system, *J.Physiol. (London)* **255**: 231-247.

Borsotto, M., Barhanin, J., Norman, R.I., and Lazdunski, M. (1984) Purification of the dihydropyridine receptor of the voltage-dependent Ca^{2+} channel from skeletal muscle transverse tubules using ^{3}H-PN 200-110, *Biochem.Biophys.Res.Comm.* **122**: 1357-1366.

Breckenridge, L.J., and Warner, A.E. (1982) Intracellular sodium and the differentiation of amphibian embryonic neurones, *J.Physiol. (London)* **332**: 393-413.

Catterall, W.A. (1991) structure and function of voltage-gated sodium and calcium channels, *Curr. Op. Neurobiol.* **1**: 5-13.

Collett, J.W., and Steele, R.E. (1992) Identification and developmental expression of *Src*[+] mRNAs in Xenopus laevis, *Dev.Biol.* **152**: 194-198.

Collett, J.W., and Steele, R.E. (1993) Alternative splicing of a neural-specific Src mRNA (*Src*[+]) is a rapid and protein synthesis-independent response to neural induction in *Xenopus laevis*, *Dev.Biol.* **158**: 487- 495.

Curtis, B.M., and Catterall, W.A. (1984) Purification of the calcium receptor of the voltage-sensitive calcium channel from skeletal muscle transverse tubules, *Biochemistry* **23**: 2113-2118.

Dale, L., and Slack, J.M.W. (1987) regional specification within the mesoderm of early embryos of *Xenopus laevis*, *Development* **99**: 279-295.

Davids, M., Loppnow, B., Tiedemann, H., and Tiedemann, H. (1987) Neural differentiation of amphibian gastrula ectoderm exposed to phorbol ester, *Roux's Arch.Dev.Biol.* **196**: 137-140.

Duprat, A.M., Saint-Jeannet, J.P., Pituello, F., Huang, S., Boudannaoui, S., Kan, P., and Gualandris, L. (1990) From presumptive ectoderm to neural cells in an amphibian, *Int.J.Dev.Biol.* **34**: 149-156.

Fukui, A. and Ashima, M. (1994) Control of cell differentiation and morphogenesis in amphibian development, *Int. J. Dev. Biol.* **38**: 257-266.

Gallien, L., and Durocher, M. (1957) Table chronologique du développement chez Pleurodeles waltl (Michah). *Bull. Biol. Fr. Belg.* **91**: 97-114.

Gerhart, J.G. (1980) Mechanisms regulating pattern formation in the amphibian egg and the early embryo. *In* "Biological regulation and development" (R.F. Goldberger ed.) pp133-315. Plenum Press, New York.,

Greenberg, D., Carpenter, C.L., and Messing, R.O. (1987) Lectin-induced enhancement of voltage-dependent calcium flux and calcium channel antagonist binding, *J.Neurochem.* **48**: 888-894.

Grunz, H.(1984), early embryonic induction: the ectodermal target cells, *In* "the role of cell interactions in early neurogenesis. Serie A" (A.M. Duprat, A.C. Kato, and M. Weber, Ed.), Life sci. Ed. Ed., Vol. 77, pp. 21-38. Plenum Press, New York.

Grunz, H. (1985a) Information transfer during embryonic induction in amphibians, *J.Embryol.exp.Morph.* **89**: 349-364.

Grunz, H. (1985b) effects of concanavalin A and vegetalizing factor on the outer and inner ectoderm layers of early gastrulae of *Xenopus laevis* after treatment with cytochalasin B, *Cell Diff.* **16**: 83-92.

Gualandris, L., Rouge, P., and Duprat, A.M. (1985) Target cell surface glycoconjugates and neural induction in an amphibian, *J.Embryol.exp.Morph.* **86**: 39-51.

Gurdon, J.B. (1987) embryonic induction. Molecular prospects, *Development* **99**: 285-306.

Hemmati-Brivanlou, A., and Melton, D.A.(1994) Inhibition of activin receptor signaling promotes neuralization in *Xenopus*, *Cell* **77**: 273-281.

Hemmati-Brivanlou, A., Kelly, O.G., and Melton, D.A.(1994) Follistatin, an antagonist of activin, is expressed in the spemann organizer and displays direct neuralizing activity, *Cell* **77**: 283-295.

Kao, J.P.Y., Harootunian, A.T., and Tsien, R.Y. (1989) Photochemically generated cytosolic calcium pulses and their detection by fluo-3, *J.Biol.Chem.* **264**: 8179-8184.

Lamb, T.M., Knecht, A.K., Smith, W.C., Stachel, S.E., Economides, A.N., Stahl, N., Yancopolous, G.D., and Harland, R.M. (1993) Neural induction by the secreted polypeptide noggin, *Science* **262**: 713-718.

Leikola, A. (1963) The mesodermal and neural competence of isolated gastrula ectoderm studied by heterogenous inductors, *Ann. Zool. Soc.* **25**: 2-50.

McPherson, P.S., and Campbell, K.P. (1993) The ryanodine receptor/Ca^{2+} release channel, *J.Biol.Chem.* **268**: 13765-13768.

Messenger, E.A., and Warner, A.E. (1979) The function of the sodium pump during differentiation of amphibian embryonic neurons, *J.Physiol. (London)* **292**: 85-105.

Moreau, M., Guerrier, P., and Vilain, J.P. (1984) Ionic regulation of oocyte maturation. *In* "Biology of Fertilization" (Metz,C.B., Monroy, A. Eds.) pp. 299-345. Vol. 1. Acad. Press, New York

Moreau, M., Leclerc, C., Gualandris-Parisot, L., and Duprat, A.-M. (1994) Increased internal Ca^{2+} mediates neural induction in the amphibian embryo, *Proc.Natl.Acad.Sci.(USA)*: (in press).

Newport, J. and Kishner, M. (1982a) A major developmental transition in early embryos: I. Characterization and timing of cellular changes at the midblastula stage, *Cell* **30**: 675-686.

Newport, J. and Kishner, M. (1982b) A major developmental transition in early embryos: II. Control of the onset of transcription, *Cell* **30**: 687-696.

Nieuwkoop, P.D., and Faber, J. (1967) Normal table of *Xenopus laevis:* a systematic and chronological survey of the development of the fertilized egg till the end of metamorphosis. North Holland, Amsterdam.

Otte, A.P., Koster, C.H., Snoek, G.T., and Durston, A.J. (1988), Protein kinase C mediates neural induction in *Xenopus laevis*, *Nature* **334**: 618-620.

Otte, A.P., van Run, P., Heideveld, M., van Driel, R., and Durston, A.J. (1989) Neural induction is mediated by cross-talk between the protein kinase C and cyclic AMP pathways, *Cell* **58**: 641-648.

Pituello, F., Homburger, V., Saint-Jeannet, J.P., Audigier, Y., Bockaert, J., and Duprat, A.M. (1991) Expression of the guanine nucleotide-binding protein G_0 correlates with the state of neural competence in the amphibian embryo, *Dev.Biol.* **145**: 311-322.

Saint-Jeannet, J.P., Huang, S., and Duprat, A.M. (1990) Modulation of neural commitment by changes in target cell contacts in *Pleurodeles waltl*, *Dev.Biol.* **141**: 93-103.

Sater, A.K., Alderton, J.M., and Steinhardt, R.A. (1994) An increase in intracellular pH during neural induction in *Xenopus*, *Development* **120**: 433-442.

Saxén, L. (1989) Neural induction, *Int.J.Dev.Biol.* **33**: 21-48.

Schmalzing, G., Eckard, P., Kröner, S., and Passow, H. (1990) Down regulation of surface sodium pump by endocytosisduring meiotic maturation of *Xenopus* oocytes, *J.membr.Biol.* **79**: 203-210.

Schmalzing, G., and Kröner, S. (1991) Micromolar free calcium exposes ouabain-binding sites in digitonin-permeabilized *Xenopus laevis* oocytes, *Biochem.J.* **269**: 757-766.

Shearman, M.S., Sekiguchi, K., and Nishizuka, Y.(1989) Modulation of ion channel activity: a key function of the protein kinase C enzyme family, *Pharmacol. Rev.* **41**: 211-237.

Sheng, M., and Greenberg, M.E. (1990), The regulation and function of c-fos and other immediate early genes in the nervous system, *Neuron* **4**: 477-485.

Sheng, M., McFadden, G., and Greenberg, M.E. (1990) Membrane depolarization and calcium induce c-fos transcription via phosphorylation of transcription factor CREB, *Neuron* **4**: 571-582.

Slack, C., Warner, A.E., and Warren, R.L. (1973) The distribution of sodium and potassium in amphibian embryos during early development, *J.Physiol. (London)* **232**: 297-312.

Slack, J.M.W. (1994) Inducing factors in *Xenopus* early embryos, *Curr. Biol.* **4**: 116-126.

Soula, C., Sagot, Y., Cochard, P., and Duprat, A.M.(1990) Astroglial differentiation from neuroepithelial precursor cells of amphibian embryos: an *in vivo* and *in vitro* analysis, *Int.J.Dev.Biol.* **34**: 351-364.

Strong, J.A., Fox, A.P., Tsien, R.W., and Kaczmarek, L.K. (1987), Stimulation of protein kinase C recruits covert calcium channels in *Aplysia* bag cell neurons, *Nature* **325**: 714-717.

Takata, K., Yamamoto, K., and Ozawa, R. (1981) Use of lectins as probes for analyzing embryonic induction, *Roux's Arch.Dev.Biol.* **190**: 92-96.

Tiedemann, H., and Born, J.(1978) Biological activity of vegetalizing and neuralizing producing factors after binding to BAC-cellulose and CNBr-Sepharose, *Roux's Arch.Dev.Biol.* **184**: 285-299.

Westenbroeck, R.E., Ahlijanian, M.K., and Catterall, W.A. (1990) Clustering of L-type Ca^{2+} channels at the base of major dendrites in hippocampal pyramidal neurons, *Nature* **347**: 281-284.

Woodland, H.R. (1989) Mesoderm formation in *Xenopus*, *Cell* **59**: 767-770.

Yang, J., and Tsien, R.W. (1993) Enhancement of N-and L-type calcium channel currents by protein kinase C in frog sympathetic neurons, *Neuron* **10**: 127-136.

DETERMINATION OF GLIAL LINEAGES DURING EARLY CENTRAL NERVOUS SYSTEM DEVELOPMENT

Philippe Cochard, Cathy Soula, Marie-Claude Giess,
Françoise Trousse, Françoise Foulquier and Anne-Marie Duprat

Centre de Biologie du Développement
CNRS UMR 9925, affiliée à l'INSERM
Université Paul Sabatier
118 route de Narbonne, Bât. 4R3
31062 Toulouse Cedex - France

INTRODUCTION

The rapidly cycling neuroepithelial cells, located in the ventricular and subventricular zones of the neural tube, give rise to most of the neurons and macroglial cells, astrocytes and oligodendrocytes, in the vertebrate central nervous system (CNS). In most CNS areas, neurons are the first to develop, followed by astrocytes, and at later stages by oligodendrocytes. An important issue in the study of early CNS development is to understand the lineage relationships of the various CNS cell types and the mechanisms by which these lineages segregate and differentiate. One aspect of this question is to define when and how precursor cells become committed to a specific differentiation pathway. For example, neuroepithelial cells could initially all be endowed with equivalent differentiation capabilities. The specification of these multipotential precursor cells towards a defined phenotype could be controlled by instructive and selective cues arising progressively from their immediate environment. Alternatively, neuroepithelial progenitors could be, from early stages in nervous system ontogeny, already segregated into subpopulations with differing potentialities. In this case, environmental cues would be less critical than intrinsic developmental programs in regulating phenotypic choices.

In this brief review, we will present and discuss data obtained in our laboratory on two fundamental questions. The first one concerns the fate and developmental repertoire of the earliest neural progenitors, i.e. of precursor cells populating the neural plate, the initial neural territory which is specified immediately after neural induction. Are they all pluripotential, endowed with equivalent differentiation capabilities or are they, at that time, already segregated into specific subtypes exhibiting restricted developmental fates ? This question was adressed specifically to the segregation of neural and astroglial phenotypes, using the amphibian embryo as a model system. The second point concerns the determination of the oligodendrocyte lineage. We have examined, in the early chick embryo, oligodendrocyte potentialities within specific CNS areas. In particular, we have sought to determine whether oligodendrocyte progenitors are distributed homogeneously within the neuroepithelium. In addition, we have analyzed the influence of environmental cues on the determination and differentiation of oligodendrocytes.

SEGREGATION OF NEURONAL AND ASTROGLIAL LINEAGES IN THE EARLY AMPHIBIAN NEURAL PLATE

We have chosen to study the segregation of neural lineages in the amphibian embryo. This choice was dictated by several advantages of this model: neuroepithelial cells are initially of large size and easily accessible at the earliest stages in nervous system ontogeny, i.e. immediately after neural induction. Moreover, the nervous system of the hatched larvae is relatively simple, especially in the posterior rhombencephalon and spinal cord, permitting an easy identification of glial and neuronal cell types. The *Pleurodeles* embryo was preferred to the more widely used *Xenopus* for the following reasons: i) neurulation is much slower in the former species, thus permitting manipulations at critical developmental stages over longer periods of time; ii) the *Xenopus* neural plate is initially made up of two cell layers, whereas that of *Pleurodeles* is constituted of a single cell layer, and therefore can more easily be compared to that of higher vertebrates.

Expression of Neuronal and Astroglial Phenotypes in Amphibia *In Vivo* and *In Vitro*

Development of Astroglial Cells in the Amphibian CNS *In Vivo*. Before undertaking the study of lineage segregation in the amphibian nervous sytem, it was necessary to document the normal *in vivo* and *in vitro* development of astroglial cells in amphibians (Soula et al., 1990). Since the pioneering work of Bignami et al. (1972), glial fibrillary acidic protein (GFAp), an intermediate filament protein, has been widely recognized in a vast number of species as a highly specific marker for astrocytes and for a limited number of related cell types in the nervous system. We therefore used anti-GFAp antibodies to trace the development of astroglial cells in *Pleurodeles* nervous system. On western blots of *Pleurodeles* brain extracts, anti-GFAp antibodies stained a band migrating with an apparent molecular mass of 62 kDa, which was identified as a GFAp-like protein (Soula et al., 1990). Thus, *Pleurodeles* astroglial cells can be readily recognized by antibodies directed against GFAp (see also Zamora and Mutin, 1988), as is also the case in *Xenopus* (Godsave et al., 1986; Szaro and Gainer, 1988).

In *Pleurodeles* (Soula et al., 1990), as in *Xenopus* (Szaro and Gainer, 1988), expression of GFAp is a precocious event in astroglial cell differentiation. The development of GFAp-immunoreactive cells in *Pleurodeles* CNS is schematized in Fig. 1. GFAp is initially detectable as early as at stage 24, i.e. about 2 days after neural plate formation, which in *Pleurodeles* occurs at stage 13. At this stage and until stage 32, astrogliocytes appear as bipolar cells, extending two long radially oriented processes which span the entire thickness of the neural tube wall. Interestingly, at these early stages, immunoreactivity is extremely polarized, initially detectable only at the pial pole of the cell, suggesting an active and oriented transport mechanism for the glial intermediate filament. Around stage 38, astroglial cells undergo a striking morphological change : they become monopolar, with a cell body localized in the ventricular zone of the neural tube, from which a thick radial process emerges, running trough the intermediate zone. This unique process then ramifies, within the now well developed marginal zone, into several thinner processes which end up against the pial surface, forming enlarged and highly immunoreactive endfeet. GFAp-immunoreactive cell somata were rarely observed in the gray matter, away from the ventricular zone, thus confirming the conclusions of Naujoks-Manteuffel and Roth (1989) in Salamandra, stating that in urodeles "displaced gliocytes" are a rare occurrence. It is important to mention here that mitotic figures displaying a high level of GFAp immunoreactivity were frequently observed in the ventricular zone, thus suggesting that such radial astrogliocytes are not post-mitotic, a property which also seems to apply to the radial glia of higher vertebrates (Misson et al., 1988). In conclusion, most astroglial cells in the *Pleurodeles* CNS appears to develop early and retain throughout life the characters of a typical radial glia.

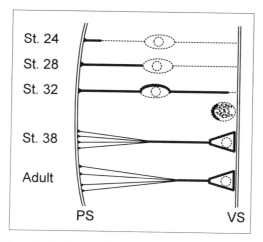

Figure 1. Schematic drawing recapitulating the localization of GFAp-immunoreactivity in astroglial cells, from transverse sections of *Pleurodeles* neural tube at the various developmental stages indicated on the left. GFAp staining is represented as thick and thin solid lines. Dotted lines delineate cell structure. Note that GFAp-immunoreactive mitotic cells, located in the ventricular zone, have been found from stage 30 to 38. See text for morphological description. PS: pial surface. VS: ventri-cular surface.

Differentiation of Neurons and Astroglial Cells in Cultures of Neural Plate Cells. Neural plate formation is triggered *in vivo* by the chordamesoderm, through the process of neural induction (see Leclerc et al., this volume). The crucial role of the chordamesoderm in neural development is not restricted to this early step. It has been clearly demonstrated that the notochord is also responsible, after neural tube formation, for the establishment of dorsoventral polarity of the neural tube and for the specification of a number of neuronal types (see Placzek et al., 1991, for review). Therefore it was of interest to determine whether the emergence of astroglial cells from neural plate precursor cells would also depend upon epigenetic cues derived from the chordamesoderm.

Stage 13 neural plates were dissociated in Ca2+/Mg2+-free culture medium, and the resulting cell suspension was cultivated either alone or in combination with chordamesoderm cells in a simple, completely defined saline medium. Glial cell differentiation was then documented at various times, using GFAp immunocytochemistry.

Cells expressing the glial marker developed rapidly in all cultures, irrespective of the presence or absence of chordamesoderm cells (Soula et al., 1990). This demonstrates that mesodermal cues beyond neural induction are not essentially required for the specification of cells of the astroglial lineage. Experiments in chick embryos, discussed in the second part of this review, indicate that this is probably not true for the development of oligodendroglia.

Clonal Analysis of the Fate of Neural Plate Cells

The rapid differentiation of neurons and astroglial cells in cultures of precursors isolated from the neural plate raised the possibility that at least some of these cells could, at this early stage, already be segregated into separate neuronal and astroglial lineages. A direct approach to address the lineage issue is to follow the fate of individual progenitors, i.e. to study phenotypes emerging from neural progenitors cloned *in vivo* and *in vitro*. Several cell cloning strategies have been recently developed : retrovirus-mediated transfer of a genetic marker (Price, 1987; Cepko, 1988; Sanes, 1989), intracellular injections of enzymes such as peroxidase (Holt et al., 1988) or vital fluorescent dyes (Bronner-Fraser and Fraser, 1988) and physical isolation of single CNS blast cells *in vitro* (Temple and Raff, 1985; Temple, 1989). Each of these techniques present particular advantages and limitations. The retroviral approach is a powerful technique, since clonally-related cells are permanently labeled, up to adult stages, whereas injected enzymes or fluorescent tracers will dilute as development proceeds as a result of cell division. However, infection of progenitors of neural progenitors by a retrovirus is a random process, whereas with intracellular labeling, the location of the injected cell can be recorded precisely, allowing the possibility of reconstructing fate maps of individual precursor cells. Another difficulty in the retroviral approach is that insertion of the retroviral gene occurs in only one of the daughter cells of the dividing progenitor (Hajihosseini et al., 1993). Therefore, phenotypes that can be recorded within the clone do not necessarily represent the full range of fates of the infected progenitor, but only those of one of its daughter cells. In case of an asymmetrical division, the fate of one of the daughter cells will be ignored entirely.

In spite of this limitation to the retroviral technique, data obtained using retroviral markers and other clonal strategies have clearly demonstrated that the segregation of neural cell lineages does not follow a simple and universal scheme throughout the vertebrate CNS. For instance, in the retina of amphibians and rodents, precursor cells are multipotential, generating both neurons and glial cells, until late in development (Turner and Cepko, 1987; Holt et al., 1988; Wetts and Fraser, 1988; Wetts et al., 1989; Turner et al., 1990). In this particular CNS region, therefore, lineage decisions do not appear as a critical mechanism in defining cell phenotypes, thus suggesting that instructive cues from the environment may play decisive roles in cell determination. Similarly, the early chick embryonic optic tectum and spinal cord contain multipotential precursors, giving rise to mixed clones of neurons and astrocytes (Leber et al., 1990; Galileo et al., 1990; Gray and Sanes, 1991, 1992) which, in the tectum, also include radial glia (Gray and Sanes, 1992). These results contrast with those obtained in the analysis of cortical lineages. In the E16 cerebral cortex of rodents, although mixed clones composed of both neurons and oligodendrocytes have been observed (Williams et al., 1991), most cloned precursors give rise to homogeneous populations of either neurons, astrocytes or oligodendrocytes, suggesting that precursors determined to a single phenotype coexist in the ventricular zone (Luskin et al., 1988; Price and Thurlow, 1988). On the other hand, elegant *in vitro* cloning experiments of E14 rat septum reveal that about one fourth of septal progenitor cells can generate mixed clones composed of both neurons and astrocytes (Temple, 1989). Such a discrepancy between results obtained in *in vivo* and *in vitro* experiments discloses a fundamental issue worth mentioning here: lineage analysis *in situ* reveals the actual fates of clonally-related cells, but not necessarily the full range of potentialities of the ancestor progenitor. Conversely, in appropriate culture conditions, cells cloned *in vitro* may give rise to progeny exhibiting a much broader range of phenotypes than would actually develop *in vivo*.

To obtain direct information regarding neural plate cell lineages, we have labeled individual neural plate precursor cells with a fluorescent tracer and followed the development of their progeny *in vivo* (Soula et al., 1993). The tracer, lysinated rhodamine dextran (LRD) was injected by iontophoresis as previously described for neural crest and retinal lineage studies (Bronner-Fraser and Fraser, 1988, 1989; Wetts and Fraser, 1988; Wetts et al., 1989).

About 120 cells were labelled at the neural plate stage in the presumptive area of the rhombencephalon and of the spinal cord, and 15 were injected during neural tube closure in the trunk neural tube. Analysis of these clones was carried out at different developmental stages, up to stage 38, i.e. about 12 days after injection. Due to a rather slow rate of division of neural tube cells in *Pleurodeles* embryos (Chibon, 1973), clone sizes never exceeded 24 cells, a number representing a maximum of 5 mitoses. Most clones at stages 34-38 contained between 6 and 16 cells (maximum 3 to 4 mitoses). Therefore the concentration of the tracer within clonally related cells was sufficiently high to permit their easy detection. Phenotypes of labeled cells were deduced from morphological, positional and immunocytochemical criteria, again using anti-GFAp antibodies as a specific marker for astroglial cells.

The main finding obtained during this study was that most (80%) neural plate and early neural tube precursors are bipotential, i.e. they give rise *in vivo* to mixed clones, constituted of both neurons and astroglial cells (Fig. 2). In some cases, clones contained only these two cell types, but generally some of the clonally-related cells could not be positively identified. Therefore it is possible that such clones contained other cell types, including undifferentiated precursors and oligodendrocytes. In addition, a small but significant number (20%) of clones were strictly constituted of neurons. Interestingly, the range of neuronal clone sizes (3 to 13 cells) was smaller than that of mixed clones (6 to 24 cells). Finally, purely astroglial clones were never observed.

Since the position of each injected cell was recorded precisely at the time of injection, a fate map was reconstructed (Fig. 2). To our surprise, we observed that although precursors for mixed clones were distributed homogeneously in the neural plate, those giving rise to neuronal clones were not. They were never found in the medial and lateral regions of the neural plate, but were exclusively located along the intermediate axes. The significance of this regional restriction of neuronal precursors is unclear at the present time.

This lineage study thus reveals that in urodele amphibians, most early neural plate precursors generate both neurons and astroglial cells. The remaining ones, which are not randomly distributed, have a restricted neuronal fate. In any case, if our observations can be generalized to other species, which remains to be established, the coexistence, within the CNS ventricular zone, of precursor cells with varying fates, already described for later developmental stages (see above), may be traced back to the earliest stage in nervous system ontogeny. But the important question raised by this study is to determine whether such an early heterogeneity in cell fate actually represents a similar heterogeneity in the programming of neural progenitors. We cannot entirely rule out the possibility that all precursors could in fact be bipotential at the origin, some becoming restricted to a neuronal fate later on by local environmental cues. To answer this question, we are currently developing an *in vitro* clonal assay system which will allow us to follow the potentialities of individual neural plate precursor cells grown in various environmental conditions.

Preliminary results indicate that when cell contacts among precursor cells were prevented, as a result of their physical isolation, purely neuronal clones became predominant, whereas clones of mixed neuronal/astroglial phenotypes were only rarely observed. Moreover, these neuronal clones were always of smaller size than that of mixed clones (Trousse, Soula, Duprat and Cochard, unpublished results). These interesting observations may indicate a direct relationship between cell division and the emergence of the astroglial phenotype. This clonal assay will be further employed to study the fate of neural plate precursor cells cultivated in the presence of defined growth and differentiation factors.

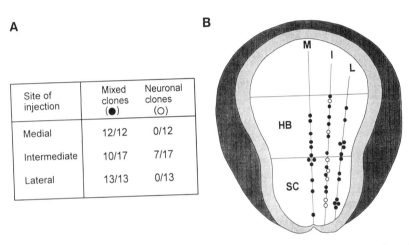

A

Site of injection	Mixed clones (●)	Neuronal clones (○)
Medial	12/12	0/12
Intermediate	10/17	7/17
Lateral	13/13	0/13

B

Figure 2. Fate map of mixed and neuronal progenitors in the neural plate (stage 13), deduced from the coordi-nates of 42 LRD-labeled cells, recorded at the time of injection, and from phenotypic analysis performed at stages 34 and 38. (**A**) Roughly similar numbers of cells were analyzed along the medial (n = 12), intermediate (n = 17) and lateral (n = 13) axes. (**B**) The fate map clearly shows the exclusive origin of neuronal precursors along the intermediate axis (I) of the neural plate, whereas mixed progenitors were found in all three areas considered. For the purpose of clarity, all progenitors have been represented on only one side of the neural plate, although they were injected on either side. Note that neuronal progenitors were found in presumptive regions of hindbrain (HB) and spinal cord (SC).

ENVIRONMENTAL INFLUENCES ON THE DETERMINATION AND DIFFERENTIATION OF OLIGODENDROCYTE PRECURSORS

The lineage experiments described and discussed in the chapter above suggest that early in development, at least some CNS precursor cells may have a restricted developmental repertoire. The important question, therefore, is to understand the mechanisms responsible for the restriction of their differentiation potentialities which control their ultimate fate *in situ*. Our own experiments in the amphibian neural plate indicate that cells with a restricted neuronal fate are not randomly distributed. Other types of progenitors with a restricted developmental repertoire may also be located precisely. In this chapter, we will review work performed in the oligodendrocyte lineage which establishes that oligodendroglial precursors are not uniformly distributed in the CNS, but restricted into defined CNS areas. Such a precise distribution may reveal the existence of specific environmental conditions prevailing in these areas and responsible for the determination of these precursors.

One of the best characterized CNS progenitors is the O-2A cell, a bipotential neural precursor generating oligodendrocytes and a subclass of astrocytes in culture (see Raff, 1989, for a review). O-2A cells were initially identified in the developing rat optic nerve (Raff et al., 1983), but there is now evidence that similar cells are also present in several CNS regions (Levi et al., 1986; Behar et al., 1988). It has clearly been demonstrated that O-2A cells generate oligodendrocytes in the CNS *in situ*, but there are still some doubts regarding the actual generation of astrocytes by O-2A cells *in vivo* (Skoff, 1990; Skoff et al., 1991). *In vitro* studies have brought important insights into the factors controlling mitosis, survival and phenotypic choices in the O-2A lineage (see Raff, 1989; Price, 1994 for reviews, and also Barres et al., 1992, 1993, 1994; Mayer et al., 1994). In addition, they have revealed that the O-2A cell exhibits extensive migratory capabilities (Small et al., 1987), a property which is shared *in situ* by immature (Gumpel et al., 1983) and mature (Wolf et al., 1986) oligodendrocytes. Investigations on the origin of 0-2A cells in the optic nerve have suggested that they may not originate from the optic nerve itself, but rather migrate into the nerve from extrinsic sources (Small et al., 1987). These results have led to the assumption that the developmental repertoire of cells intrinsic to the embryonic optic nerve may be limited to the production of type-1 astrocytes (Small et al., 1987).

Another apparent restriction in oligodendroglial potentialities within specific CNS areas has recently been demonstrated in the spinal cord. Experiments in which the ventral and dorsal halves of the embryonic (E14) rat spinal cord are cultivated separately indicate that oligodendroglial potentialities are initially almost entirely restricted to the ventral part of the spinal cord (Warf et al., 1991; Noll and Miller, 1993). Furthermore, at this precise developmental stage, cells expressing the α receptor for platelet-derived growth factor (PDGF-αR), which, in the CNS, is expressed by cells of the oligodendrocyte lineage (Pringle et al., 1992) are localized only in the ventral part of the spinal cord ventricular zone (Pringle and Richardson, 1993), thus suggesting a ventral localization of oligodendroglial precursors. This study has recently been extended by a strikingly similar localization of another marker of the oligodendrocyte lineage, the enzyme CNP (2',3'-cyclic nucleotide phosphohydrolase) (Yu et al., 1994).

Differentiation Potentialities of Cells in the Optic Nerve Primordium

We have studied the developmental repertoire of optic nerve cells in the chick embryo, using explant cultures of optic stalks carefully microdissected at various developmental stages (Giess et al., 1990, 1992). Neurons, astrocytes and oligodendrocytes were identified using antibodies directed against neurofilaments, GFAp and galactocerebrosides (GalC), respectively. In addition, Ol-1, a monoclonal antibody directed against sulfatides, was used to label immature oligodendrocytes (Ghandour and Nussbaum, 1990). Unfortunately, the A2B5 antibody, which recognizes tetrasialogangliosides on the surface of neurons and O-2A cells (Eisenbarth et al., 1979), was not a reliable marker for oligodendrocyte precursors in the chick embryo.

The results of these experiments are summarized in Fig. 3. As expected, optic nerve cells gave rise to astrocytes throughout development. In addition, they also appeared capable of generating neurons, although this potential diminished progressively during embryonic development (Giess et al., 1990). A similar potential for neuron generation was

Table 1. Oligodendrocyte differenciation in cultures of chick optic nerve explants

Age at explantation	Time in culture	Oligodendrocytes
4 days	≤ 21 days	0[1]
5 days	≤ 21 days	0
6 days	≤ 21 days	0
7 days	> 3 days	+++
8-13 days	> 2 days	++++

[1]0 denotes total absence of oligodendrocytes

also demonstrated in the fetal (Giess et al., 1990) and newborn (Omlin and Waldmeyer, 1989) rat optic nerve. Consequently, although the optic nerve is one of the few CNS areas which in the adult are entirely devoid of neuronal cell bodies, neuronal potentialities are initially present in this neuroepithelial territory. The reason why neuronal potentialities are not expressed during normal *in vivo* development is unclear. An interesting hypothesis would be that retinal axons running through the optic nerve are responsible for repressing this potential among optic nerve cells. Surprisingly, however, optic stalks explanted between 4 and 6 days of incubation (E4 - E6) were totally unable to produce oligodendrocytes, even after more than 3 weeks *in vitro* (Table 1). In contrast, numerous immature (Ol-1+/GalC-) and mature (GalC+) oligodendrocytes differentiated very rapidly (within 3 days) in cultures from E7 and older optic nerves (Giess et al., 1992).

These results thus confirm the observations of Small et al. (1987) in the rat optic nerve. They indicate that up to a precise developmental stage (E7 in chicks, E17 in rats), optic stalk cells are unable to generate oligodendrocytes. Cells intrinsic to the optic nerve may be, from the onset, restricted in their potential to generate neurons and astrocytes. In this case, oligodendrocyte precursors populating the optic nerve after E7 in chicks (or after

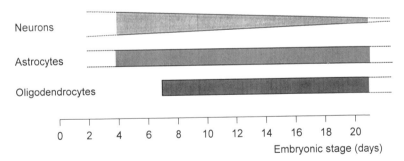

Figure 3. Summary of neural differentiation potentialities in the cultivated embryonic chick optic nerve. Optic stalk cells explanted *in vitro* at all stages can differentiate into neurons and astrocytes. Neuronal potentialities decrease, however, as development proceeds. In contrast, optic stalk cells from E4 to E6 embryos are totally unable to generate oligodendrocytes. The oligodendroglial potential appears abruptly at E7 in the optic nerve.

E17 in rats) may be of extrinsic origin, as proposed by Small et al. (1987). Alternatively, optic stalk precursor cells may receive, around E6-E7 in chick embryos, a developmental signal of critical importance for their subsequent orientation towards the oligodendroglial phenotype.

Oligodendroglial Differentiation Potentialities in the Spinal Cord and in other CNS Regions

We then sought to analyze oligodendroglial potentialities in the chick spinal cord, as well as in other CNS regions. Dorsal and ventral microexplants were dissected from E4 and E7 embryos and cultivated separately. Cultures were analyzed at various times using the anti-sulfatide (Ol-1) and anti-GalC antibodies. The results indicate that E4 dorsal cord tissue produced few or no oligodendrocytes, whereas well differentiated oligodendrocytes were already identifiable in 7-day cultures of ventral explants. This difference in oligodendrogenesis between the dorsal and ventral spinal cord regions is transient and restricted to an early stage in spinal cord ontogeny, since E7 dorsal explants produced numerous oligodendrocytes. Similar results were obtained in cultures of E4 and E7 dorsal and ventral rhombencephalon, indicating that a ventral origin for oligodendrocyte precursors is not limited to the spinal cord. In contrast, the entire E4 mesencephalon, dorsal and ventral, was unable to generate oligodendrocytes in similar culture conditions.

Notochord Effect On Oligodendrocyte Differentiation In Dorsal Spinal Cord Cultures

The marked ventral predominance in oligodendroglial potentialities summarized above could indicate that in the spinal cord and rhombencephalon, oligodendrogenesis is controlled by ventrally-derived cues, and, in particular, by signals produced by the notochord, an axial mesodermal structure which extends along the ventral aspect of the rhombencephalic and truncal neural tube. The establishment of the dorsoventral polarity of the neural tube is controlled, at least in part, by diffusible signals produced by the notochord and by a specialized subset of cells, located along the ventral midline of the neural tube, termed the floor plate (see Placzek et al., 1991, for review). To test this possibility, E4 dorsal explants were cocultivated with segments of notochord dissected from E2.5 and E4 chick embryos. In 80% of cases (n=25), notochord tissue stimulated expression of the oligodendrocyte phenotype , as evidenced by the differentiation of numerous (from 50 to several hundreds) Ol-1+/GalC+ oligodendrocytes, whereas dorsal explants cultivated alone in control experiments contained few, if any, Ol-1+ cells and no GalC-immunoreactive cell. However, the effect of notochord tissue was short range, since oligodendrocytes were produced only by dorsal explants located in contact with the notochord, or in its immediate vicinity (Fig. 4).

Similar experiments were repeated, using the E4 optic stalk instead of the dorsal spinal cord, to see whether early optic nerve cells would also be sensitive to notochord derived cues. Thus, optic stalk explants were associated in culture with notochord segments. In all cases (n=12), these cocultures were entirely negative for Ol-1 and GalC immunoreactivity.

These results have two important implications: i) E4 dorsal spinal cord tissue, but not E4 optic stalk, contains cells with the ability, upon appropriate stimulation, to produce oligodendroglia; whether oligodendrocyte precursors which populate later on this CNS region are induced locally, between E4 and E7, by an appropriate developmental signal, or arise from determined progenitors migrating from the ventral cord, as suggested by the group of Miller (Warf et al., 1991; Noll and Miller, 1993) remains to be determined; ii) notochord can promote expression of the oligodendroglial phenotype in the dorsal spinal cord. This raises the interesting possibility that oligodendroglial determination in the ventral spinal cord could be elicited, directly or indirectly, by the notochord, as already evidenced for a number of neuronal types (Placzek et al. 1991).

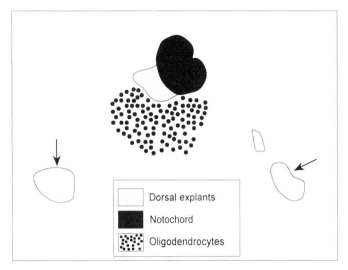

Figure 4. Oligodendrocyte differentiation after 16 days in E4 dorsal spinal cord/notochord cocultures. Camera lucida drawing of the area containing the notochord segment. The notochord has rounded up on the culture substratum. Regions containing well differentiated (Ol-1+/GalC+) oligodendrocytes are indicated by black dots. All oligodendrocytes were located in close vicinity to notochord tissue. In contrast, immunoreactive cells were not observed around dorsal explants located away from notochord tissue (arrows).

In any case, several points still await further study, before definitive conclusions can be drawn about the role of the notochord in this process *in vivo*. In particular, it will be important to define the specificity of the notochord effect, and the influence of early notochordectomy upon oligodendrocyte differentiation in the ventral spinal cord region. It will also be necessary to determine the mechanisms involved in notochord stimulation of oligodendrogenesis.

CONCLUSIONS

Experiments described in the first part of this article clearly indicate that the segregation of neuronal and astroglial lineages is not an early event in CNS ontogeny. In the amphibian neural plate, even if some cells have a restricted neuronal fate, most precursors generate neurons and astroglial cells and are therefore at least bipotential. Iontophoresis injections of a fluorescent tracer in ventricular cells at progressively older developmental stages will help elucidate the lineage relationships of neurons and astroglial cells. A similar analysis will also be carried out for the oligodendroglial lineage, using specific oligodendrocyte markers in amphibians (Steen et al., 1989). In particular, it will be of interest to determine the possible existence in amphibians of a common progenitor for both neurons and oligodendrocytes, as suggested by retroviral cloning experiments in the rat cerebral cortex *in vivo* and *in vitro* (Williams et al., 1991). In addition, *in vitro* clonal experiments will allow us to determine the full range of potentialities of neural plate precursor cells.

This *in vitro* approach may also give important insights into the mechanisms regulating early lineage decisions and phenotypic choices. Bulk cultures of isolated neural plate have already demonstrated that signals produced by the chordamesoderm are not decisive in specifying the astroglial phenotype. Preliminary results obtained in clonal cultures of neural plate cells (Trousse, Soula and Cochard, unpublished data) suggest that the proportion of neuronal clones over mixed neuronal/astroglial clones is much greater *in vitro* than that observed *in vivo*, indicating that cell differentiation does not follow a rigid program, imposed early in development. The decision for a bipotential progenitor to become a neuron or an astrocyte may therefore be controlled among neuroepithelial cells themselves, through cellular interactions which remain to be defined. One of these control mechanisms may involve cell-cell communication through a variety of signalling molecules, including growth and differentiation factors.

In contrast, expression of the oligodendrocyte phenotype may be induced experimentally in the dorsal spinal cord by the notochord, suggesting that notochord-derived factors may be of importance in defining the location and timing of oligodendroglial determination. Obviously, further experiments are required to confirm whether the determination of the oligodendrocyte lineage obeys to the same signals than those which appear responsible for the patterning of neuronal phenotypes in the spinal cord (e.g. see Price, 1994). For instance, it will be important to determine the effect of the floor plate on oligodendrocyte determination and control whether notochord or floor plate tissue can induce the oligodendrocyte lineage *in vivo*. In addition, it will be necessary to identify the relevant inducing factors. Cells of the O-2A lineage are sensitive to a growing number of soluble factors, which affect selectively, or pleiotropically, their proliferation, differentiation and survival. Future experiments, using cultures of dorsal spinal cord cells, will probably help defining environmental factors responsible in the initial specification of the oligodendrocyte lineage.

REFERENCES

Barres BA, Hart IK, Coles HSR, Burne JF, Voyvodic JT, Richardson WD, Raff MC (1992): Cell death and control of cell survival in the oligodendrocyte lineage. *Cell* 70:31-46.

Barres BA, Schmid R, Sendtner M, Raff M (1993): Multiple extracellular signals are required for long-term oligodendrocyte survival. *Development* 118:283-295.

Barres BA, Lazar MA, Raff MC (1994): A novel role for thyroid hormone, glucocorticoids and retinoic acid in timing oligodendrocyte development. *Development* 120:1097-1108.

Behar T, McMorris FA, Novotny EA, Barker JL, Dubois-Dalcq M (1988): Growth and differentiation properties of O-2A progenitors purified from rat cerebral hemispheres. *J. Neurosci. Res.* 21:168-180.

Bignami A, Eng LF, Dahl D, Uyeda CT (1972): Localization of the glial fibrillary acidic protein in astrocytes by immunofluorescence. *Brain Res.* 43:429-435.

Bronner-Fraser M, Fraser SE (1988): Cell lineage analysis reveals multipotency of some avian neural crest cells. *Nature* 335:161-164.

Bronner-Fraser M, Fraser SE (1989): Developmental potential of avian trunk neural crest cells *in situ*. *Neuron* 3:755-766.

Cepko C (1988): Retrovirus vectors and their applications in neurobiology. *Neuron* 1:345-353.

Chibon P (1973): Cell proliferation in late embryos and young larvae of the newt *Pleurodeles waltlii* Michah. In Balls M, Billett FS (eds): "*The Cell Cycle in Development and Differentiation.*" London/New York: Cambridge Univ. Press, pp. 257-277.

Eisenbarth GS, Walsh FS, Nirenberg M (1979): Monoclonal antibody to a plasma membrane antigen of neurons. *Proc Natl Acad Sci USA* 76:4913-4917.

Galileo DS, Gray GE, Owens GC, Majors J, Sanes JR (1990): Neurons and glia arise from a common progenitor in chicken optic tectum: demonstration with two retroviruses and cell type-specific antibodies. *Proc. Natl. Acad. Sci. USA* 87:458-462.

Ghandour MS, Nussbaum JL (1990): Oligodendrocyte cell surface recognized by a novel monoclonal antibody specific to sulfatide. *Neuroreport* 1:13-16.

Giess MC, Cochard P, Duprat AM (1990): Neuronal potentialities of cells in the optic nerve of the chicken embryo are revealed in culture. *Proc. Natl. Acad. Sci. USA* 87:1643-1647.

Giess MC, Soula C, Duprat AM, Cochard P (1992): Cells from the early chick optic nerve generate neurons but not oligodendrocytes *in vitro*. *Dev. Brain res.* 70:163-171.

Godsave SF, Anderton BH, Wylie CC (1986): The appearance and distribution of intermediate filament proteins during differentiation of the central nervous system, skin and notochord of *Xenopus* laevis. *J. Embryol. exp. Morphol.* 97:201-223.

Gray GE, Sanes JR (1991): Migratory paths and phenotypic choices of clonally related cells in the avian optic tectum. *Neuron* 6:211-225.

Gray GE, Sanes JR (1992): Lineage of radial glia in the chicken optic tectum. *Development* 114:271-283.

Gumpel M, Baumann N, Raoul M, Jacque C (1983): Survival and differentiation of oligodendrocytes from neural tissue transplanted into new-born mouse brain. *Neurosci. Lett.* 37:307-311.

Hajihosseini M, Iavachev L, Price J (1993): Evidence that retroviruses integrate into post-replication host DNA. *EMBO J.* 12:4969-4974.

Holt CE, Bertsch TW, Ellis HM, Harris WA (1988): Cellular determination in the *Xenopus* retina is independent of lineage and birth date. *Neuron* 1:15-26.

Leber SM, Breedlove SM, Sanes JR (1990): Lineage, arrangement, and death of clonally related motoneurons in chick spinal cord. *Neurosci.* 10:2451-2462.

Levi G, Gallo V, Ciotti MT (1986): Bipotential precursors of putative fibrous astrocytes and oligodendrocytes in rat cerebellar cultures express distinct surface features and "neuron-like"-aminobutyric acid transport. *Proc. Natl. Acad. Sci. USA* 83:1504-1508.

Luskin MB, Pearlman AL, Sanes JR (1988): Cell lineage in the cerebral cortex of the mouse studied *in vivo* and *in vitro* with a recombinant retrovirus. *Neuron* 1:635-647.

Mayer M, Bhakoo K, Noble M (1994): Ciliary neutrophic factor and leukemia inhibitory factor promote the generation, maturation and survival of oligodendrocytes *in vitro*. *Development* 120:143-153.

Misson JP, Edwards MA, Yamamoto M, Caviness VS (1988): Mitotic cycling of radial cells of the foetal murine cerebral wall: a combined autoradiographic and immunocytochemical study. *Dev. Brain res.* 38:183-190.

Naujoks-Manteuffel C, Roth G (1989): Astroglial cells in a salamander brain (Salamandra salamandra) as compared to mammals: a glial fibrillary acidic protein immunohistochemistry study. *Brain Res.* 487:397-401.

Noll E, Miller RH (1993): Oligodendrocyte precursors originate at the ventral ventricular zone dorsal to the midline region in the embryonic rat spinal cord. Development 118:563-573.

Noll E, Miller RH (1994): Regulation of oligodendrocyte differentiation: a role for retinoic acid in the spinal cord. *Development* 120:649-660.

Omlin FX, Waldmeyer J (1989): Differentiation of neuron-like cells in cultured rat optic nerves: a neuron or common neuron-glia progenitor? *Dev. Biol.* 133:247-253.

Placzek M, Yamada T, Tessier-Lavigne M, Jessell TM, Dodd J (1991): Control of dorsoventral pattern in vertebrate neural development: induction and polarizing properties of the floor plate. *Development 113 (Suppl. 2)*:105-122.

Price J (1987): Retroviruses and the study of cell lineage. Development 101:409-419.

Price J (1994): Glial cell lineage and development. *Current Opinion Neurobiol.* 4:680-686.

Price J, Thurlow L (1988): Cell lineage in the rat cerebral cortex: a study using retroviral-mediated gene transfer. *Development* 104:473-482.

Pringle NP, Richardson WD (1993): A singularity of PDGF alpha-receptor expression in the dorsoventral axis of the neural tube may define the origin of the oligodendrocyte lineage. *Development* 117:525-533.

Pringle NP, Mudhar HS, Collarini EJ, Richardson WD (1992): PDGF receptors in the rat CNS - During late neurogenesis, PDGF alpha-receptor expression appears to be restricted to glial cells of the oligodendrocyte lineage. *Development* 115:535-551.

Raff MC (1989): Glial cell diversification in the optic nerve. *Science* 243:1450-1455.

Raff MC, Miller RH, Noble M (1983): A glial progenitor cell that develops *in vitro* into an astrocyte or an oligodendrocyte depending on culture medium. *Nature* 303:390-395.

Sanes JR (1989): Analysing cell lineage with a recombinant retrovirus. *Trends Neurosci.* 12:21-28.

Skoff RP (1990): Gliogenesis in rat optic nerve - Astrocytes are generated in a single wave before oligodendrocytes. *Dev. Biol.* 139:149-168.

Skoff RP, Knapp PE (1991): Division of astroblasts and oligodendroblasts in postnatal rodent brain - Evidence for separate astrocyte and oligodendrocyte lineages. *Glia* 4:165-174.

Small RK, Riddle P, Noble M (1987): Evidence for migration of oligodendrocyte-type-2 astrocyte progenitor cells into the developing rat optic nerve. *Nature* 328:155-157.

Soula C, Sagot Y, Cochard P, Duprat AM (1990): Astroglial differentiation from neuroepithelial precursor cells of amphibian embryos: an *in vivo* and *in vitro* analysis. *Int. J. Dev. Biol.* 34:351-364.

Soula C, Foulquier F, Duprat AM, Cochard P (1993): Lineage analysis of early neural plate cells: cells with purely neuronal fate coexist with bipotential neuroglial progenitors. *Dev. Biol.* 159:196-207.

Steen P, Kalghatgi L, Constantine-Paton M (1989): Monoclonal antibody markers for amphibian oligodendrocytes and neurons. *J. Comp. Neurol.* 289:467.

Szaro BG, Gainer H (1988): Immunocytochemical identification of non-neuronal intermediate filament proteins in the developing *Xenopus* laevis nervous system. *Dev. Brain res.* 43:207-224.

Temple S (1989): Division and differentiation of isolated CNS blast cells in microculture. *Nature* 340:471-473.

Temple S, Raff MC (1985): Differentiation of a bipotential glial progenitor cell in single cell microculture. *Nature* 313:223-225.

Turner DL, Cepko CL (1987): A common progenitor for neurons and glia persists in rat retina late in development. *Nature* 328:131-136.

Turner DL, Snyder EY, Cepko CL (1990): Lineage-independent determination of cell type in the embryonic mouse retina. *Neuron* 4:833-845.

Warf BC, Fok-Seang J, Miller RH (1991): Evidence for the ventral origin of oligodendrocyte precursors in the rat spinal cord. *Cell Diff.* 11:2477-2488.

Wetts R, Fraser SE (1988): Multipotent precursors can give rise to all major cell types of the frog retina. *Science* 239:1142-1145.

Wetts R, Serbedzija GN, Fraser E (1989): Cell lineage analysis reveals multipotent precursors in the ciliary margin of the frog retina. *Dev. Biol.* 136:254-263.

Williams BP, Read J, Price J (1991): The generation of neurons and oligodendrocytes from a common precursor cell. *Neuron* 7:685-693.

Wolf MK, Brandenberg MC, Billings-Cagliardi S (1986): Migration and myelination by adult glial cells: reconstructive analysis of tissue culture experiments. *Cell Diff.* 6:3731-3738.

Yu W-P, Collarini EJ, Pringle NP, Richardson WD (1994): Embryonic expression of myelin genes: evidence for a focal source of oligodendrocyte precursors in the ventricular zone of the neural tube. *Neuron* 12: 1353-1362.

Zamora AJ, Mutin M (1988): Vimentin and glial fibrillary acidic protein filaments in radial glia of the adult urodele spinal cord. *Neurosci.* 27:279-288.

IN VIVO AND *IN VITRO* STUDIES ON THE DIFFERENTIATION OF THE NEURAL CREST IN THE AVIAN MODEL

Catherine Ziller and Nicole M. Le Douarin

Institut d'Embryologie Cellulaire et Moléculaire
du CNRS et du Collège de France
49bis, Avenue de la Belle-Gabrielle
94736 Nogent-sur-Marne cedex, France

INTRODUCTION

The neural crest of the vertebrate embryo is a population of cells that arises from the lateral ridges of the neural tube at late neurula stage. While the neural tube is closing, the neural crest cells leave their original position and undergo extensive migrations through the developing embryo, settle to various homing sites where they give rise to a large diversity of different cell types. Therefore, the ontogeny of the neural crest constitutes a remarkable model for studies on cell migration and cell lineage segregation.

The derivatives of the neural crest include most constituents of the peripheral nervous system (PNS), i.e. sensory and autonomic neurons, ganglionic satellite cells and Schwann cells; pigment cells; endocrine and paraendocrine cells; mesectodermal derivatives in the head (bones, cartilage, connective tissues, meninges).

The migration pathways and phenotypic fates of neural crest cells have been studied with several labeling techniques. The quail-chick labeling system devised by N. Le Douarin (1982 for a review) relies on a histological difference between chick and quail cells; in the nucleus of the latter, a mass of heterochromatic DNA is associated with the nucleolus which thus reacts positively with several DNA stainings. Embryonic tissues from one species of birds can be grafted into the other at early stages of development and the fate of the implanted cells can thus be followed in the chimeric embryos. Isotopic transplantations of defined segments of the neural primordium along the neural axis have allowed precise identification of the migration pathways and the derivatives of the neural crest. These experiments also show that the neural crest is regionalized : different levels of the neural crest give rise to different sets of derivatives. However, heterotopic transplantations show that when they are translocated from one level of the body axis to another, neural crest cells behave according to the new environment in which they are placed experimentally, and not according to their level of origin. In fact, all the levels of the neural crest have the potential to give rise to all types of derivatives (except the cephalic mesenchyme). This demonstrates the pluripotentiality of the neural crest as a population and the important role of the embryonic environment in determining the differentiation of neural crest cells.

Organization of the Early Vertebrate Embryo
Edited by N. Zagris *et al.*, Plenum Press, New York, 1995

I HETEROGENEITY OF THE NEURAL CREST. ANALYSIS OF THE PROGRESSIVE DETERMINATION OF NEURAL CREST CELLS

The role of factors from the microenvironment may be either to select among differently committed precursors or to exert an instructive action on pluripotent precursors; an intermediate possibility is that the neural crest could be composed of a mixture of pluripotent and of more or less restricted progenitors.

In fact, the latter hypothesis is confirmed by a number of data, obtained mainly *in vitro*, which show that the neural crest is a heterogeneous population when migration starts. The apparent homogeneity of the early neural crest was challenged by the finding that several antigenic markers are not uniformly expressed by migrating crest cells (see for instance Barbu et al., 1986). Some of these markers are related to particular cell types: the melanocytic lineage for example is identified early in neural crest development by a number of specific markers (Steel et al., 1992; Kitamura et al., 1992; Nataf et al., 1993). The heterogeneity of the neural crest is also supported by *in vitro* explantation experiments, in which a subset of neuronal precursors appears to be specified early in development (Ziller et al., 1983).

The state of commitment of neural crest cells has been precisely evaluated in cloning experiments, performed in our laboratory and in others, *in vivo* and *in vitro*. (Sieber-Blum and Cohen, 1980; Baroffio et al., 1988; Bronner-Fraser and Fraser, 1988,1989; Dupin et al. 1990; Frank and Sanes, 1991; Fraser and Bronner-Fraser 1991; Ito and Sieber-Blum, 1991; Ito et al., 1993). In our experiments, neural crest cells from the cephalic region of quail embryos at 2 days of incubation (E2) were dissociated into a suspension, seeded one by one under microscopic control onto a feeder layer of mitomycin arrested 3T3 mouse fibroblasts; this substrate and the culture medium provided optimal conditions for attachment, survival and differentiation. After 7 to 13 days in culture, the cells had generated clones that could be distinguished from mouse fibroblasts by Hoechst nuclear staining. The phenotypes obtained in the clones were analysed with several markers, allowing the main neural crest derived phenotypes to be recognized : antibodies to neurofilament proteins, tyrosine hydroxylase, substance P, VIP (vasoactive intestinal polypeptide), SMP (Schwann cell myelin protein), HNK1; presence of melanin; toluidine blue staining of cartilage.

The main result of these experiments was that most clones (over 80%) were composed of various combinations of several different phenotypes, demonstrating the pluripotency of the founder cell. Few clones only (about 15%) were composed of cells expressing all the same phenotype, for instance Schwann cells, indicating that the ancestor cell was a committed precursor. Sometimes, all the phenotypes that could be characterized, including cartilage, a mesectodermal derivative,were found in the same clone, suggesting the existence in the early migrating cephalic crest of cells exhibiting some characteristics of stem cells (Baroffio et al., 1991). True pluripotent self-renewing stem cells were evicenced in clonal cultures of rat neural crest cells (Stemple and Anderson, 1992). In the mammalian as in the avian neural crest, pluripotent progenitors coexist with oligopotent and unipotent cells. Such a diversity among neural crest cells when migration starts suggests that these cells are generated through progressive restrictions of the developmental potentials of initially totipotent stem cells.

This was confirmed by cloning experiments performed with quail neural crest-derived cells taken at later stages of development, during migration or at their sites of arrest : in the somitic mesoderm a E3, in dorsal root ganglia (DRG), in sympathetic ganglia, in the skin, in the visceral arches, in the digestive tract (Duff et al., 1991; Sextier-Sainte-Claire Deville et al., 1992; Ito and Sieber-Blum, 1993; Richardson and Sieber-Blum, 1993; Sextier-Sainte-Claire Deville et al., 1994). These experiments showed that with increasing embryonic age the proliferation and differentiation potentials of the neural crest cells became more and more restricted: less phenotypes were found in the clones and the size of the clones was reduced as compared to the clones derived from early migrating neural crest cells.

However, in clonal and in mass cultures of neural crest derivatives, pluripotent cells are found together with cells that have achieved their terminal differentiation. The DRG and the enteric plexuses, for instance, contain precursors which differentiate *in vitro* into adrenergic cells, while *i*the adrenergic phenotype is never expressed *in vivo* in DRG nor in the enteric nervous system (Xue et al., 1985; Pomeranz et al., 1993; Sextier-Sainte-Claire Deville et al., 1994).

II ENVIRONMENTAL FACTORS CONTROLLING THE DIFFERENTIATION OF NEURAL CREST DERIVATIVES

The fact that such resting pluripotent precursors persist during ontogeny in neural crest derived organs had been demonstrated previously by back-transplantation experiments. A piece of a developing peripheral ganglion from a quail embryo was implanted into the neural crest migration pathways of a chick host embryo at E2. Cells from the implanted ganglion behaved like neural crest cells: they migrated along pathways corresponding to the level of implantation, settled in appropriate homing sites and differentiated accordingly. For instance, cells from the ciliary ganglion, which is cholinergic, can colonize the sympathetic ganglia and the adrenal medulla of the host embryo and in these locations they differentiate into catecholaminergic cells (Le Douarin et al., 1978).

In back-transplantations,sensory and autonomic ganglia behaved differently. Cells from sensory DRG colonized the host's DRG, sympathetic ganglia and adrenal medulla, whereas autonomic ganglion cells migrated to sympathetic ganglia, adrenal medulla and enteric plexuses but did not colonize the DRG of the host and did not differentiate into sensory neurons. Apparently sensory precursors can survive only in DRG, which develop in close contact with the neural tube, i.e. to the central nervous system (CNS) (Le Douarin, 1986, for a review). This observation lead to the hypothesis that a CNS derived factor could be necessary for the development of DRG. This was demonstrated by an experiment (Kalcheim and Le Douarin, 1986), in which a silastic membrane was placed between neural tube and somites of a chick embryo at E3. The neural crest derived cells that were aggregating to form the DRG were thus separated from the developing spinal chord. The result was that the DRG cells died but when the silastic membrane was impregnated with neural tube extract or with brain derived neurotrophic factor (BDNF) and laminin, the DRG cells survived, suggesting that BDNF is the factor from CNS that is necessary for the survival of sensory neuron precursors (Kalcheim et al., 1987). Later on, when the DRG neurons differentiate, they require other target derived factors for growth and survival, such as for example NGF.

Other factors probably produced by the CNS have been shown to act on neural crest cells : neurotrophin 3 (NT3) as a mitogen *in vitro* (Kalcheim et al., 1992), basic fibroblast growth factor (bFGF) as a survival factor on non-neuronal trunk crest cells *in vivo* and *in vitro* (Kalcheim, 1989). In clonal cultures of neural crest cells, BDNF caused an increase in the number of sensory neuron precursors, suggesting a role of this factor in cell type specification (Sieber-Blum, 1991).

The fact that the neurotrophins -NGF, BDNF, NT3, NT4/5- act on neuronal development has been documented mainly *in vitro*, and, so far, only limited information was available on the role of endogenous neurotrophic factors *in vivo*. Recently, important progress has been made in this respect through targeted mutations in mouse of the genes encoding neurotrophins and their receptors of the Trk family of tyrosine kinases. Indeed, disruption of these genes causes severe deficiencies in the nervous system and the mice usually die during the perinatal period. Mice lacking NGF (Crowley et al., 1994) or the Trk/NGF receptor (Smeyne et al., 1994), or NT3 (Ernfors et al., 1994; Farinas et al., 1994) display loss of sensory and sympathetic neurons. Disruption of the BDNF gene (Ernfors et al., 1994; Jones et al., 1994l) leads mainly to sensory defects, while inactivation of the gene for Trk B, the receptor of BDNF and NT4 affects both sensory and motor neurons (Klein et al., 1993).

Many attempts have been made to try and identify the factors which are responsible for the adrenergic differentiation of the autonomic neuronal precursors. *In vitro* studies have shown that chick embryo extract (CEE) promotes the adrenergic phenotype in neural crest cultures (Le Douarin and Smith, 1988, for a review). It was found (Dupin et al., 1993) that the positive response of crest cells *in vitro* to CEE, as evidenced by an increased number of cells expressing tyrosine hydroxylase immunoreactivity, involves cAMP as a second messenger and the activation of β adrenergic receptors which are present on neural crest cells. The promoting effect of CEE on neural crest cultures is inhibited by antagonists to β adrenergic receptors or by eliminating catecholamines from the CEE. This suggests that, *in vivo*, exogenous catecholamines contained in the embryonic environment may trigger the onset of the adrenergic differentiation of the autonomic precursors.

It was shown that adrenergic differentiation *in vitro* is also significantly stimulated by insulin and insulin-like growth factor (IGF1). In cultures of neural crest derived cells from E2.5-E3 quail embryos, these factors acted in a dose dependent manner to increase the number of cells expressing tyrosine-hydroxylase immunoreactivity and/or containing catecholamines (Nataf and Monier, 1992). A positive effect of insulin and IGF1 was also observed in DRG cell cultures. DRG cells normally never express adrenergic features *in vivo*. If the medium is supplemented with insulin, IGF1 or CEE, which contains insulin, cells expressing tyrosine hydroxylase immunoreactivity or containing catecholamines differentiate in the cultures (Xue et al., 1987, 1988, 1992). These cells are probably the cryptic precursors that undergo adrenergic differentiation in the back-transplantation experiments. It is likely that peptides of the insulin family are among the environmental factors which are involved in sympatho-adrenal development.

The melanogenic lineage is another neural crest derivative which depends upon the environment for its development. The molecular mechanisms underlying this process involve the steel factor and its tyrosine kinase receptor c-kit, which play a role in the survival and differentiation of two other classes of migratory cells, primordial germ cells and hemopoietic cells (Williams et al., 1992, for a review). *In vivo*, c-kit is required for the development of normal coat colour in mouse (Nishikawa et al., 1991). *In vitro*, steel and c-kit have a positive effect on survival and possibly on proliferation and differentiation of mouse melanocyte precursors (Morrison-Graham and Weston, 1993) and the production of differentiated pigment cells in quail trunk neural crest cultures is stimulated by chick steel factor (Lahav et al., 1994, in press).

III ROLE OF THE ENVIRONMENT IN THE DIVERSIFICATION OF THE NEURAL CREST-DERIVED GLIAL LINEAGES

The differentiation of PNS glial cells provides another example of the influence of the environment on the development of neural crest-derived cells and of their plasticity. A marker for Schwann cells has been discovered in our laboratory by means of a monoclonal antibody recognizing a cell surface protein, SMP (for Schwann cell myelin protein, Dulac et al., 1988), a glycoprotein of the immunoglobulin-like superfamily (Dulac et al., 1992) present on all Schwann cells, myelinating or not (and on oligodendrocytes in the CNS). Enteric glial cells and intraganglionic neuron-associated satellite cells are SMP-negative. But when dissociated and cultured *in vitro*, enteric glial cells and satellite cells rapidly start to express SMP on their surface, suggesting that the gut mesenchyme and the ganglion are environments that inhibit SMP expression (Dulac and Le Douarin, 1991; Sextier-Sainte-Claire-Deville et al., 1994). This was confirmed by co-culture experiments, in which

Schwann cells from peripheral nerves, expressing SMP, and neural crest cells, that are able to become SMP positive in culture, were associated with gut mesenchyme. In these associations, neural crest cells did not acquire the SMP marker and the differentiated Schwann cells lost it. In control cultures with skin and limb bud muscle, i.e. tissues in which Schwann cells are normally present, SMP expression was acquired by neural crest cells, and maintained on Schwann cells (Dulac and Le Douarin, 1991). Therefore, the SMP positive phenotype is the constitutive, or default, pathway of peripheral glial cell differentiation. Schwann cells, enteric glial cells and satellite cells belong to the same lineage, in which the commitment to a particular glial cell type is reversible (Le Douarin et al., 1993, for a review).

CONCLUSION

In summary, the data obtained from both *in vivo* and *in vitro* studies demonstrate the crucial role of the environment in the determination of developmental fates of neural crest precursor cells. Some of the growth factors acting on survival, proliferation and/or differentiation that are involved in neural crest ontogeny have been identified.

Of great importance also are investigations devoted to the molecular mechanisms by which cell-to-cell and cell-to-matrix interactions occur in neural crest development. Such interactions are essential in determining the migration routes, hence also in selecting the differentiation pathways taken by the neural crest cells (see Bronner-Fraser, 1993; Le Douarin and Ziller, 1993, for reviews).

REFERENCES

Barbu, M., Ziller, C., Rong, P.M., and Le Douarin, N.M. (1986) Heterogeneity in migrating neural crest cells revealed by a monoclonal antibody. *J. Neurosci.* 6:2215-2225.

Baroffio, A., Dupin, E., and Le Douarin, N.M. (1988) Clone-forming ability and differentiation potential of migratory neural crest cells. *Proc. Natl. Acad. Sci. U. S. A.* 85:5325-5329.

Baroffio, A., Dupin, E., and Le Douarin, N.M. (1991) Common precursors for neural and mesectodermal derivatives in the cephalic neural crest. *Development* 112:301-305.

Bronner-Fraser, M., and Fraser, S. (1989) Developmental potential of avian trunk neural crest cells *in situ*. *Neuron* 3:755-766.

Bronner-Fraser, M. (1993) Environmental influences on neural crest cell migration. *J. Neurobiol.* 24:233-247.

Crowley, C., Spencer, S.D., Nishimura, M.C., Chen, K.S., Pittsmeek, S., Armanini, M.P., Ling, L.H., McMahon, S.B., Shelton, D.L., Levinson, A.D., and Phillips, H.S. (1994) Mice Lacking Nerve Growth Factor Display Perinatal Loss of Sensory and Sympathetic Neurons Yet Develop Basal Forebrain Cholinergic Neurons. *Cell* 76:1001-1011.

Duff, R.S., Langtimm, C.J., Richardson, M.K., and Sieber-Blum, M. (1991) *In vitro* clonal analysis of progenitor cell patterns in dorsal root and sympathetic ganglia of the quail embryo. *Dev. Biol.* 147:451-459.

Dulac, C., Cameron-Curry, P., Ziller, C., and Le Douarin, N.M. (1988) A surface protein expressed by avian myelinating and nonmyelinating Schwann cells but not by satellite or enteric glial cells. *Neuron* 1:211-220.

Dulac, C., and Le Douarin, N.M. (1991) Phenotypic plasticity of Schwann cells and enteric glial cells in response to the microenvironment. *Proc. Nat. Acad. Sci.* 88:6358-6362.

Dulac, C., Tropak, M.B., Cameron-Curry, P., Rossier, J., Marshak, D.R., Roder, J., and Le Douarin, N.M. (1992) Molecular characterization of the Schwann cell myelin protein, SMP ; structural similarities within the immunoglobulin superfamily. *Neuron* 8:323-334.

Dupin, E., Baroffio, A., Dulac, C., Cameron-Curry, P., and Le Douarin, N.M. (1990) Schwann-cell differentiation in clonal cultures of the neural crest, as evidenced by the anti-Schwann cell myelin protein monoclonal antibody. *Proc. Nat. Acad. Sci. USA* 87:1119-1123.

Dupin, E., Maus, M., and Fauquet, M. (1993) Regulation of the quail tyrosine hydroxylase gene in neural crest cells by cAMP and β-adrenergic ligands. *Dev. Biol.* 159:75-86.

Ernfors, P., Lee, K.F., and Jaenisch, R. (1994) Mice lacking brain-derived neurotrophic factor develop with sensory deficits. *Nature* 368:147-150.

Ernfors, P., Lee, K.F., Kucera, J., and Jaenisch, R. (1994) Lack of neurotrophin-3 leads to deficiences in the peripheral nervous system and loss of limb proprioceptive afferents. *Cell* 77:503-512.

Farinas, I., Jones, K.R., Backus, C., Wang, X.Y., and Reichardt, L.F. (1994) Severe sensory and sympathetic deficits in mice lacking neurotrophin-3. *Nature* 369:658-661.

Frank, E., and Sanes, J.R. (1991) Lineage of neurons and glia in chick dorsal root ganglia: analysis *in vivo* with a recombinant retrovirus. *Development* 111:895-908.

Fraser, S.E. and Bronner-Fraser, M. (1991) Migrating neural crest cells in the trunk of the avian embryo are multipotent. *Dev.* 112:913-920.

Ito, K., and Sieber-Blum, M. (1991) *In vitro* clonal analysis of quail cardiac neural crest development. *Dev. Biol.* 148:95-106.

Ito, K., and Sieber-Blum, M. (1993) Pluripotent and developmentally restricted neural crest-derived cells in posterior visceral arches. *Dev. Biol.* 156:191-200.

Jones, K.R., Farinas, I., Backus, C., and Reichardt, L.F. (1994) Targeted disruption of the BDNF gene perturbs brain and sensory neuron development but not motor neuron development. *Cell* 76:989-999.

Kalcheim, C., and Le Douarin, N.M. (1986) Requirement of a neural tube signal for the differentiation of neural crest cells into dorsal root ganglia. *Dev. Biol.* 116:451-466.

Kalcheim, C., Barde, Y.A., Thoenen, H., and Le Douarin, N.M. (1987) In vivo effect of brain-derived neurotrophic factor on the survival of developing dorsal root ganglion cells. *EMBO J.* 6:2871-2873.

Kalcheim, C. (1989) Basic fibroblast growth factor stimulates survival of nonneuronal cells developing from trunk neural crest. *Dev. Biol.* 134:1-10.

Kalcheim, C., Carmeli, C., and Rosenthal, A. (1992) Neurotrophin-3 is a mitogen for cultured neural crest cells. *Proc. Nat. Acad. Sci.* 89:1661-1665.

Kitamura, K., Takiguchi-Hayashi, K., Sezaki, M., Yamamoto, H., and Takeuchi, T. (1992) Avian neural crest cells express a melanogenic trait during early migration from the neural tube - observations with the new monoclonal antibody, MELB-1. *Development* 114:367-378.

Klein, R., Smeyne, R., Wurst, W., Long, L.K., Auerbach, B.A., Joyner, A., and Barbacid, M. (1993) Targeted disruption of the trkB neurotrophin receptor gene results in nervous system lesions and neonatal death. *Cell* 75:113-122.

Lahav, R., Ziller, C., Lecoin, L., Nataf, V., Martin, F.H., Carnahan, J.F., Langley, K.E., Boone, T.C., and Le Douarin, N.M. (1993) Effect of *steel* gene product on melanogenesis in avian neural crest cell cultures. *Differentiation* (in press).

Lallier, T. and Bronner-Fraser, M. (1993) Inhibition of neural crest cell attachment by integrin antisense oligonucleotides. *Science* 259:692-695.

Le Douarin, N.M., Teillet, M.A., Ziller, C., and Smith, J. (1978) Adrenergic differentiation of cells of the cholinergic ciliary and Remak ganglia in avian embryo after *in vivo* transplantation. *Proc. Natl. Acad. Sci. U. S. A.* 75:2030-2034.

Le Douarin, N.M. (1982) "The Neural Crest" Cambridge University Press, Cambridge.

Le Douarin, N.M. (1986) Cell line segregation during peripheral nervous system ontogeny. *Science* 231:1515-1522.

Le Douarin, N.M. and Smith, J. (1988). Development of the peripheral nervous system from the neural crest. *Annu. Rev. Cell Biol.* 4:375-404.

Le Douarin N.M., and Ziller, C. (1993) Plasticity in neural crest cell differentiation. *Curr. Op. Cell Biol.* 5:1036-1043.

Le Douarin, N.M., Ziller, C., and Couly, G.F. (1993) Patterning of neural crest derivatives in the avian embryo: in vivo and in vitro studies. *Dev. Biol.* 159:24-49.

Morrison-Graham, K., and Weston, J.A. (1993) Transient steel factor dependence by neural crest-derived melanocyte precursors. *Dev. Biol.* 159:346-352.

Nataf, V., and Monier, S. (1992) Effect of insulin and insulin-like growth factor-I on the expression of the catecholaminergic phenotype by neural crest cells. *Dev. Brain Res.* 69:59-66.

Nataf, V., Mercier, P., Ziller, C., and Le Douarin, N.M. (1993) Novel markers of melanocyte differentiation in the avian embryo. *Exp. Cell Res.* 207:171-182.

Nishikawa, S., Kusakabe, M., Yoshinaga, K., Ogawa, M., Hayashi, S., Kunisada, T., Era, T., Sakakura, T., and Nishikawa, S. (1991) *In utero* manipulation of coat color formation by a monoclonal anti-c-kit antibody: two distinct waves of c-kit-dependency during melanocyte development. *EMBO J.* 10:2111-2118.

Pomeranz, H.D., Rothman, T.P., Chalazonitis, A., Tennyson, M., and Gershon, M.D. (1993) Neural crest-derived cells isolated from the gut by immunoselection develop neuronal and glial phenotypes when cultured on laminin. *Dev. Biol.* 156:341-361.

Richardson, M.K., and Sieber-Blum, M. (1993) Pluripotent neural crest cells in the ceveloping skin of the quail embryo. *Dev. Biol.* 157:348-358.

Sextier-Sainte-Claire Deville, F., Ziller, C., and Le Douarin, N. (1992) Developmental potentialities of cells derived from the truncal neural crest in clonal cultures. *Dev. Brain. Res.* 66:1-10.

Sextier-Sainte-Claire Deville, F., Ziller, C., and Le Douarin, N.M. (1994) Developmental potentials of enteric neural crest-derived cells in clonal and mass cultures. *Dev. Biol.* 163:141-151.

Sieber-Blum, M., and Cohen, A.M. (1980) Clonal analysis of quail neural crest cells: they are pluripotent and differentiate in vitro in the absence of non-crest cells. *Dev. Biol.* 80:96-106.

Sieber-Blum, M. (1991) Role of the neurotrophic factors BDNF and NGF in the commitment of pluripotent neural crest cells. *Neuron* 6:949-955.

Smeyne, R.J., Klein, R., Schnapp A, Long, L.K., Bryant, S., Lewin, A., Lira, S.A., and Barbacid, M. (1994) Severe sensory and sympathetic neuropathies in mice carrying a disrupted Trk/NGF receptor gene. *Nature* 368:246-248.

Steel, K.P., Davidson, D.C., and Jackson, I.J. (1992) TRP-2/DT, a new early melanoblast marker, shows that steel growth factor (c-kit ligand) is a survival factor. *Development* 115:1111-1119.

Stemple, D.L., and Anderson, D.J. (1992) Isolation of a stem cell for neurons and glia from the mammalian neural crest. *Cell* 71:973-985.

Williams, D.E., de Vries, P., Namen, A.E., Widmer, M.B., and Lyman, S.D. (1992) The Steel factor. *Dev. Biol.* 151:368-376.

Xue, Z.G., Smith, J., and Le Douarin, N.M. (1985) Expression of the adrenergic phenotype by dorsal root ganglion cells of the quail in culture *in vitro*. *C. R. Acad. Sci. Paris.* 300:483-488.

Xue, Z.G., Smith, J., and Le Douarin, N.M. (1987) Developmental capacities of avian embryonic dorsal root ganglion cells: neuropeptides and tyrosine hydroxylase in dissociated cell cultures. *Brain. Res.* 431:99-109.

Xue, Z.G., Le Douarin, N.M., and Smith, J. (1988) Insulin and insulin-like growth factor-1 can trigger the differentiation of catecholaminergic precursors in cultures of dorsal root ganglia. *Cell. Differ. Dev.* 25:1-10.

Xue, Z.G., Xue, X.J., Fauquet, M., Smith, J., and Le Douarin, N. (1992) Expression of the gene encoding tyrosine hydroxylase in a subpopulation of quail dorsal root ganglion cells cultured in the presence of insulin or chick embryo extract. *Developmental Brain Research* 69:23-30.

Ziller, C., Dupin, E., Brazeau, P., Paulin, D., and Le Douarin, N.M. (1983) Early segregation of a neuronal precursor cell line in the neural crest as revealed by culture in an chemically defined medium. *Cell* 32:627-638.

RETINOIDS AND AXIAL PATTERNING
IN THE EARLY VERTEBRATE EMBRYO

A.J. Durston, J. van der Wees, W.W.M. Pijnappel,
J.G. Schilthuis, and S.F. Godsave

Hubrecht Laboratory
Netherlands Institute for Developmental Biology
Uppsalalaan 8
3584 CT Utrecht
The Netherlands

INTRODUCTION

The purpose of this article is to discuss some of the evidence concerning a possible role for retinoids (vitamin A metabolites) as morphogens regulating axial patterning in the early vertebrate embryo. We do not intend to review all of the extensive literature bearing on this point, but to concentrate instead on some aspects which are especially relevant for axial patterning in very early developmental stages, and, particularly, on evidence obtained using embryos of the Amphibian Xenopus laevis.

RETINOIDS AND AXIAL PATTERNING

It is well known that treating early vertebrate embryos with the active retinoid all-trans retinoic acid (tRA) causes teratogenesis. Many defects are found, most notably in the heart, head, neural crest and limbs. It was pioneering investigations of the teratogenic effects of tRA on the developing chicken limb which first triggered strong interest in the idea that tRA might function as a morphogen during embryogenesis. The experiments showed that implanting a bead soaked in tRA could mimic a natural organiser region (the zone of polarising activity: ZPA) in determining the axial polarity of the limb (Tickle et al., 1982). Recent findings demonstrating the availability and action of other relevant signal molecules (sonic hedgehog, BMP-2, FGF-4), now complicate the picture considerably (reviewed in Maden, 1994), but it is still rather probable that an endogenous retinoid is involved in limb patterning.

The relevance of retinoids for axial patterning in the very early embryo was first demonstrated by the discovery that Xenopus embryos have a responsive period around gastrulation when they are particularly sensitive to tRA (Durston et al., 1989; Sive et al.,

1990). Treatment during this period disturbs formation of the antero-posterior (a-p) axis. Most obvious is a concentration dependent loss of head structures, but there are also other abnormalities, including defects in tail formation at high concentrations (Sive et al., 1990). The gastrula specific posteriorising effect was interesting because of the mass of evidence, initiated by the pioneering experiments by Spemann and his collaborators, and extended by many others, that the main a-p axis of the Amphibian embryo is patterned by intercellular signals acting in the gastrula. Spemann's experiments revealed that the dorsal lip of the blastopore in the gastrula is an organiser region which emits head and tail specific signals at sequential stages during gastrulation (Spemann and Mangold, 1924; Spemann, 1931), and further experiments revealed signals which specify a-p identity both in the neural plate and in the axial mesoderm (Mangold, 1933; Nieuwkoop, 1952; Toivonen and Saxén, 1968; Hama, 1978). Recent transplantation experiments have shown that organisers from different vertebrates (chickens and mice) have general properties, being able to exert at least some of their organiser functions when transplanted to a Xenopus gastrula or combined with Xenopus embryonic tissue (Kintner and Dodd, 1991; De Robertis et al., 1992), while the early posteriorising effect of tRA is also general in vertebrates, having now been demonstrated in mice, chickens and fish. It is also notable that the organising ZPA region in the chicken limb bud can be replaced not only by a tRA soaked bead, but also by an organiser (Hensen's node) from the chicken gastrula (Hornbruch and Wolpert, 1986). These findings make it interesting to speculate that a retinoid has a conserved role in early vertebrate patterning.

Experiments by several authors now show clearly that tRA has effects in the gastrula both on the axial mesoderm and on the neural plate. Its action on the neural plate resembles that of a neural transformation signal proposed in the classical literature in that it posteriorises induced neural tissue but is not, itself, a neural inducer (Nieuwkoop et al., 1952). Treating whole Xenopus embryos with tRA thus increases the volume of the hindbrain at the expense of a decrease in the volume of the forebrain (Durston et al., 1989). It also induces the expression of posterior neural markers both in the presumptive forebrain region of the intact embryo (Ruiz i Altaba and Jessell, 1991b), and in explants of presumptive forebrain tissue from the late gastrula (Dekker et al., 1992). Interestingly, early tRA treatments also appear to contract or delete the most anterior part of the developing hindbrain as well as disturbing hindbrain segmentation in Xenopus and zebrafish embryos (Papalopulu et al., 1991; Holder and Hill, 1991). In the axial mesoderm, tRA has been shown to reduce the capacity of organiser (dorsal mesoderm) tissue to induce the formation of head structures in Einsteck experiments (Sive and Cheng, 1991; Ruiz i Altaba and Jessell, 1991a; Cho et al., 1991). tRA also clearly ventralises mesoderm in growth factor induced explants (Ruiz i Altaba and Jessell, 1991a), but this tendency was not obvious in whole embryos (Sive and Cheng, 1991). These results make it interesting to consider whether an endogenous retinoid might function as a posteriorising morphogen, both for the neural plate and the axial mesoderm in the early embryo.

ACTIVE RETINOIDS *IN VIVO*

If a retinoid does, indeed, function as a morphogen during embryogenesis, one would expect that it should be available *in vivo* and probably localised, in the embryo. It would not be surprising, considering the posteriorising effect of tRA, and its inducing effect on Hox gene expression, if an active retinoid became localised posteriorly around the blastopore during gastrulation. Since the original discovery that all-trans retinoic acid (tRA) is a naturally occurring active retinoid, a number of other natural retinoid metabolites have been shown to be biologically active. These are: the 9-cis stereoisomer of retinoic acid (9-cis-RA), the ligand for the RXR family of receptors described below (Heyman et al., 1992;

Levin et al., 1992), as well as all-trans-3,4-didehydro-retinoic acid (ddRA) (Thaller and Eichele, 1990), all-trans-4-oxo-retinoic acid (4-oxo-RA) (Pijnappel et al., 1993), 14-hydroxy-4,14-retroretinol (14-HRR) (Buck et al., 1991) and anhydroretinol (AR) (Buck et al., 1993). Investigations using HPLC analysis have now reported that four of these active retinoids, tRA (Thaller and Eichele, 1987), 9-cis-RA (Creech Kraft et al., 1994b), 4-oxo-RA (Pijnappel et al., 1993), and ddRA (Thaller and Eichele, 1990), are available *in vivo*, in developing embryos. Two of them have also already been reported as being localised, or synthesised locally, during embryogenesis. tRA has been reported as a posterior to anterior gradient in the developing limb bud (Thaller and Eichele, 1987). tRA and 9-cis-RA have also been reported as being localised in Xenopus gastrula and neurula embryos; 9-cis-RA as anterior dorsal and posterior dorsal; tRA as anterior only, and ventral rather than dorsal (Creech Kraft et al., 1994b). tRA synthesis from retinol has, on the other hand, been reported to be localised in the posterior region (Hensen's node organiser), in the early mouse embryo (Hogan et al., 1992). A different approach, using *in vivo* activation of a luciferase reporter construct driven by a retinoic acid responsive element has now also reported a posterior (organiser) localisation of active retinoids in the Xenopus neurula (Chen et al., 1994), and a study using F9 teratocarcinoma cells transfected with a retinoid responsive lacZ construct reported localised active retinoid secretion in the posterior part of the developing central nervous system of the later rat embryo (Wagner et al., 1992). Taken together, these investigations clearly leave some uncertainty about the localisation of active retinoids in the early vertebrate embryo. Further investigation is required. It should be borne in mind that HPLC investigation of the localisation of endogenous retinoids is made difficult by the existence of overlapping retinoid peaks which hinder definitive identification, and that we do not yet know which are the most important active retinoids *in vivo*. Some of the studies using reporter constructs also employ the endogenous retinoid receptors in the embryo, and these can have localised availability (see below).

An interesting question concerning active retinoids concerns their specificities for different retinoid sensitive processes *in vivo*. It is notable that tRA and 4-oxo-RA both apparently have qualitatively similar effects on Xenopus embryogenesis (inducing microcephaly if applied at the gastrula stage), but that 4-oxo-RA (which was characterised as an inactive tRA catabolite on the basis of its relatively weak activity in regulating growth and differentiation of cultured cells (Surekha Rao et al., 1972; Frolik et al., 1979; Williams et al., 1987), is very potent in disturbing patterning in the early embryo (5 x more than tRA) (Pijnappel et al., 1993). The basis of this biological specificity has not yet been accounted for, in terms of receptor specificity or other molecular differences. It is also notable that 9-cis-RA, which can work via a totally different receptor pathway than RA or 4-oxo-RA (see below), apparently induces qualitatively similar morphological defects in Xenopus embryos as these ligands (Creech Kraft et al., 1994a; and own unpublished observations). Explaining and characterising such biological specificities will clearly require much more investigation, both at the embryological and at the molecular level.

RECEPTORS AND RETINOID SIGNAL TRANSDUCTION

The present consensus is that at least some, and probably all of the biological effects of acidic retinoids are mediated by members of the nuclear receptor super family of ligand-inducible transcription factors. There are presently two identified types of nuclear retinoid receptors: the RARs and RXRs. Each receptor type comprises three closely related subtypes (α, β, γ), encoded by separate genes and each receptor subtype gene also encodes several isoforms, via the use of different promoters and differential splicing. These receptor subtypes are all available in mammals, birds and amphibians, although the gene for the RARβ subtype has not yet been cloned in Xenopus (see below) (reviewed in Mangelsdorf et al., 1994). The RAR receptor type is recognised with high affinity by several naturally occurring acidic retinoid ligands, namely tRA (Petkovitch et al., 1987; Giguère et al., 1987), 9-cis-RA (Levin et al., 1992; Heyman et al., 1992), ddRA (Allenby et al., 1993), and 4-oxo-RA (Pijnappel et al., 1993). The RXR receptor type has only one known natural ligand: 9-cis-RA (Levin et al., 1992; Heyman et al., 1992). It is by no means certain that the complexity sketched above ends with RARs and RXRs, because two new families of retinoid related orphan receptors (RORs/RZRs (Giguère et al., 1994; Carlberg et al., 1994), and RVR/Rev-erbβ (Retnakaran et al., 1994; Forman et al., 1994) which closely resemble RARs and RXRs in their DNA sequence, have now also been cloned. We note, however, that a ligand for RZRβ has now very recently been identified as the non-retinoid hormone melatonin (Becker-André et al., 1994). Another class of nuclear orphan receptors (COUP-TFs) are also relevant because they probably act as negative regulators of retinoid signalling (Kliewer et al., 1992a; Tran et al., 1992; Cooney et al., 1993; Jonk et al., 1994).

At the molecular level, the RARs and RXRs function as transactivators or silencers of transcription by binding to response elements in the promoters of, or enhancers for, target genes. The most important role for RARs is probably to act, in RAR-RXR heterodimers, by binding to response elements consisting of a direct repeat (DR) of AGGTCA or a closely related half site motif, spaced by one, two or five nucleotides (DR1, DR2 or DR5). The DR2 and DR5 response elements mediate transactivation via activated RAR-RXR heterodimers. The DR1 response element mediates silencing via activated RAR-RXR heterodimers, probably via competition with activated RXR homodimers (Mangelsdorf et al., 1991; reviewed in Mangelsdorf et al., 1994), see also below.

Unlike the RARs, RXRs can function as homodimers, which activate gene expression via the DR1 response element. They also function very widely as promiscuous partners in heterodimers with several different nuclear receptors, including RARs, thyroid hormone receptors (TRs), vitamin D receptors (VDRs), peroxisome proliferator-activated receptors (PPARs), and COUP TFs, with each heterodimer acting via its appropriate specific direct repeat response element (reviewed in Green, 1993, and Mangelsdorf et al., 1994).

The nature of this mechanism makes it clear that retinoid signalling will interact with other signalling pathways. RXRs can function explicitly in heterodimers to transduce signals due to other non retinoid hormones. RARs have also been reported to form heterodimers with TRs (Glass et al., 1989) and VDRs (Schräder et al., 1993). They may also interact with other pathways indirectly by competing for the available pool of RXR receptors. It is possible that COUP-TFs inhibit retinoid signalling via RARs by the same mechanism, or more likely by binding of COUP-TF homodimers (Cooney et al., 1993) or COUP-RXR heterodimers (Kliewer et al., 1992a; Widom et al., 1992; Cooney et al., 1993) to the DR1 retinoid response element in competition with RXR homodimers. This connectedness obviously has far reaching implications. It can not absolutely be excluded, for example, that retinoids and their receptors are not directly relevant for early pattern formation at all, but that the developmental effects observed from retinoid treatment are indirect, due to interference with a so far unidentified endogenous hormonal morphogen system. The existence of localised retinoid ligands with high specificity (above) and of localised retinoid receptors (below) and, indeed, the very existence of complex retinoid signalling systems in the early embryo do, however, make this possibility unattractive, both from a functional and from an evolutionary point of view. The existence of this complex molecular machinery raises other interesting questions. One concerns the relationship between molecular and biological specificities, and the expectation that the different retinoid ligands may work via different receptor types and subtypes to regulate different biological processes and different target genes. The investigations in this area are still in their infancy, but are promising.

Studies using synthetic retinoids have identified ligands with quite high specificity for particular receptor subtypes (Hashimoto et al., 1990; Hashimoto, 1991; Lehman et al., 1991, 1992; Graupner et al., 1991) and some specificities have also been identified for natural ligands, apparently depending on the assay used (Allenby et al., 1994). The most striking specificity so far identified for a natural ligand is that 9-cis-RA is the only known natural ligand for the RXRs, and this fact has now been given added significance by the revealing discovery that only the RAR ligand binds to RAR-RXR heterodimers when these are bound to DNA (Kurokawa et al., 1994). It is likely that the RXR functions as a passive cofactor in RAR-RXR heterodimers and in some other heterodimers between RXR and hormone receptors, and that a main function of the RXR ligand is to activate RXR homodimers. 9-cis RA is known to induce RXR homodimerisation (Zhang et al., 1992). It is interesting to note that 9-cis-RA binding is able to activate RXR-PPAR heterodimers (Keller et al., 1993, Kliewer et al., 1992b), and that these, like RXR homodimers act via the DR1 response element. RXRs thus participate in two very different types of retinoid signal transduction pathway, mediating signals from RAR and RXR ligands respectively. A second aspect of retinoid receptor and ligand specificity concerns the point that, besides being transcription factors, which regulate gene expression specifically via retinoid responsive DNA elements, RARs regulate gene expression via protein - protein interactions which have their own specificities. One such interaction which is rather well known is a retinoid dependent inhibitory interaction with the AP-1 transcription complex (Schüle et al, 1991). Some very recent exciting findings using synthetic retinoids now show that this protein - protein interaction has its own unique retinoid ligand specificity (Fanjul et al., 1994). Different functions of the same retinoid receptor may thus be regulated separately by different ligands. Relating these different molecular specificities to different biological functions of the retinoid signalling system in the embryo is thus an important challenge for the future. There is, so far, very little progress in this area. For example, the basis of the high specificities of 4-oxo-RA and 9-cis-RA for teratogenesis in the early embryo remains obscure.

RETINOID RECEPTORS AND EMBRYOGENESIS

Considering the molecular complexity sketched briefly above, it is hardly surprising that the importance of different retinoid ligands and receptors for pattern formation in the early embryo is still obscure. The relevance of known RARs and RXRs for embryogenesis has been investigated by characterising their expression patterns, and by functional analysis (via gain and loss of function experiments).

Much is now known, from *in situ* hybridisation studies and other approaches, about the mRNA expression patterns of different RAR and RXR subtypes in the mouse embryo. There are also limited data available in other vertebrates chicken (Smith and Eichele, 1991), zebrafish (Joore et al., 1994, White et al., 1994) and Xenopus (Ellinger-Ziegelbauer and Dreyer, 1991; Sharpe, 1992)). The data show that different RAR and RXR subtypes are widely expressed, both in the adult organism and during embryogenesis and that they each have individual expression patterns, which may be overlapping or exclusive with each other and with the expression patterns of cytoplasmic retinoid binding proteins (CRABPs and CRBPs). These data suggest that individual receptor subtypes have specific, and probably multiple, functions. Among the murine RARs, the RARα subtype is rather generally expressed during embryogenesis while RARβ and RARγ are more restricted, sometimes being expressed in non-overlapping patterns (Dollé et al., 1990; Ruberte et al., 1990, 1991, 1993; Mendelsohn et al., 1994). An interesting example for axis formation is that RARβ is available only in the closed neural tube, and RARγ only in the open neural folds during murine neurulation (Ruberte et al., 1991). Among the RXRs, murine RXRβ is rather generally available, while murine RXRα and especially γ have more restricted expression patterns during embryogenesis (Mangelsdorf et al., 1992; Dollé et al., 1994; reviewed in Mangelsdorf et al., 1994). Another aspect, which complicates the picture further, is that different isoforms of a receptor subtype can be expressed differently (Mendelsohn et al., 1994a and see below). For details of these receptor expression patterns, the reader is referred to the original literature.

There is relatively little information available about the expression of retinoid receptor subtypes in Xenopus. The literature indicates that at least the RARα and γ and RXRα, β and γ subtypes are expressed with subtype specific timing in the early embryo (Ellinger-Ziegelbauer and Dreyer, 1991; Blumberg et al., 1992; Sharpe, 1992; Marklew et al., 1994). Xenopus RARβ has not yet been cloned. RARγ is, so far, the only subtype for which localised expression has been reported during gastrulation. This subtype shows posterior dorsal expression, around the dorsal lip of the blastopore, as well as a second anterior expression zone (Ellinger-Ziegelbauer and Dreyer, 1991). These anterior and posterior zones persist into the neurula and tailbud stages, where isoform specific patterns (γ1, γ2) are demonstrable (Pfeffer and De Robertis, 1994). Another RAR isoform, RARα2, shows localised expression within the neural tube by the late neurula stage (Sharpe, 1992). These data are thus fragmentary, but they and the data from other vertebrates, are consistent with specific functions for individual retinoid receptor subtypes and even isoforms during embryogenesis.

The functional significance of the different retinoid receptor subtypes, and of retinoid signalling, has been investigated by genetic manipulation in a number of different vertebrates. The most significant progress so far has been made by knocking out murine RARs and RXRs via germ-line homologous recombination in transgenic mice. Single gene knock-outs reveal that, whereas knocking out the RARα1 or RARβ2 or RARγ2 isoforms delivers an apparently normal phenotype (Lufkin et al., 1993; Mendelsohn et al., 1994c; Lohnes et al., 1993), knocking out the RARα or γ or RXRα genes delivered quite severe developmental abnormalties and that crossing these mutants (RARα, RARγ, RARβ2, RARα1, RXRα) to make double mutants delivered a much more severe phenotype in many cases (Kastner et al., 1994; Lohnes et al., 1994; Mendelsohn et al., 1994b). It is not appropriate to describe all of the findings made in these experiments here; the reader is referred to the original publications for details. We note, however, that the defects obtained are complex, as would be expected from the complex expression patterns of the different RARs and RXRs, and that some of them duplicate defects observed in the offspring of vitamin A deficient mice. We also note that the defects obtained include axial homeotic transformations, as could be expected if endogenous retinoids have a role in axial patterning. RARγ knock-outs thus generate anteriorising homeotic transformations of the cervical axial skeleton (Lohnes et al., 1993), and other, more or less severe axial transformations are observed in several of the double mutants (Lohnes et al., 1994). The results from these homologous recombination knock-outs are thus consistent with multiple roles for retinoids in embryogenesis, including a role in axial patterning. The increased severity of the double mutant phenotypes also indicates functional redundancy among retinoid receptors. Elucidating the functions of the individual receptors and the extent of the functional redundancy will require much more investigation. A very interesting *in vitro* study in RARγ deficient F9 murine teratocarcinoma cells suggested a form that this redundancy might take. These cells now failed to express Hoxa-1 in response to RA, but overexpressed its paralogue Hoxb-1 (Boylan et al., 1993).

The functional studies in Xenopus are, so far, not extensive and have delivered little insight into a possible role for a retinoid in axial patterning. They show that overexpression in pre-gastrula stages of intact Xenopus RARγ2, or of a truncated dominant negative version of RARγ2, by mRNA injection into the zygote, causes no obvious disturbance of embryogenesis, even though these treatments enhance tRA induced teratogenesis and suppress the tRA-induced expression of a retinoid responsive receptor construct (Smith et al., 1994). Overexpression of v-erb A (a truncated version of the thyroid hormone receptor which can be predicted to interfere with retinoid signalling) did, however, induce developmental defects (Schuh et al., 1993), but no obvious axial defects. These studies are clearly still in their infancy. The future will be interesting.

In conclusion, expression studies and gain and loss of function studies presently lead to no clear conclusions regarding the functioning of RARs and RXRs in early embryogenesis. The expression studies show patterns which are compatible with conserved specialised functions for particular receptor subtypes. The functional studies in the mouse indicate redundancy of function, and suggest multiple functions for retinoid signalling in embryogenesis. The studies in Xenopus are not extensive.

HOX GENES AS RETINOID TARGET GENES

Some of the effects of retinoids in modulating the embryonic main body axis are likely to be mediated via regulation of Hox gene expression. There are 4 clusters of Hox genes in vertebrates, which were originally identified because of their homology with Drosophila homeotic genes in the antennapedia and bithorax complexes (Duboule & Dollé, 1989; Graham et al., 1989, reviewed in McGinnis & Krumlauf, 1992). During early development, the Hox genes, encoding transcription factors, are expressed in sequential zones along the a-p axis in the hindbrain and trunk regions. They show a-p colinearity of expression, such that progressively more 5' genes are expressed in progressively more posterior zones (McGinnis & Krumlauf, 1992; Krumlauf, 1994; Dekker et al., 1992; Godsave et al., 1994). As with the homeotic genes in Drosophila, these expression zones appear to be involved in providing a code for position along the a-p axis.

The Hox genes begin to be expressed during gastrulation, when a-p patterning is being established (Gaunt, 1987; Gaunt, 1988; Wilkinson et al., 1989; Dekker et al., 1992; Deschamps & Wijgerde, 1993) and in Xenopus, the characteristic sequence of Hoxb gene expression zones seen at tailbud stages appear to develop very early, by the late gastrula/early neurula stage (Godsave et al., 1994). These data suggest that the establishment of localised Hox gene expression is regulated by factors active in the gastrula, which may include retinoids.

There is substantial evidence to suggest that retinoids do indeed play a role in regulating Hox gene expression. The Hox genes are sensitive to retinoic acid both *in vitro*, in embryocarcinoma cells, and in early embryos. tRA induces transcription of 3' Hox genes first and to a greater extent than that of progressively more 5' genes (Simeone et al., 1990; 1991; Conlon & Rossant, 1992; Dekker et al., 1992; Leroy & De Robertis, 1992). This colinearity in the response to tRA is what would be predicted if a gradient of a retinoid produced at the posterior end of the embryo was responsible for a sequential activation of Hox genes along the a-p axis. A gradient developing from a posterior source would also be expected to initiate Hox gene expression posteriorly with the expression spreading to an anterior border. In mouse embryos, spreading of Hoxb gene expression zones from the posterior end of late primitive streak stage embryos has been described. By the head fold/early somite stages, the expression reaches an anterior expression border and from then on appears to be clonally transmitted (Wilkinson et al., 1989; Deschamps & Wijgerde, 1993). The 3' Hox genes, which are most sensitive to retinoids would be expected to spread to more anterior borders than the less sensitive 5' genes, as is the case. The idea that retinoids may be regulating this process is supported by the finding that tRA treatment of early mouse embryos results in a more anterior expression of 3' Hoxb genes in the embryo (Conlon and Rossant, 1992: Wood et al., 1994). Hoxb gene expression can also be induced by RA in explants of anterior neural tissue from late gastrulae of Xenopus

(Dekker et al., 1992). In the mesoderm, tRA also causes transformations of Hox gene expression domains in murine prevertebrae leading to homeotic transformations of vertebrae and to the idea of a Hox code for vertebral identity (Kessel & Gruss, 1991).

An involvement of retinoids in the activation of Hox genes during embryogenesis is most directly supported by the recent finding of an enhancer containing a RARE downstream of the Hoxb-1 gene, the most 3' gene in the Hoxb cluster. This RARE is required for the early expression of Hoxb-1 RNA in the neuroectoderm in mouse embryos (Marshall et al., 1994). The Hoxa-1 gene also contains a downstream enhancer with a RARE (Langston & Gudas, 1992) and a RARE has been identified upstream of the murine and human Hoxd-4 genes (Moroni et al., 1993; Pöpperl & Featherstone, 1993).

An interesting finding concerning RA-induced Hox gene expression in human embryonal carcinoma cells is that 3' Hox gene expression may be required for the expression of more 5' Hox genes. Inhibition of HOXB1 or HOXB3 causes a reduction in mRNA levels of more 5' HoxB genes, and this effect increases towards the 5' end of the cluster. In the case of HOXB3 inhibition, it was shown additionally that more 5' Hox genes in other clusters were also affected (Faiella et al., 1994). The cluster organisation and regulation may therefore be important in the response to retinoids. However, in vivo, the picture is more complicated. Although the expression domain of Hoxb-1 was found to be substantially reduced in Hoxa-1 knock-out mice, the expression of a number of other Hox genes appeared to be unaffected (Dollé et al., 1993). It will be interesting to examine the expression of 5' Hox genes in double knock-out experiments.

The effects of tRA on Hox gene expression are actually rather complex. For example, RA has different effects on Hoxb-1 expression before and after the onset of somitogenesis in mouse embryos. Hoxb-1 expression is normally restricted to rhombomere 4 in the hindbrain. Treatment of embryos with tRA before somites start forming, results in expression of Hoxb-1 throughout the pre-otic (anterior) hindbrain, whereas treatment of older embryos leads to a duplication of the hindbrain segments r4,5 and a second stripe of Hoxb-1 expression in the pre-otic hindbrain (Marshall et al., 1992; Wood et al., 1994). These findings suggest that retinoids may play several distinct roles in the patterning of the a-p axis, including a relatively late role in subdivision of the hindbrain. In further support of this idea is the finding of a RARE in a repressor responsible for restricting the expression of Hoxb-1 to r4 in the hindbrain in later development (Studer et al., 1994).

These data support the idea that Hox gene regulation is mediated partly by positional information supplied by retinoids and that RA effects on axial patterning are partly mediated via effects on Hox gene expression.

CONCLUDING REMARKS

The findings above lead to a complex and inconclusive picture concerning the functioning of retinoids in early embryogenesis. The rather specific teratogenic effects of retinoids, the availability and localisation of sufficient concentrations of active retinoid ligands in the gastrula embryo and the specific connections between retinoids and important developmental genes, like the Hox genes, and, particularly the availability of a retinoid response element in the Hoxa-1, Hoxb-1 and Hoxd-4 genes suggest that a retinoid may act as an early developmental signal, possibly acting as a posterior to anterior gradient in the gastrula. The literature concerning the molecular mechanisms of retinoid signal transduction indicates complexity, the existence of multiple retinoid dependent pathways, and interactions with other signal transduction pathways. The relevance of these aspects for embryogenesis is still obscure. The gain and loss of function experiments with particular retinoid receptors so far show relatively mild effects on early embryogenesis. The results clearly indicate redundancy between different receptor subtypes (as demonstrated by the more severe phenotype of double receptor knock-outs). More extensive analysis is obviously required to elucidate the relevance of retinoid signalling for early embryogenesis.

ACKNOWLEDGMENTS

We thank Paul van der Saag for critical comments on the manuscript. This work was supported by the Netherlands Science Foundation (NWO), via MW projects 900-582-082 and 900-28-061, and SLW project # 417 442. We also acknowledge support via the Koningin Wilhelmina Fonds project # HUBR 93-677 and via EEC Biotech program # PL 920060.

REFERENCES

Allenby, G., Bocquel, M.T., Saunders, M., Kazmer, S., Speck, J., Rosenberger, M., Lovey, A., Kastner, P., Grippo, J.F., Chambon, P., and Levin, A.A., 1993, Retinoic acid receptors and retinoid X receptors: Interactions with endogenous retinoic acids, *Proc. Natl. Acad. Sci. USA* 90:30

Allenby, G., Janocha, R., Kazmer, S., Speck, J., Grippo, J.F., and Levin, A.A., 1994, Binding of 9-*cis*-retinoic acid and all-*trans*-retinoic acid to retinoic acid receptors α, β, and γ, *J. Biol. Chem.* 269:16689

Becker-André, M., Wiesenberg, I., Schaeren-Wiemers, N., André, E., Missbach, M., Saurat, J.H., and Carlberg, C., 1994, Pineal gland hormone melatonin binds and activates an orphan of the nuclear receptor superfamily, *J. Biol. Chem.* 269:28531

Blumberg, B., Mangelsdorf, D.J., Dyck, J.A., Bittner, D.A., Evans, R.M., and De Robertis, E.M., 1992, Multiple retinoid-responsive receptors in a single cell: Families of the retinoid "X" receptors and retinoic acid receptors in the *Xenopus* egg, *Proc. Natl. Acad. Sci. USA* 89:2321

Boylan, J.F., Lohnes, D., Taneja, R., Chambon, P., and Gudas, L.J., 1993, Loss of retinoic acid receptor γ function in F9 cells by gene disruption results in aberrant *Hoxa-1* expression and differentiation upon retinoic acid treatment, *Proc. Natl. Acad. Sci. USA* 90:9601

Buck, J., Derguini, F., Levi, E., Nakanishi, K., and Hammerling, U., 1991, Intracellular signaling by 14-hydroxy-4,14-retro-retinol, *Science* 254:1654

Buck, J., Grun, F., Derguini, F., Chen, Y., Kimura, S., Noy, N., and Hammerling, U., 1993, Anhydroretinol - a naturally occurring inhibitor of lymphocyte physiology, *J. Exp. Med.* 178:675

Carlberg, C., van Huijsduijnen, R.H., Staple, J.K., DeLamarter, J.F., and Becker-André, M., 1994, RZRs, a new family of retinoid-related orphan receptors that function as both monomers and homodimers, *Mol. Endocrinol.* 8:757

Chen, Y.P., Huang, L., and Solursh, M., 1994, A concentration gradient of retinoids in the early *Xenopus laevis* embryo, *Dev. Biol.* 161:70

Cho, K.W.Y., and De Robertis, E.M., 1990, Differential activation of *Xenopus* homeobox genes by *** mesoderm-inducing growth factors and retinoic acid, *Genes Dev.* 4:1910

Cho, K.W.Y., Morita, E.A., Wright, C.V.E., and De Robertis, E.M., 1991, Overexpression of a homeodomain protein confers axis-forming activity to uncommitted Xenopus embryonic cells, *Cell* 65:55

Conlon, R.A., and Rossant, J., 1992, Exogenous retinoic acid rapidly induces ectopic expression of murine Hox-2 genes *in vivo*, *Development* 116:357

Cooney, A.J., Leng, X., Tsai, S.Y., O'Malley, B.W., and Tsai, M.-J., 1993, Multiple mechanisms of Chicken Ovalbumin Upstream Promoter Transcription Factor-dependent repression of transactivation by the vitamin D, thyroid hormone, and retinoic acid receptors, *J. Biol. Chem.* 268:4152

Creech Kraft, J., Schuh, T., Juchau, M., and Kimelman, D., 1994a, The retinoid-X receptor ligand, 9-*cis*-retinoic acid, is a potential regulator of early *Xenopus* development, *Proc. Natl. Acad. Sci. USA* 91:3067

Creech Kraft, J., Schuh, T., Juchau, M.R., and Kimelman, D., 1994b, Temporal distribution, localization and metabolism of all-*trans*-retinol, didehydroretinol and all-*trans*-retinal during *Xenopus* development, *Biochem. J.* 301:111

Dekker, E.J., Pannese, M., Houtzager, E., Timmermans, A., Boncinelli, E., and Durston, A., 1992, *Xenopus* Hox-2 genes are expressed sequentially after the onset of gastrulation and are differentially inducible by retinoic acid, *Development suppl.*:195

De Robertis, E. M., Blum, M., Niehrs, C., and Steinbeisser, H., 1992, Goosecoid and the Organiser, *Development Suppl.*:167

Deschamps, J., and Wijgerde, M., 1993, Two phases in the establishment of HOX expression domains, *Dev. Biol.* 156:473

Dollé, P., Ruberte, E., Leroy, P., Morriss-Kay, G., and Chambon, P., 1990, Retinoic acid receptors and cellular retinoid binding proteins. I. A systematic study of their differential pattern of transcription during mouse organogenesis, *Development* 110:1133

Dollé, P., Lufkin, T., Krumlauf, R., Mark, M., Duboule, D., and Chambon, P., 1993, Local alterations of *Krox-20* and *Hox* gene expression in the hindbrain suggest lack of rhombomeres-4 and rhombomere-5 in homozygote null *Hoxa-1* (*Hox-1.6*) mutant embryos, *Proc. Natl. Acad. Sci. USA* 90:7666

Dollé, P., Fraulob, V., Kastner, P., and Chambon, P., 1994, Developmental expression of murine retinoid X receptor (RXR) genes, *Mech. Dev.* 45:91

Duboule, D., and Dollé, P., 1989, The structural and functional organization of the murine HOX gene family resembles that of Drosophila homeotic genes, *EMBO J.* 8:1497

Durston, A.J., Timmermans, J.P.M., Hage, W.J., Hendriks, H.F.J., de Vries, N.J., Heideveld, M., and Nieuwkoop, P.D., 1989, Retinoic acid causes an anteroposterior transformation in the developing central nervous system, *Nature* 340:140

Ellinger-Ziegelbauer, H., and Dreyer, C., 1991, A retinoic acid receptor expressed in the early development of *Xenopus laevis*, *Genes Dev.* 5:94

Faiella, A., Zappavigna, V., Mavilio, F., and Boncinelli, E., 1994, Inhibition of retinoic acid-induced activation of 3' human *HOXB* genes by antisense oligonucleotides affects sequential activation of genes located upstream in the four *HOX* clusters, *Proc. Natl. Acad. Sci. USA* 91:5335

Fanjul, A., Dawson, M.I., Hobbs, P.D., Jong, L., Cameron, J.F., Harlev, E., Graupner, G., Lu, X.P., and Pfahl, M., 1994, A new class of retinoids with selective inhibition of AP-1 inhibits proliferation, *Nature* 372:107

Forman, B.M., Chen, J., Blumberg, B., Kliewer, S.A., Henshaw, R., Ong, E.S., and Evans, R.M., 1994, Cross-talk among RORα1 and the Rev-erb family of orphan nuclear receptors, *Mol. Endocrinol.* 8:1253

Frolik, C.A., Roberts, A.B., Tavela, T.E., Roller, P.P., Newton, D.L., and Sporn, M.B., 1979, Isolation and identification of 4-hydroxy- and 4-oxoretinoic acid. In vitro metabolites of all-*trans*-retinoic acid in hamster trachea and liver, *Biochemistry* 18:2092

Gaunt, S.J., 1987, Homoeobox gene Hox-1.5 expression in mouse embryos: earliest detection by in situ hybridization is during gastrulation, *Development* 101:51

Gaunt, S.J., 1988, Mouse homeobox gene transcripts occupy different but overlapping domains in embryonic germ layers and organs: a comparison of *Hox-3.1* and *Hox-1.5*, *Development* 103:135

Giguère, V., Ong, E.S., Segui, P., and Evans, R.M., 1987, Identification of a receptor for the morphogen retinoic acid, *Nature* 330:624

Giguère, V., Tini, M., Flock, G., Ong, E., Evans, R.M., and Otulakowski, G., 1994, Isoform-specific amino-terminal domains dictate DNA-binding properties of RORα, a novel family of orphan hormone nuclear receptors, *Genes Dev.* 8:538

Glass, C.K., Lipkin, S.M., Devary, O.V., and Rosenfeld, M.G., 1989, Positive and negative regulation of gene transcription by a retinoic acid-thyroid hormone receptor heterodimer, *Cell* 59:697

Godsave, S., Dekker, E.J., Holling, T., Pannese, M., Boncinelli, E., and Durston, A., 1994, Expression patterns of *Hoxb* genes in the *Xenopus* embryo suggest roles in anteroposterior specification of the hindbrain and in dorsoventral patterning of the mesoderm, *Dev. Biol.* 166:465

Graham, A., Papalopulu, N., and Krumlauf, R., 1989, The murine and Drosophila homeobox gene complexes have common features of organization and expression, *Cell* 57:367

Graupner, G., Malle, G., Maignan, J., Lang, G., Pruniéras, M., and Pfahl, M., 1991, 6'-Substituted naphtalene-2-carboxylic acid analogs, a new class of retinoic acid receptor subtype-specific ligands, *Biochem. Biophys. Res. Comm.* 179:1554

Green, S., 1993, Promiscuous liaisons, *Nature* 361:590

Hama, T., 1978, Dynamics of the organizer. B. New findings on the regionality and morphogenetic movement of the organizer, in:"Organizer - A Milestone of a Half-Century from Spemann," O. Nakamura and S. Toivonen, eds., Elsevier/North-Holland Biomedical Press, Amsterdam.

Hashimoto, Y., 1991, Retinobenzoic acids and nuclear retinoic acid receptors, *Cell Struct. Funct.* 16:113

Hashimoto, Y., Kagechika, H., and Shudo, K., 1990, Expression of retinoic acid receptor genes and the ligand-binding selectivity of retinoic acid receptors (RARs), *Biochem. Biophys. Res. Comm.* 166:1300

258

Heyman, R.A., Mangelsdorf, D.J., Dyck, J.A., Stein, R.B., Eichele, G., Evans, R.M., and Thaller, C., 1992, 9-*cis*-Retinoic acid is a high affinity ligand for the retinoid X receptor, *Cell* 68:397

Hogan, B.L.M., Thaller, C., and Eichele, G., 1992, Evidence that Hensen's node is a site of retinoic acid synthesis, *Nature* 359:237

Holder, N., and Hill, J., 1991, Retinoic acid modifies development of the midbrain-hindbrain border and affects cranial ganglion formation in zebrafish embryos, *Development* 113:1159

Hornbruch, A., and Wolpert, L., 1986, Positional signalling by Hensen's node when grafted to the chick limb bud, *J. Embryol. exp. Morph.* 94:257

Jonk, L.J.C., Dejonge, M.E.J., Pals, C.E.G.M., Wissink, S., Vervaart, J.M.A., Schoorlemmer, J., and Kruijer, W., 1994, Cloning and expression during development of three murine members of the COUP family of nuclear orphan receptors, *Mech. Dev.* 47:81

Joore, J., van der Lans, G.B.L.J., Lanser, P.H., Vervaart, J.M.A., Zivkovic, D., Speksnijder, J.E., and Kruijer, W., 1994, Effects of retinoic acid on the expression of retinoic acid receptors during zebrafish embryogenesis, *Mech. Dev.* 46:137

Kastner, P., Grondona, J.M., Mark, M., Gansmuller, A., LeMeur, M., Decimo, D., Vonesch, J.L., Dollé, P., and Chambon, P., 1994, Genetic analysis of RXRα developmental function: Convergence of RXR and RAR signaling pathways in heart and eye morphogenesis, *Cell* 78:987

Keller, H., Dreyer, C., Medin, J., Mahfoudi, A., Ozato, K., and Wahli, W., 1993, Fatty acids and retinoids control lipid metabolism through activation of peroxisome proliferator-activated receptor- retinoid X receptor heterodimers, *Proc. Natl. Acad. Sci. USA* 90:2160

Kessel, M., and Gruss, P., 1991, Homeotic transformations of murine vertebrae and concomitant alteration of Hox codes induced by retinoic acid, *Cell* 67:89

Kintner, C.R., and Dodd, J., 1991, Hensen's node induces neural tissue in *Xenopus* ectoderm - implications for the action of the organizer in neural induction, *Development* 113:1495

Kliewer, S.A., Umesono, K., Heyman, R.A., Mangelsdorf, D.J., Dyck, J.A., and Evans, R.M., 1992a, Retinoid X receptor-COUP-TF interactions modulate retinoic acid signalling, *Proc. Natl. Acad. Sci. USA* 89:1448

Kliewer, S.A., Umesono, K., Noonan, D.J., Heyman, R.A., and Evans, R.M., 1992b, Convergence of 9-*cis* retinoic acid and peroxisome proliferator signalling pathways through heterodimer formation of their receptors, *Nature* 358:771

Krumlauf, R., 1994, Hox genes in vertebrate development, *Cell* 78:191

Kurokawa, R., Direnzo, J., Boehm, M., Sugarman, J., Gloss, B., Rosenfeld, M.G., Heyman, R.A., and Glass, C.K., 1994, Regulation of retinoid signalling by receptor polarity and allosteric control of ligand binding, *Nature* 371:528

Langston, A.W., and Gudas, L.J., 1992, Identification of a retinoic acid responsive enhancer 3' of the murine homeobox gene Hox-1.6, *Mech. Dev.* 38:217

Lehmann, J.M., Dawson, M.I., Hobbs, P.D., Husmann, M., and Pfahl, M., 1991, Identification of retinoids with nuclear receptor subtype-selective activities, *Cancer Research* 51:4804

Lehmann, J.M., Jong, L., Fanjul, A., Cameron, J.F., Lu, X.P., Haefner, P., Dawson, M.I., and Pfahl, M., 1992, Retinoids selective for retinoid X receptor response pathways, *Science* 258:1944

Leroy, P., and De Robertis, E.M., 1992, Effects of lithium chloride and retinoic acid on the expression of genes from the *Xenopus* laevis Hox 2 complex, *Developmental Dynamics* 194:21

Levin, A.A., Sturzenbecker, L.J., Kazmer, S., Bosakowski, T., Huselton, C., Allenby, G., Speck, J., Kratzeisen, C., Rosenberger, M., Lovey, A., and Grippo, J.F., 1992, 9-*cis*-Retinoic acid stereoisomer binds and activates the nuclear receptor RXRα, *Nature* 355:359

Lohnes, D., Kastner, P., Dierich, A., Mark, M., LeMeur, M., and Chambon, P., 1993, Function of retinoic acid receptor γ in the mouse, *Cell* 73:643

Lohnes, D., Mark, M., Mendelsohn, C., Dollé, P., Dierich, A., Gorry, P., Gansmuller, A., and Chambon, P., 1994, Function of the retinoic acid receptors (RARs) during development (I) Craniofacial and skeletal abnormalities in RAR double mutants, *Development* 120:2723

Lufkin, T., Lohnes, D., Mark, M., Dierich, A., Gorry, P., Gaub, M.P., LeMeur, M., and Chambon, P., 1993, High postnatal lethality and testis degeneration in retinoic acid receptor α mutant mice, *Proc. Natl. Acad. Sci. USA* 90:7225

Maden, M., 1994, Developmental biology - The limb bud .2, *Nature* 371:560

Mangelsdorf, D.J., Umesono, K., Kliewer, S.A., Borgmeyer, U., Ong, E.S., and Evans, R.M., 1991, A direct repeat in the cellular retinol-binding protein type II gene confers differential regulation by RXR and RAR, *Cell* 66:555

Mangelsdorf, D.J., Borgmeyer, U., Heyman, R.A., Zhou, J.Y., Ong, E.S., Oro, A.E., Kakizuka, A., and Evans, R.M., 1992, Characterization of three RXR genes that mediate the action of 9-*cis* retinoic acid, *Genes Dev.* 6:329

Mangelsdorf, D.J., Umesono, K., and Evans, R.M., 1994, The retinoid receptors, *in*:"The Retinoids: Biology, Chemistry and Medicine," M.B. Sporn, A.B. Roberts, D.S. Goodman, eds., Raven Press Ltd., New York.

Mangold, O., 1933, Über die Induktionsfähigkeit der verschiedenen Bezirke der Neurula von Urodelen, *Naturwissenschaften* 21:761

Marklew, S., Smith, D.P., Mason, C.S., and Old, R.W., 1994, Isolation of a novel RXR from *Xenopus* that most closely resembles mammalian RXR beta and is expressed throughout early development, *Biochem. Biophys. Acta* 1218:267

Marshall, H., Nonchev, S., Sham, M.H., Muchamore, I., Lumsden, A., and Krumlauf, R., 1992, Retinoic acid alters hindbrain *Hox* code and induces transformation of rhombomeres 2/3 into a 4/5 identity, *Nature* 360:737

Marshall, H., Studer, M., Pöpperl, H., Aparicio, S., Kuroiwa, A., Brenner, S., and Krumlauf, R., 1994, A conserved retinoic acid response element required for early expression of the homeobox gene *Hoxb-1*, *Nature* 370:567

McGinnis, W., and Krumlauf, R., 1992, Homeobox genes and axial patterning, *Cell* 68:283

Mendelsohn, C., Larkin, S., Mark, M., LeMeur, M., Clifford, J., Zelent, A., and Chambon, P., 1994a, RARβ isoforms: distinct transcriptional control by retinoic acid and specific spatial patterns of promoter activity during mouse embryonic development, *Mech. Dev.* 45:227

Mendelsohn, C., Lohnes, D., Décimo, D., Lufkin, T., LeMeur, M., Chambon, P., and Mark, M., 1994b, Function of the retinoic acid receptors (RARs) during development (II) Multiple abnormalities at various stages of organogenesis in RAR double mutants, *Development* 120:2749

Mendelsohn, C., Mark, M., Dollé, P., Dierich, A., Gaub, M.P., Krust, A., Lampron, C., and Chambon, P., 1994c, Retinoic acid receptor β2 (RARβ2) null mutant mice appear normal, *Dev. Biol.* 166:246

Moroni, M.C., Viganó, M.A., and Mavilio, F., 1993, Regulation of the human HOXD4 gene by retinoids, *Mech. Dev.* 44:139

Nieuwkoop, P.D., Boterenbrood, E.C., Kremer, A., Bloemsma, F.F.S.N., Hoessels, E.L.M.J., Meyer, G., and Verheyen, F.J., 1952, Activation and organization of the central nervous system in amphibians, *J. Exp. Zool.* 120:1

Papalopulu, N., Clarke, J.D.W., Bradley, L., Wilkinson, D., Krumlauf, R., and Holder, N., 1991, Retinoic acid causes abnormal development and segmental patterning of the anterior hindbrain in *Xenopus* embryos, *Development* 113:1145

Petkovich, M., Brand, N.J., Krust, A., and Chambon, P., 1987, A human retinoic acid receptor which belongs to the family of nuclear receptors, *Nature* 330:444

Pfeffer, P.L., and De Robertis, E.M., 1994, Regional specificity of RARγ isoforms in *Xenopus* development, *Mech. Dev.* 45:147

Pijnappel, W.W.M., Hendriks, H.F.J., Folkers, G.E., van den Brink, C.E., Dekker, E.J., Edelenbosch, C., van der Saag, P.T., and Durston, A.J., 1993, The retinoid ligand 4-oxo-retinoic acid is a highly active modulator of positional specification, *Nature* 366:340

Pöpperl, H., and Featherstone, M.S., 1993, Identification of a retinoic acid response element upstream of the murine *Hox-4.2* gene, *Mol. Cell. Biol.* 13:257

Ruberte, E., Dolle, P., Krust, A., Zelent, A., Morriss-Kay, G., and Chambon, P., 1990, Specific spatial and temporal distribution of retinoic acid receptor gamma transcripts during mouse embryogenesis, *Development* 108:213

Ruberte, E., Dollé, P., Chambon, P., and Morriss-Kay, G., 1991, Retinoic acid receptors and cellular retinoid binding proteins. II. Their differential pattern of transcription during early morphogenesis in mouse embryos, *Development* 111:45

Ruberte, E., Friederich, V., Chambon, P., and Morriss-Kay, G., 1993, Retinoic acid receptors and cellular retinoid binding proteins. III. Their differential transcript distribution during mouse nervous system development, *Development* 118:267

Ruiz i Altaba, A., and Jessell, T., 1991a, Retinoic acid modifies mesodermal patterning in early *Xenopus* embryos, *Genes Dev.* 5:175

Ruiz i Altaba, A., and Jessell, T.M., 1991b, Retinoic acid modifies the pattern of cell differentiation in the central nervous system of neurula stage *Xenopus laevis* embryos, *Development* 112:945

Schräder, M., Bendik, I., Becker-André, M., and Carlberg, C., 1993, Interaction between retinoic acid and vitamin D signaling pathways, *J. Biol. Chem.* 268:17830

Schüle, R., Rangarajan, P., Yang, N., Kliewer, S., Ransone, L.J., Bolado, J., Verma, I.M., and Evans, R.M., 1991, Retinoic acid is a negative regulator of AP-1 responsive genes, *Proc. Natl. Acad. Sci. USA* 88:6092

Schuh, T.J., Hall, B.L., Creech Kraft, J., Privalsky, M.L., and Kimelman, D., 1993, v-erbA and citral reduce the teratogenic effects of all-*trans* retinoic acid and retinol, respectively, in *Xenopus* embryogenesis, *Development* 119:785

Sharpe, C.R., 1992, Two isoforms of retinoic acid receptor α expressed during *Xenopus* development respond to retinoic acid, *Mech. Dev.* 39:81

Simeone, A., Acampora, D., Arcioni, L., Andrews, P.W., Boncinelli, E., and Mavilio, F., 1990, Sequential activation of HOX2 homeobox genes by retinoic acid in human embryonal carcinoma cells, *Nature* 346:763

Simeone, A., Acampora, D., Nigro, V., Faiella, A., D'Esposito, M., Stornaiuolo, A., Mavilio, F., and Boncinelli, E., 1991, Differential regulation by retinoic acid of the homeobox genes of the four hox loci in human embryonal carcinoma cells, *Mech. Dev.* 33:215

Sive, H.L., and Cheng, P.F., 1991, Retinoic acid perturbs the expression of *Xhox.lab* genes and alters mesodermal determination in *Xenopus laevis*, *Genes Dev.* 5:1321

Sive, H.L., Draper, B.W., Harland, R.M., and Weintraub, H., 1990, Identification of a retinoic acid-sensitive period during primary axis formation in *Xenopus laevis*, *Genes Dev.* 4:932

Smith, S.M., and Eichele, G., 1991, Temporal and regional differences in the expression pattern of distinct retinoic acid receptor-β transcripts in the chick embryo, *Development* 111:245

Smith, D.P., Mason, C.S., Jones, E.A., and Old, R.W., 1994, Expression of a dominant negative retinoic acid receptor γ in *Xenopus* leads to partial resistance to retinoic acid, *Roux's Arch. Dev. Biol.* 203:254

Spemann, H., 1931, Über den Anteil von Implantat und Wirtskeim an der Orientierung und Beschaffenheit der induzierten Embryonalanlage, *Roux's Arch. Entw. mech. Org.* 123:389

Spemann, H., and Mangold, H., 1924, Über Induktion von Embryonalanlagen durch Implantation artfremder Organisatoren, *Roux's Arch. Entw. mech. Org.* 100:599

Studer, M., Pöpperl, H., Marshall, H., Kuroiwa, A., and Krumlauf, R., 1994, Role of a conserved retinoic acid response element in rhombomere restriction of *Hoxb-1*, *Science* 265:1728

Surekha Rao, M.S., John, J., and Cama, H.R., 1972, Studies on vitamin A2: Preparations, properties, metabolism and biological activity of 4-oxoretinoic acid, *Int. J. Vit. Nutr. Res.* 42:368

Thaller, C., and Eichele, G., 1987, Identification and spatial distribution of retinoids in the developing chick limb bud, *Nature* 327:625

Thaller, C., and Eichele, G., 1990, Isolation of 3,4-didehydroretinoic acid, a novel morphogenetic signal in the chick wing bud, *Nature* 345:815

Tickle, C., Alberts, B., and Lee, J., 1982, Local application of retinoic acid to the limb bud mimics the action of the polarizing region, *Nature* 296:564

Toivonen, S., and Saxén, L., 1968, Morphogenetic interactions of presumptive neural and mesodermal cells mixed in different ratios, *Science* 159:539

Tran, P., Zhang, X.-K., Salbert, G., Hermann, T., Lehmann, J.M., and Pfahl, M., 1992, COUP orphan receptors are negative regulators of retinoic acid response pathways, *Mol. Cell. Biol.* 12:4666

Wagner, M., Han, B., and Jessell, T.M., 1992, Regional differences in retinoid release from embryonic neural tissue detected by an in vitro reporter assay, *Development* 116:55

White, J.A., Boffa, M.B., Jones, B., and Petkovich, M., 1994, A zebrafish retinoic acid receptor expressed in the regenerating caudal fin, *Development* 120:1861

Widom, R.L., Rhee, M., and Karathanasis, S.K., 1992, Repression by ARP-1 sensitizes Apolipoprotein AI gene responsiveness to RXRα and retinoic acid, *Mol. Cell. Biol.* 12:3380

Wilkinson, D.G., Bhatt, S., Cook, M., Boncinelli, E., and Krumlauf, R., 1989, Segmental expression of Hox-2 homoeobox-containing genes in the developing mouse hindbrain, *Nature* 341:405

Williams, J.B., Shields, C.O., Brettel, L.M., and Napoli, J.L., 1987, Assessment of retinoid induced differentiation of F9 embryonal carcinoma cells with an enzyme linked immunoadsorbent assay for laminin: statistical comparison of dose response curves, *J. Anal. Biochem* 160:267

Wood, H., Pall, G., and Morriss-Kay, G., 1994, Exposure to retinoic acid before or after the onset of somitogenesis reveals separate effects on rhombomeric segmentation and 3' *HoxB* gene expression domains, *Development* 120:2279

Zhang, X.-K., Lehmann, J., Hoffmann, B., Dawson, M.I., Cameron, J., Graupner, G., Hermann, T., Tran, P., and Pfahl, M., 1992, Homodimer formation of retinoid X receptor induced by 9-*cis* retinoic acid, *Nature* 358:587

THE Pax FAMILY OF TRANSCRIPTION FACTORS IN EMBRYONIC DEVELOPMENT

Georges Chalepakis and Peter Gruss

Max-Planck institute for Biophysical Chemistry
Department of Molecular Cell Biology
D-37077 Gottingen, Germany

The Pax gene family consists of nine genes identified in the mouse genome by their homology to the **Drosophila** segmentation gene, paired. The Pax gene products have the biochemical properties of transcription factors and share the conserved paired domain that exhibits DNA binding activity. Pax genes are expressed in temporally and spatially restricted patterns during embryogenesis and control key aspects of embryonic development. Pax gene mutations are semidominantly inherited, and homozygous mouse mutants are usually lethal.

Pax PROTEIN STRUCTURE

All Pax proteins share a conserved paired domain which is 128 amino acids in length and represents the primary DNA binding motif of the Pax proteins (Burri et al., 1989). The paired domain is located at the amino-terminal end of and is the longest conserved structural motif of the Pax proteins whose total lenght varies between 361 (Pax1) and 479 (Pax3) amino acids (Fig. 1). Carboxy-terminal to the paired domain, an octapeptide sequence of unknown function is also conserved among the Pax proteins except Pax4 and Pax6. Pax3, Pax4, Pax6 and Pax7 also contain a paired-type homeodomain carboxy-terminal to the octapeptide. Pax2, Pax5 and Pax8 contain only part of the homeodomain including the first α-helix (Fig. 1). The carboxy-terminal region is rich in prolines, serines and threonines and mediates the transcriptional activity of these proteins (Glaser et al., 1994; Chalepakis et al., 1994c). Pax

Organization of the Early Vertebrate Embryo
Edited by N. Zagris *et al.*, Plenum Press, New York, 1995

GENE	CHROMO-SOME	STRUCTURE		PROTEIN	MUTANTS	
		PD	HD		MOUSE	HUMAN
Pax-1	2	N[////] o ────────]C		361 aa	undulated (un)	
Pax-2	19	N[/ // o] ────]C		392 / 415 aa		
Pax-3	1	N[/ // o ▮▮▮]C		479 aa	splotch (Sp)	Waardenburg Syndrome I
Pax-4	6	N[▓▓ // ▮▮▮]C		?		
Pax-5	4	N[/ // o]]C		391 aa		
Pax-6	2	N[▓▓ // ▮▮▮]C		422 / 436 aa	small eye (Sey)	Aniridia
Pax-7	4	N[/ // o ▮▮▮]C		>300 aa		
Pax-8	2	N[/ // o]]C		457 aa		
Pax-9	12	N[//// o ────]C		?		

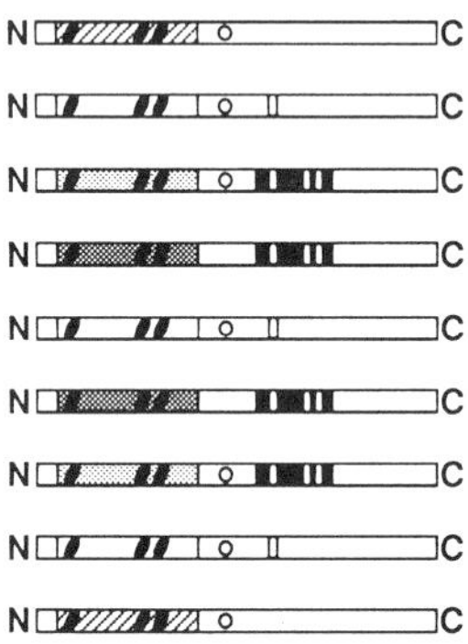

Figure 1. The Pax gene family. The paired domain (PD) and homeodomain (HD) are shown as boxes. α-helices in the paired domain are indicated by black ovals and in the homeodomain by white ovals. The presence of the conserved first α-helix of the homeodomain in Pax2, Pax5 and Pax8 is indicated. The octapeptide (when present) is indicated by a circle carboxy-terminal to the paired domain. Paired domains which are represented with the same hatching belong to the same Pax subfamily.

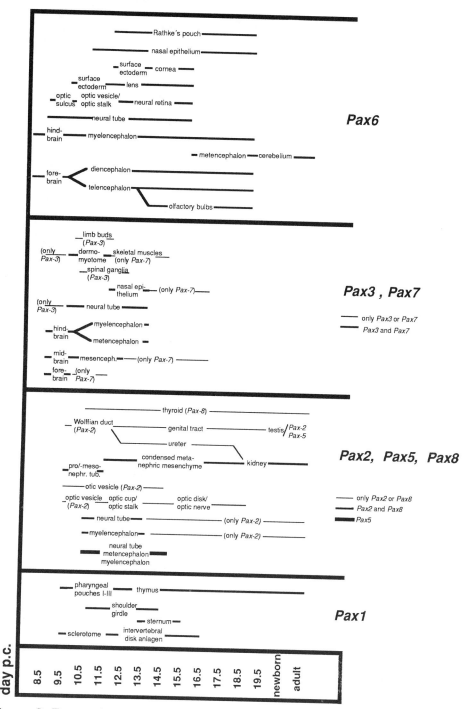

Figure 2. Expression domains of Pax genes from data which are so far available.

genes can be classified into four subfamilies according to their genomic organisation (similar intron-exon boundaries) and the paired domain sequences (Walther et al., 1991). The paralogous groups include Pax1 and Pax9; Pax2, Pax5 and Pax8; Pax3 and Pax7; Pax4 and Pax6.

EXPRESSION PATTERN OF Pax GENES

All Pax genes except Pax1 and Pax9 are expressed in the developing nervous system. The onset of expression is detectable between days 8 and 9.5 of embryogenesis and expression precedes along the entire antero-posterior axis in a tissue specific manner (Fig. 2).

Pax1

Expression of Pax1 starts at day-9 p.c. in the sclerotomal part of the differentiating somites (Deutsch et al., 1988). Later on, Pax1 expression is confined to the intervertebral disk anlagen. Expression has been also detected in the sternum and in the thymus.

Pax2, Pax5 and Pax8

Pax2 (Nornes et al., 1990; Dressler et al., 1990) and Pax8 (Plachov et al., 1990) transcripts have been first identified in the neural tube and in the hindbrain. Expression in the neural tube is confined to a specific subset of postmitotic differentiating cells on both sides of the dorso-ventral midine, the sulcus limitans. Pax8 transcripts are no longer detectable after day-13.5 p.c., whereas Pax2 continues to be expressed throughout development and in adult tissues. Around day-10.5 p.c. both genes are expressed in the developing kidney. Furthermore, Pax8 expression is detected in the thyroid from the earliest developmental stages to the adult stage and Pax2 transcripts were also found in the developing ear and eye.

Pax5 is expressed in the neural tube and in the developing brain, predominantly in the midbrain-hindbrain boundary. In addition, Pax5 is also expressed in the pre-B, pro-B and mature B-cells, as well as in the testis (Adams et al., 1992; Asano et al., 1992).

Pax3 and Pax7

Both genes exhibit a similar expression pattern during development. At day-8.5 p.c. Pax3 is expressed in the dorsal aspect of the neural tube, including the roof plate and neural crest cells and their derivatives (Goulding et al., 1991). Pax7 is expressed similarly to Pax3, but transcripts are not detectable in the rood plate and in the neural crest cells (Jostes et al., 1990). Both genes are also detected in the dermomyotome part of the developing somites. Accordingly, Pax3 expression has also been detected in the myogenic precursor cells which migrate from myotomes into the limb buds to form muscles (Fig. 2).

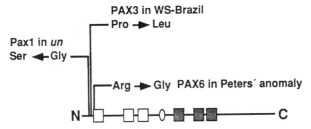

Figure 3. Shown is the position of point mutations around the first α-helix of the paired domain which were identified in Pax1 in the undulated mouse mutant (un), or in the human PAX3 in a Brazilian family with Waardenburg syndrome, or in the human PAX6 gene in patients with Peter's anomaly. In the lower part are shown the three α-helices of the paired domain (amino-terminal) or the homeodamain (carboxy-terminal) in a linear representation of a Pax protein.

Pax6

Pax6 is almost exclusively expressed in the developing central nervous system (Walther and Gruss, 1991). Expression in the neural tube is restricted to the ventral ventricular zone. In the brain, Pax6 transcripts are first detected around day-8.5 p.c. with a gap in expression in the roof of the mesencephalon and metencephalon.

Expression is also detected in the nose and in all developing structures in the eye starting from day-8.5 p.c.

Pax MUTANTS

One of the most interesting features of the Pax family is that mutations in some of the Pax genes have been correlated with certain mouse mutant phenotypes and human diseases (for reviews see: Deutsch and Gruss 1991; Gruss and Walther 1992; Chalepakis et al., 1992; Chalepakis et al., 1993; Noll, 1993; Fritsch and Gruss, 1993; Strachan and Read, 1994; Stuart et al., 1994).

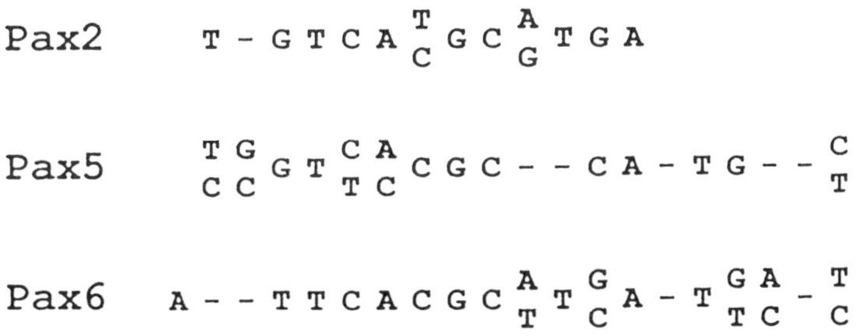

$$\text{Pax2} \quad \text{T - G T C A }^{\text{T}}_{\text{C}}\text{ G C }^{\text{A}}_{\text{G}}\text{ T G A}$$

$$\text{Pax5} \quad ^{\text{T G}}_{\text{C C}}\text{ G T }^{\text{C A}}_{\text{T C}}\text{ C G C - - C A - T G - - }^{\text{C}}_{\text{T}}$$

$$\text{Pax6} \quad \text{A - - T T C A C G C }^{\text{A}}_{\text{T}}\text{ T }^{\text{G}}_{\text{C}}\text{ A - T }^{\text{G A}}_{\text{T C}}\text{ - }^{\text{T}}_{\text{C}}$$

Figure 4. Consensus DNA binding sequences for paired domains.

Pax1

Mutations in the Pax1 gene cause the mouse undulated (un) phenotype. Undulated is characterised by a kinky-tail (Balling et al., 1988; Balling et al., 1992) and malformations are detected in the sternum and along the vertebral axis where the intervertebral disks are larger and the vertebral centres are reduced in size.

Pax3

Alterations in the Pax3 locus have been associated with the mouse splotch (sp) phenotype (Epstein et al., 1991a, 1991b, 1993; Goulding et al., 1993) and the human Waardenburg syndrome type I (WSI) (Tassabehji et al., 1992, 1993; Baldwin et al., 1992; Morell et al., 1992; Tsukamoto et al., 1992). Heterozygous splotch mutants display pigmentation defects which are visible as white spotting on the belly, the back, the tail and the limbs. Homozygous splotch mice are embryonic lethal and embryos die around day-14 p.c. Developmental defects include spina bifida, exencephaly, neural tissue overgrowth, deficiency in the development of certain skeletal muscles and neural crest cell- associated deficiency. Waardenburg syndrome is a hereditary disease which occurs in around 1 to 2 of 100,000 individuals. Phenotypic features of the Waardenburg syndrome include deafness, pigmentation disturbances (heterochromia irides, white forelock and eyelash), and dystopia canthorum (lateral displacement of the inner eye corner).

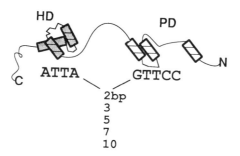

Figure 5. Interaction of Pax3 with the ATTA motif (through the homeodomain) and the GTTCC sequence (through the paired domain) of the e5 binding site. The distance between the two DNA binding submotifs can vary between 2 and 10 base pairs without affecting significantly the binding of Pax3.

Pax6

The small eye phenotype in mouse and the human aniridia syndrome result from mutations in the Pax6 gene (Hill et al., 1991; Ton et al., 1991; Jordan et al., 1992; Glaser et al., 1992).

Heterozygous small eye animals show reduced body size, small eyes, cataracts, abnormal folding of the retina and reduction of the pigment layer. Homozygous small eye mice are characterised by distortion and degeneration of the optic vesicle and absence of development of the nasal cavities and of the olfactory bulbs. By the time of birth, animals have neither eye nor nasal tissue and die soon because they cannot breathe. Aniridia is a congenital, panocular disorder which occurs in around 1 of 100,000 individuals of the human population. It is characterised by the complete or partial absence of the iris and also affects the cornea, lens, retina and optic nerve.

DNA BINDING OF Pax PROTEINS

Computer structure analysis of the paired domain suggests the presence of three α- helices. THe first α-helix is located around position 23-31 in the paired domain, the second at position 80-89, and the third at position 94-106. A detailed structure-functional analysis of the Pax5 paired domain revealed that it is composed of two subdomains which interact independently with the DNA (Czerny et al., 1993). One subdomain includes the first α-helix and the other includes the second and the third α-helices of the paired domain. The

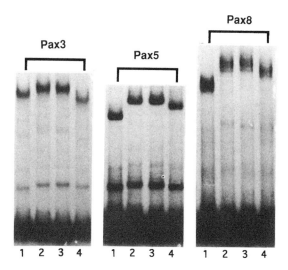

Figure 6. Gel shift analysis of circularly permuted DNA fragments. The four DNA fragments (numbered in the lower part of the figure) have the same length and nucleotide composition but contain the Pax binding site at different positions within the fragments. The different migration of the protein -DNA complexes (upper part of the gel shift) is indicative for different conformations of the protein-bound DNA fragments. The conformational changes are induced upon binding of the Pax protein. The protein-free DNA fragments (lower part of the gel shift) are running equally.

importance of the protein region around the first α-helix for the interaction with DNA can be evaluated from the localisation of individual point mutations within this region which were found in Pax1, Pax3 and Pax6 and are responsible for the phenotypes of undulated (Balling et al., 1988), Waardenburg syndrome I found in a Brazilian family (Baldwin et al., 1992) and Pater's anomaly (Glaser et al., 1994) respectively (Fig. 3). The above Pax1 (Chalepakis et al., 1991) and Pax3 (Chalepakis et al., 1994a) mutations have been analysed and were found to diminish the DNA binding activity of the paired domain. Recent reports describe the DNA binding specificities of different paired domains. Consensus binding sites for the Pax2 and Pax6 paired domains were deduced from in vitro selection of randomly synthesised oligonucleotide sequences and subsequent amplification of the selected sequences (Epstein et al., 1994). A consensus binding sequence for Pax5 was derived from the binding site in the CD19 promoter by introducing various mutations and subsequent alignment (Czerny et al., 1993). The consensus binding sites are listed in Fig. 4.

Pax3, which contains both a paired domain and a paired type homeodomain, binds to the so called e5 sequence found in the promoter of the **Drosophila** even-skipped gene. This sequence consists of an ATTA motif, recognized from the Pax3 homeodomain, and of a GTTCC site recognized from the paired domain (Goulding et al., 1991; Chalepakis et al., 1994). Efficient binding of Pax3 to this sequence requires the presence of both motifs. Mutation of one of these sites dramatically reduces its DNA binding affinity. Interestingly, the distance between the two DNA binding submotifs can vary from 2 to 10 base pairs

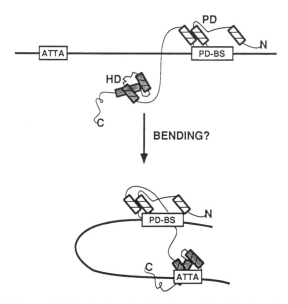

Figure 7. Hypothetical model illustrating the binding of the Pax3 paired domain with the appropriate DNA binding site (PD-BS), the induction of DNA conformational changes in terms of DNA bending, the interaction of the homeodomain with an ATTA recognition sequence and the subsequent stabilisation of a DNA loop structure.

(Chalepakis et al., 1994b) and Pax3 is still able to interact efficiently with these sequences (Fig. 5). Furthermore, the protein spacing between the paired domain and the homeodomain can be shortenerd by 45 amino acids without changing the flexibility of the truncated Pax3 protein for interaction with the various spacing derivatives shown in Fig. 5 (Chalepakis et al., 1994b).

Recently it has been shown that binding of paired domain proteins can change the conformation of the protien-bound DNA (Chalepakis et al., 1994b). These conformational changes can be interpreted in terms of paired domain induced DNA-bending (Fig. 6). Thus, combining the findings that the distance between the homeodomain (e.g. ATTA) and the paired domain (e.g. GTTCC) recognition sequences can vary considerably without affecting the nature of Pax3-DNA interaction, and that binding of the paired domain could bend the DNA, we propose the following model for the action of Pax3 (Fig. 7). In the first step Pax3 interacts with its cognate suquence and bends the DNA; in a second step a homeodomain lying far upstream approaches the interacting angle of the Pax3 homeodomain which then binds to this sequence thereby stabilising a DNA loop structure.

Acknowledgement

The authors are greatly indebted to H. Boger for critical comments on the manuscript.

REFERENCES

Adams, B., Dorfler, P., Aguzzi, A., Kozmik, Z., Urbanek, P., Maurer-Fogy, I., and Busslinger, M. (1992) Pax-5 encodes the transcription factor BSAP and is expressed in B lymphocytes, the developing CNS, and adult testis. Genes Dev. **6**: 1889-1607.

Asano, M., and Gruss, P. (1992). Pax-5 is expressed at the midbrain-hindbrain boundary during mouse development. Mech. Dev. **39**: 29-39.

Baldwin, C.T., Hoth, C.F., Amos, J.A., da-Silva, E.O., and Milunsky, A. (1992) An exonic mutation in the HuP2 paired domain gene causes Waardenburg's syndrome. Nature **355**: 637-638.

Balling, R., Deutsch, U., and Gruss, P. (1988) **Undulated**, a mutation affecting the development of the mouse skeleton, has a point mutation in the paired box of Pax1. Cell **55**: 531-535.

Balling, R., Lau, C.F., Dietrich, S., Wallin, J., and Gruss, P. (1992) Development of the skeletal system. Ciba Foundation Symposium **165**: 132-140.

Burri, M., Tromvoukis, Y., Bopp, D., Frigerio, G., and Noll, M. (1989) Conservation of the paired domain in metazoans and its structure in three isolated human genes. EMBO J. **8**: 1183-1190.

Chalepakis, G., Fritsch, R., Fickenscher, H., Deutsch, U., Goulding, M., and Gruss, P. (1991) The molecular basis of the **undulated**/Pax-1 mutation. Cell **66**: 873-884.

Chalepakis, G., Tremblay, P., and Gruss, P. (1992) Pax genes, mutants and molecular function. J.Cell Sci. **16** (Suppl.): 61-67.

Chalepakis, G., Stoykova, A., Wijnholds J., Tremblay, P., and Gruss, P. (1993) Pax: gene regulators in the developing nervous system. J.Neurobiol. **24**: 1367-1384.

Chalepakis, G., Goulding, M., Read, A., Strachan, T., and Gruss, P. (1994a). Molecular basis of splotch and Waardenburg Pax-3 mutations. Proc. Natl. Acad. Sci. USA **91**: 3685-3689.

Chalepakis, G., Wijnholds J., and Gruss, P. (1994b) Pax-3-DNA interaction: flexibility in the DNA binding and induction of DNA conformational changes by paired domains. Nucleic Acids Res. 22: 3131-3137.

Chalepakis, G., Jones, F.S., Edelman, G.M., and Gruss, P. (1994c) Pax-3 contains domains for transcription activation and transcription inhibition. Proc. Natl. Acad. Sci. USA, in press.

Czerny, T., Schaffner, G., and Busslinger, M. (1993) DNA-sequence recognition by Pax proteins-Bipartite structure domain and its binding-site. Genes Dev. 7: 2048-2061.

Deutsch, U., Dressler, G.R., and Gruss, P. (1988) Pax1, a member of a paired box homologous murine gene family, is expressed in segmented structures during development. Cell 53: 617-625.

Deutsch. U., and Gruss, P. (1991) Murine paired domain proteins as regulatory factors of embryonic development. Sem. Dev. Biol. 2: 413-424.

Dressler, G.R., Deutsch, U., Chowdhury, K., Nornes, H.O., and Gruss, P. (1990) Pax2, a new murine paired box containing gene and its expression in the developing excretory system. Development 109: 787-795.

Epstein, D.J., Vekemans, M., and Gros P. (1991a) Splotch (Sp^{2H}), a mutation affecting development of the mouse neural tube, shows a deletion within the paired homeodomain of Pax-3. Cell 67: 767-774.

Epstein, D.J., Malo, D. Vekemans, M., and Gros, P. (1991b) Molecular characterization of a deletion encompassing the Splotch mutation on chromosome 1. Genomics 10: 89-93.

Epstein, D.J., Vogan, K.J., Trasler, D.G., and Gros, P. (1993) A mutation within intron- 3 of the Pax-3 gene produces aberrantly spliced messenger-RNA transcripts in the Splotch (Sp) mouse mutant. Proc. Natl. Acad. Sci. USA 90: 532-536.

Epstein, J., Cai, J., Glaser, T., Jepeal, L., and Maas, R. (1994) Identification of a Pax paired domain recognition sequence and evidence for DNA-dependent conformational changes. J.Biol. Chem. 269: 8355-8361.

Fritsch, R. and Gruss, P. (1993) In Robertson, E., Maxfield, F.R. and Vogel, H.J. (ed), Cell-cell signaling in vertebrate development. Academic press, New York (in press). Foundation Symposium, 165: 132-140.

Glaser, T., Walton, D.S., and Maas, R.L. (1992) Genomic structure, evolutionary conservation and aniridia mutations in the human PAX6 gene. Nature Genet. 2:232-239.

Glaser, T., Jepeal, L., Edwards, J. G., Young, S.R., Favor, J., and Maas, R.L. (1994) Pax6 gene dosage effect in a family with congenital cataracts, aniridia, anophthalmia and central nervous system defects. Nature Genet. 7: 463-471.

Goulding, M.D., Chalepakis, G., Deutsch, U., Erselius, J., and Gruss, P. (1991) Pax-3, a novel murine DNA binding protein expressed during early neurogenesis. EMBO J. 10: 1135-1147.

Goulding, M.D., Sterrer, S., Fleming, J., Balling, R., Nadeau, J., Moore, K.J., Brown, S.D.M., Steel, K.P., and Gruss, P. (1993) Analysis of the Pax-3 gene in the mouse mutant Splotch. Genomics 17: 355-363.

Gruss, P., and Walther, C. (1992) Pax in development. Cell 69: 719-722.

Hill, R.E., Favor, J., Hogan, B.L.M., Ton, C.C.T., Saunders, G.F., Hanson, J.M., Prosser, J., Jordan, T., Hastie, N.D., and van Hyningen, V. (1991) Mouse Small eye results from mutations in a paired-like homeobox-containing gene. Nature 354: 522-525.

Jordan, T., Hanson, I., Zaletayev, D., Hodgson, S., Prosser. J., Seawright, A., Hastie, N., and van Heyningen, V. (1992) The human PAX6 gene is mutated in two patients with aniridia. Nature Genet. 1: 328-332.

Jostes, B., Walther, C., and Gruss, P. (1990) The murine paired box gene, Pax7, is expressed specifically during the development of the nervous and muscular system. Mechanisms of Development 33: 27-38.

Morell, R., Friedman, T.B., Moeljopawiro, S., Hartono, Soewito, and Asher, Jr. J.H. (1992) A Frameshift mutation in the HuP2 paired domain of the probable human homolog of murine **Pax-3** is responsible for Waardenburg syndrome type 1 in an indonesian family. Hum. Mol. Genet. **1**: 243-247.

Noll, M., (1993) Evolution and role of Pax genes. Curr. Opin. Genet. Dev. **3**: 595-605.

Nornes, H.O., Dressler, G.R., Knapik, E.W., Deutsch, U., and Gruss, P. (1990) Spatially and temporally restricted expression of **Pax-2** during murine embryogenesis. Development **109**: 797-809.

Plachov, D., Chowdhury, K., Walther, C., Simon, D., Guenet, J-L., and Gruss, P. (1990) Pax8, a murine paired box gene expressed in the developing excretory system and thyroid gland. Development **110**: 643-651.

Strachan, T., and Read, A.P. (1994) PAX genes, Curr. Opin. Genet. Dev. **4**: 427-438.

Stuart, E.T., Kioussi, C., and Gruss, P. (1994) Mammalian Pax genes. Annu. Rev. Genet. **27**: 219-236.

Tassabehji, M., Read, A.P., Newton, V.E., Harris, R., Balling, R., Gruss, P., and Strachan, T. (1992a) Waardenburg's syndrome patients have mutations in the human homologue of the **Pax-3** paired box gene. Nature **355**: 635-636.

Tassabehji, M., Read, A.P., Newton, V.E., Patton, M., Gruss, P., Harris, R., and Strachan, T. (1993) Mutations in the **PAX3** gene causing Waardenburg syndrome Type 1 and Type 2. Nature Genet. **3**: 36-30.

Ton, C.C.T., Hirvonen, H., Miwa, H., Weil, M.M., Monaghan, P., Jordan, T., van Heyningen, V., Hastie, N.D., Meijers-Heijboer, H., Drechsler, M., Royer-Pokora, B., Collins, F., Swaroop, A., Strong, L.C., and Saunders, G.F. (1991) Positional cloning of a paired box- and homeobox-containing gene from the Aniridia region. Cell **67**: 1059-1074.

Tsukamoto, K., Tohma, T., Ohta, T., Yamakawa, K., Fukushima, Y., Nakamura, Y., and Niikawa, N. (1992) Cloning and characterization of the inversion breakpoint at chromosome 2q35 in a patient with Waardenburg syndrome type I. Hum. Mol. Genet. **1**: 315-317.

Walther, C., Guenet, J. -L., Simon, D., Deutsch, U., Jostes, B., Goulding, M.D., Plachov, D., Balling, R., and Gruss, P. (1991) Pax: A murine multigene family of paired box containing genes. Genomics **11**: 424-434.

Walther, C., and Gruss, P. (1991) **Pax-6**, a murine paired box gene, is expressed in the developing CNS. Development **113**: 1435-1449.

HOMEOBOX GENES IN THE DEVELOPING HEAD OF VERTEBRATES

Edoardo Boncinelli, Massimo Gulisano, Maria Pannese
and Maria Grazia Giribaldi

DIBIT, San Raffaele Scientific Institute, Via Olgettina
60, 20132 Milano, Italy

INTRODUCTION

The study of vertebrate homologues of regulatory genes operating in the trunk of *Drosophila* has provided invaluable information about the genetic control of positional values in development. Many of these genes contain a homeobox and it is now well established that many homeobox genes control cell identity in specific regions or segments both in invertebrates and vertebrates. *Hox* genes (McGinnins and Krumlauf, 1992) stand out among the various homeobox gene families so far identified as the vertebrate cognates of *Drosophila* homeotic genes. They control the specification of body regions along the vertebrate axis and provide positional cues for the developing neural tube from the branchial area to the tail.

Despite this progress, the development of the anteriormost body domain, the head or better the anterior head, has remained relatively obscure in invertebrates and vertebrates alike (Finkelstein and Boncinelli, 1994). In the insect embryo, the nature of anterior head segmentation has been controversial, and the genes that govern it mostly unknown. In vertebrates, the very existence of compartments or segments in the forebrain and midbrain has been contested, and underlying molecular mechanisms of pattern formation undetermined.

A recent breakthrough has come with the identification of genes in *Drosophila* and their homologues in vertebrates that appear critical to anterior head and brain specification. We will focus here primarily on four vertebrate homologues of two of these genes, called *empty spiracles* (*ems*) and *orthodenticle* (*otd*) in the fruitfly (Finkelstein and Perrimon, 1991; Cohen and Jürgens, 1991 for reviews). These four genes are *Emx1* and *Emx2* (Simeone et al., 1992a,b), related to *ems*, and *Otx1* and *Otx2* (Simeone et al., 1992a, 1993; Finkelstein and Boncinelli, 1994), related to *otd*. The two *Otx* genes code for homeoproteins containing a homeodomain of the *bicoid* class. Homeodomains of this class contain a characteristic lysine residue at position 50, corresponding to position 9 of the recognition helix. The restricted family of homeobox genes encoding a homeoprotein of the *bicoid* class only include so far *bicoid* itself, *orthodenticle*, the two vertebrate *Otx* genes and *goosecoid* (*gsc*), a regulatory gene originally isolated in *Xenopus* where it has been suggested to play a role in executing Spemann's organizer phenomenon (Cho et al., 1991).

The four *Emx* and *Otx* vertebrate genes are expressed in extended regions of the developing rostral brain of mouse midgestation embryos, including the presumptive cerebral cortex and olfactory bulbs. Here we summarize expression data and discuss a possible role of the four genes in establishing the boundaries of the various embryonic brain regions. We will focus on three main aspects of these patterns: expression of the four genes in E10 mouse embryos; expression of *Otx* genes in midgestation mouse diencephalon; expression

Organization of the Early Vertebrate Embryo
Edited by N. Zagris *et al.*, Plenum Press, New York, 1995

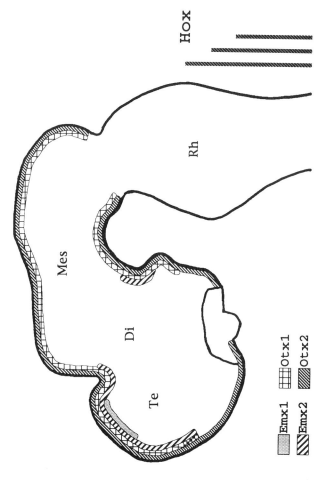

Figure 1. Summary of the expression domains of *Emx* and *Otx* genes in the developing central nervous system of E10 mouse embryos. Expression of members of the *Hox* gene family is also indicated. Di, diencephalon; Mes, mesencephalon; Rh, rhomboencephalon; Te, telencephalon.

of *Otx2* in early mouse, *Xenopus* and chick embryos. Perhaps the most interesting feature of this story is what it promises to reveal about the development and structure of the anterior head.

EXPRESSION OF THE FOUR GENES IN E10 MOUSE EMBRYOS

At day 10 of development (E10) the developing neural tube of the mouse shows recognisable presumptive regions corresponding to the future anatomical subdivisions. The entire neural tube consists of neuroepithelial cells in active proliferation and most of the specific differentiative events have not yet occurred. In E10 mouse embryos all four genes are expressed. Their expression domains (Simeone et al., 1992a) are continuous regions of the developing brain contained within each other in the sequence *Emx1<Emx2<Otx1<Otx2* (Fig. 1). The *Emx1* expression domain includes the dorsal telencephalon with a posterior boundary slightly anterior to that between presumptive diencephalon and telencephalon. *Emx2* is expressed in dorsal and ventral neurectoderm with an anterior boundary slightly anterior to that of *Emx1* and a posterior boundary within the roof of presumptive diencephalon. This boundary most probably coincides with the boundary between first and second thalamic segment (Kuhlenbeck, 1973) which will subsequently give rise to ventral thalamus and dorsal thalamus, respectively (see also below). The *Otx1* expression domain contains the *Emx2* domain. It covers a continuous region including part of the telencephalon, the diencephalon and the mesencephalon with an anterior boundary approximately coincident with that of *Emx2*. Laterally, the posterior boundary of *Otx1* domain coincides with that of the mesencephalon. In median sections a strong hybridisation signal extends only half way along the mesencephalon, dividing the mesencephalic dorsal midline in two domains. Finally, the *Otx2* expression domain contains the *Otx1* domain, both dorsally and ventrally, and practically covers the entire fore- and mid-brain, to the exclusion of the early optic area.

Expression of *Emx* and *Otx* genes identifies several regions in the forebrain (Fig. 1). Some of these regions seem to correspond to presumptive anatomical subdivisions, whereas the significance of others remains to be assessed. Dorsally, for example, it is clear that the two *Emx* genes identify a presumptive cortical region, part of which will be neocortex and archicortex. *Emx2* expression also appears to define the boundary between future dorsal and ventral thalamus. On the other hand, it is notable that expression of these genes does not offer an unambiguous cue for the boundary between presumptive ventral thalamus and posterior dorsal telencephalon.

In summary, analysis of E10 brain shows a pattern of nested expression domains of the four genes in brain regions defining an embryonic rostral, or pre-isthmic, brain as opposed to hindbrain and spinal cord. The first appearance of the four genes is also sequential (Simeone et al., 1992a): *Otx2* is already expressed in E5.5, followed by *Otx1* and *Emx2* in E8-8.5 and finally by *Emx1* in E9.5. It seems reasonable to postulate a role of the four homeobox genes in establishing or maintaining the identity of the various embryonic brain regions. In this line, the specification of the regions of the early rostral brain seems to be a discrete process with its center in the dorsal telencephalon.

EXPRESSION OF *EMX* AND *OTX* GENES IN MIDGESTATION MOUSE EMBRYOS

Emx Gene Expression

Emx1 and *Emx2* are expressed in presumptive cerebral cortex in a developmental period, between day 9.5 and day 16 of development, corresponding to major events in cortical neurogenesis (Simeone et al., 1992b). In its full extension, E12.5 to E13.5, the *Emx1* expression domain comprises cortical regions including primordia of neopallium, hippocampal and parahippocampal archipallium. *Emx1* expression seems characteristic of cortical regions, mainly but not exclusively hexalaminar in nature (Kuhlenbeck, 1973). In the same period, the *Emx2* expression domain comprises presumptive cortical regions

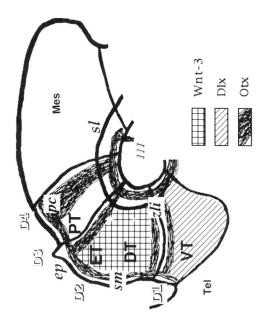

Figure 2. Schematic representation of the expression domains of *Otx1* and *Otx2*, *Wnt3*, and *Dlx* genes in diencephalon of E12.5 embryos. Within the diencephalic regions bold letters designate the columnar nomenclature: DT, dorsal thalamus; ET, epithalamus; PT, pretectum and VT, ventral thalamus. Outside the profile, the proposed new subdivision into four neuromeres, D1 to D4 (Figdor and Stern, 1993), is indicated. *ep*, epiphysis; Mes, mesencephalon; *pc*, posterior commissure; *sl*, sulcus limitans; *sm*, stria medullaris; Tel, telencephalon; *zli*, zona limitans intrathalamica; *III*, 3rd cranial nerve.

including neopallium, hippocampal and parahippocampal archipallium and selected paleopallial localisations, but no basal internal grisea. In E12.5 embryos, the hybridization signal is uniformly distributed across the cortex without major differences but starting from E13.75 it appears to be confined to the germinal neuroepithelium of the ventricular zone, excluding the intermediate zone and cortical plate. From day 14.5 on, *Emx2* cortical expression progressively declines in anterior and ventrolateral regions and at day 17 of development is confined to specific cell layers in hippocampus.

Otx Gene Expression

In midgestation mouse embryos the two *Otx* genes are expressed in specific restricted regions of the developing brain (Simeone et al., 1993). Both are expressed in basal telencephalon, in diencephalon and mesencephalon but not in spinal cord. Their expression domains in mesencephalon show a sharp posterior boundary, both dorsally and ventrally, approximately at the level of rhombic isthmus. From E9.25 onward, the expression of both genes clearly marks the posterior boundary of mesencephalon to the exclusion of presumptive anterior cerebellar domains. *Otx1* is also expressed in dorsal telencephalon, whereas *Otx2* expression has disappeared from this region at day 11.75.

Otx1 and *Otx2* are also expressed in restricted regions of diencephalon of midgestation embryos: epithalamus, dorsal thalamus and mammillary region of posterior hypothalamus. In these regions, the hybridization signal is almost exclusively confined to cells of the ventricular zone. Their expression domain does not include the ventral thalamus. A two-layered narrow stripe of expression is detectable at the level of the boundary between dorsal and ventral thalamus, that is the zona limitans intrathalamica, the precursor of lamina medullaris externa and mammillo-thalamic tract. Other localizations are: fasciculus retroflexus, the precursor of habenulo-interpenduncular tract, stria medullaris, including the region surrounding the posterior commissure, primordium of mammillotegmental tract, epiphysis, fornix and sulcus lateralis hypothalami posterioris. Posterior to diencephalon, *Otx1* and *Otx2* are expressed in mesencephalic regions of tectum and tegmentum, possibly at the level of presumptive periventricular bundles.

Both *Otx* genes are also expressed in the olfactory epithelium, as well as in the developing inner ear from early expression in the otic vesicle to epithelia in auricular ducts of sacculus and cochlea and in the devoloping eye, including the external sheaths of the optic nerve.

Areas and Boundaries in Diencephalon

Expression of *Otx* genes in diencephalon and mesencephalon of E12.5-14.5 embryos colocalizes with boundary regions and presumptive axon tracts, including anterior and posterior commissure (Fig. 2). This expression is confined to precursor cells surrounding these structures as if these cells could be used as borders of pathways for the pioneer axon tracts. This is particularly evident in posterior commissure and along the zona limitans intrathalamica. *Otx* gene expression in posterior commissure is limited to cells of ventricular epithelium, whereas primary fibers running on its surface are not labelled. Expression of *Otx* genes along the zona limitans intrathalamica might constitute a framework for the axon patterning of lamina medullaris and other structures physically separating dorsal thalamus from ventral thalamus. The existence of this barrier might account for the sharp dorsal boundary of the expression domain in ventral thalamus of *Dlx* genes (see Boncinelli, 1994 for a review; Simeone et al., 1994) and for the sharp ventral boundary of the expression domain in dorsal thalamus of other genes like *Wnt3* (Salinas and Nusse, 1992). Both *Otx* genes are also expressed around the developing optic nerve. This localization is similar to that along the zona limitans intrathalamica in providing clues to axon pathfinding and patterning. In this light, expression of *Otx* genes might provide a global framework for the primary scaffold of specific axon pathways in the early neuroepithelium of the forebrain. We are currently testing this hypothesis with in vitro analysis of axon growth and propagation on *Otx*-transfected cells.

It is of interest to consider the possibility that *Otx* genes play a different role in the development of the head in at least two different stages. They first specify territories or areas in rostral brain of E8-E10 mouse embryos and provide later on a set of positional cues required for growing axons to follow specific pathways within the embryonic central nervous system. It is not clear whether the two functions are independent. It is also of interest to consider that in flies mutant for *orthodenticle*, pioneer axons of the posterior commissures fail to develop normally as if appropriate positional cues were missing (Tessier-Lavigne, 1992).

EARLY *OTX2* EXPRESSION

Otx2 Expression in the Mouse Blastocyst

Otx2 transcripts are already detectable in embryonic structures of the implanted mouse blastocyst since the prestreak stage at about day 5.5-5.7 of development (Simeone et al., 1993). *Otx2* transcripts are present in the embryonic ectoderm, or epiblast, but not in extraembryonic ectoderm. The epiblast of the pre-gastrula mouse embryo is believed to be the sole source of all tissues of the fetus. A variety of experiments demonstrate the potency of the epiblast to form derivatives of all three germ layers in gastrulating as well as in prestreak embryos. The same expression pattern is observed in the ectoderm of E6 embryos and of early-streak E6.5 embryos. *Otx2* expression is still confined to primary ectoderm: invaginating cells of the primitive streak do not express it. Between day 7 and 7.5 of development the posterior boundary of the *Otx2* expression domain progressively recedes to anterior regions where it will remain confined. These regions correspond to the neuroectoderm of the prosencephalon and mesencephalon. *Otx2* will remain expressed in these regions until late in gestation.

The progressive confinement of *Otx2* expression from the entire epiblast to presumptive fore- and mid-brain neuroectoderm occurs concomitantly with progressive regionalization of cell fates within the epiblast. This *Otx2* progressive confinement also correlates with the expression of other early developmental genes. For example, what is known about the evolution of expression patterns of early *Hox* genes, in particular *Hoxb1* (McGinnins and Krumlauf, 1992), suggests a relationship between the progressive displacement towards anterior of both the anterior border of the *Hox* expression domain and the posterior border of the *Otx2* expression domain.

Otx2 Expression in the Early Frog Embryo

The fact that in the mouse *Otx2* expression demarcates anterior regions from pre-streak stages and rostral brain regions from the formation of the headfold until late in gestation raises the question of whether the *Otx2* homeodomain protein might play a role in specifying anterior structures, rather than being a mere marker of position. A direct approach to this problem is the microinjection of synthetic mRNA in *Xenopus* embryos. With this in mind, we looked for the frog homologue of *Otx2*, *Xotx2*. The study of its expression in normal and experimentally manipulated frog embryos suggests for this gene a role in patterning the body axis and in specifying anterior regions and their spatial relationship with trunk structures (Pannese et al., 1994).

RNase protection experiments reveal that *Xotx2* is expressed at low levels throughout early *Xenopus* development from unfertilized egg to late blastula, when its expression significantly increases. Isolated animal caps dissected from stage 8 embryos also show low expression levels of *Xotx2*. The presence of *Xotx2* transcripts in animal caps is confirmed by whole-mount in situ hybridization of stage 8 embryos. Then, *Xotx2* transcripts are progressively confined to dorsal internal regions of marginal zone where they are clearly detectable at stage 9.5. At stage 10.25 the major expression site of *Xotx2* is in migratory deep zone cells that are fated to give rise to prechordal mesendoderm. In addition, it is also expressed in dorsal bottle cells. At stage 10.5 *Xotx2* expression persists in these cell types and posteriorly, above the dorsal blastopore lip, this expression clearly respects the boundary between internal deep zone cells and external cell layers. Conversely, in more anterior regions *Xotx2* expression extends to cells of presumptive anterior neuroectoderm, where it persists throughout embryogenesis.

At stage 14, *Xotx2* expression appears to be confined to mesendoderm and ectoderm cells of anterior dorsal embryonic regions. *Xotx2* trascripts are also detectable in a more ventral position, in both layers of the ectodermal region including the stomodeal-hypophyseal anlage and the cement gland anlage. In summary, at this stage the *Xotx2* expression domain is spatially restricted in a sort of anterior stripe extending across the three germinal layers. This is highly reminiscent of the *orthodenticle* expression domain in *Drosophila* blastoderm. It is also of interest to note that the regions of the *Xenopus* embryo where *Xotx2* is expressed specifically contain those cells which never undergo convergence and extension movements during gastrulation (Keller et al., 1992).

Comparison of the expression domains of *Otx2* and *Xotx2* in early mouse and frog embryos shows several similarities and some dissimilarities. Before the onset of neurulation this gene is expressed in presumptive rostral brain in both species. In both cases the expression of this gene demarcates presumptive fore- and mid-brain. This localized neuroectodermal expression appears in *Xenopus* after a phase during which *Xotx2* is expressed in migrating anterior mesoendoderm. It is very difficult to detect expression of *Otx2* in the corresponding cells of extending head process in mouse embryos, but this cannot be excluded. Major differences between the number of cells contained in these structures in the two species and a different relative expression intensity might explain this discrepancy. A second question concerns the relationship between early expression in the entire mouse epiblast and later expression in anterior neuroectoderm exhibited by *Otx2*. From the observation of mouse embryos it is difficult to assess whether *Otx2* later expression derives directly from the first wave of expression in epiblast, simply because some posterior cells no longer express this gene while anterior cells still contain its transcripts. Comparison of mouse and *Xenopus* expression data rather suggest that restricted expression in anterior neuroectoderm is a new event, regulated independently from the previous extended expression, possibly through different control regions.

Various treatments of early embryos cause a general reorganization of *Xotx2* expression. In particular, retinoic acid treatment essentially abolishes its expression in neuroectoderm. These embryos lack most of the rostral brain including fore- and mid-brain, whereas hindbrain and spinal cord regions are not reduced. It is tempting to speculate that there is a direct correlation between the deletion of these regions and the lack of *Xotx2* expression in neuroectoderm of embryos which had been subjected to retinoic acid treatment. It is known that RA affects expression of many other genes including homeobox genes of the *Hox* family (McGinnis and Krumlauf, 1992 for a review). All these genes contribute to the actual phenotype of treated embryos, but *Xotx2* is the only gene found so far to be expressed in the presumptive rostral brain whose expression is downregulated by retinoic acid, both in cultured cells (Simeone et al., 1993) and in embryos. Lack of anterior structures in embryos derived from UV-treated zygotes, where *Xotx2* expression is also significantly reduced, is consistent with the notion of a role of this gene in the development of the head region.

Microinjection of *Xotx2* mRNA in 1-, 2- and 4-cell stage embryos produces shortened embryos with severely reduced trunk and tail structures and an expansion of internal head structures. These phenotypes can be viewed in the light of at least two interpretative schemes. According to the first scheme, the relative sizes of the body regions allocated in early embryogenesis for the development of head and trunk structures are altered in microinjected embryos. Regions specified for presumptive head structures are slightly expanded at the expense of those fated to give rise to trunk and tail structures. According to a second interpretative scheme, reduced trunk and tail structures result from interferences with the movements of convergence and extension taking place during gastrulation and neurulation and giving rise to more posterior regions. We already noticed that endogenous *Xotx2* is expressed in presumptive tissues which are not interested by these movements. An extensive misexpression of *Xotx2* transcripts might interfere with the dynamic fate of posterior regions of the axis. Of course, the two interpretative schemes need not to be mutually esclusive.

Most of the embryos overexpressing *Xotx2* also show the presence of an additional cement gland, both in anterior and posterior localizations. Cement gland is one of the most anterior structures of the developing *Xenopus* body and induction of a secondary cement gland by ectopic expression of *Xotx2* suggests for this gene a role in specifying anterior

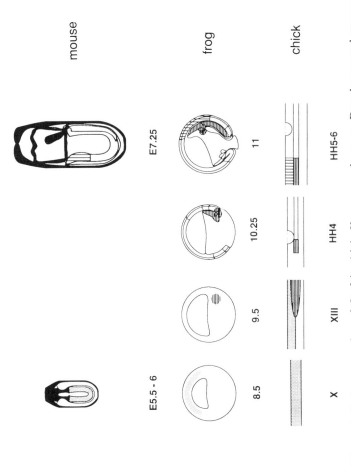

Figure 3. Early expression of *Otx2* in chick, *Xenopus* and mouse. Developmental stages are indicated. Very early *Otx2* expression in epiblast is shown as stippling, mesendodermal expression is indicated by horizontal stripes and neuroectodermal expression by vertical stripes. A notch in chick diagram indicates the node.

head structures. The presence of the *Xotx2* homeodomain is required to produce all of these effects. We have seen that this homeodomain contains a specific lysine residue at position 9 of the recognition helix. *Xotx2* constructs containing a homeodomain where this lysine is substituted by a glutamine or a glutamic acid residue also fail to cause these effects.

The expression pattern of *Xotx2* in *Xenopus* embryos suggests that this gene plays a role in events leading to preparation and execution of gastrulation, especially in connection with the specification and possibly patterning of anterior regions. Results of microinjection experiments suggest that *Xotx2* gene products are able to respecify the subdivision of the body along the anterior-posterior axis. In conclusion, both in mice and frogs, *Otx2* gene products might play a central role in the initial events of axis formation and in particular in specifying anterior head and its spatial relationship with trunk structures.

Otx2 Expression in the Early Chick Embryo

The chick embryo provides an excellent model system for the study of gastrulation because it is at early stages relatively large, translucent and flat. Furthermore, it can develop for long periods of time in vitro, where it is readily accessible to some experimental manipulations. With this in mind, we cloned the chick homologue of *Otx2*, *c-otx2*, and analyzed its expression pattern during gastrulation (Bally-Cuif et al., 1994). Two distinct phases of *c-otx2* expression, separated by the onset of primitive streak regression, were observed. *c-otx2* transcripts were first detected in the unincubated egg and up to stage XIII, in all epiblast and forming hypoblast cells. Its expression in epiblast is no surprise owing to *Otx-2* expression in mouse epiblast, whereas expression in hypoblast comes as a novelty. The hypoblast develops after stage X under the epiblast layer, and is believed to receive contributions from both epiblast cells and from cells spreading from the posterior marginal zone. The hypoblast is not fated to contribute to the embryo proper, and it may perhaps be homologous to the primitive endoderm of mammals, and to the vegetal pole of frogs.

c-otx2 is also expressed in a very early population of middle layer cells. These presumptive early mesendoderm or mesoblast cells also express the chick homologue of *goosecoid*, *c-gsc* (Izpisúa-Belmonte et al., 1993). They have been shown to contribute to Hensen's node and it has been suggested that they are already determined as inductive cells (Izpisúa-Belmonte et al., 1993). The presence of these early mesendoderm cells in chick blastoderm is difficult to notice in the absence of molecular markers like *c-gsc* and *c-otx2*, especially because they can be easily confused with residual polyinvaginating hypoblast cells. Nothing is known about the existence of corresponding early mesendoderm cell populations in the mouse, but we have seen that in *Xenopus* the cognate *Xotx2* gene is already expressed by late blastula in a restricted cell population in the internal dorsal region of the marginal zone. These cells are probably fated to give rise to anterior mesendoderm and the so-called prechordal plate. These structures subsequently express both *otx-2* and *gsc*. Comparative analysis of frog and chicken gastrulation suggests that these early mesendodermal precursors segregate very early from other cell populations.

During primitive streak formation and progression, *c-otx2* expression is progressively restricted to anterior regions in all cell layers. *c-otx2* expressing cells are also accumulating in the streak with a highest expression in its anterior portion and a decreasing intensity in more posterior regions. When the extension of the streak is maximal, transcripts are only found in Hensen's node. This extremely restricted expression lasts for a very short period of time and a second, distinct, phase of *c-otx2* expression starts at the beginning of streak regression.

c-otx2 transcripts progressively disappear from the regressing node and a strong expression is now detectable in anterior mesendoderm and neuroectoderm, with the exception of forming notochord and floor plate. The expression in anterior neuroectoderm becomes predominant with headfold formation and appears clearly spatially restricted, with a sharp posterior boundary at the level of midbrain-hindbrain constriction. Subsequently, *c-otx2* expression will remain mainly confined to these areas of the neural tube (fore- and mid-brain) and to anterior mesoderm and endoderm.

The expression of *c-otx2* in the forming neural plate is most probably due to a reinitiation of expression at the time of streak regression, rather than to the maintenance of transcription in anterior cells. The best argument on this matter is the finding of a very transient intermediate stage, at the time of maximal streak extension, when *c-otx2* is almost exclusively transcribed in nodal cells and anterior expression in epiblast cells is lost. This two-step expression profile was not previously appreciated in the mouse, but it might have gone unnoticed due to the very transient nature of the intermediate stage.

The cells positive for *c-otx2* during the two steps are not clearly related by lineage, which further suggests that its expression might be regulated independently during the two phases. A large portion of the anterior primitive streak cells, which are positive during streak progression, will contribute to notochord and somites, which are *c-otx2* negative. Conversely, the *c-otx2* negative epiblast cells which are located just anterior to the node are fated to give rise to neuroepithelial cells which will form the floor of the neural tube throughout its entire length. Expression in the floor of the anterior neural tube therefore clearly results from de novo initiation in these cells. This induction might result, at least in part, from inductive signals emanating from the underlying prechordal mesendoderm cells, which probably express *c-otx2* in a continual manner from prestreak stages. *c-otx2* might exert different functions at these different developmental stages.

The first phase of *c-otx2* expression bears strong similarity with that of *c-gsc*. Therefore, we compared the expression of the two genes. This analysis demonstrated that *c-otx2* is transcribed first and its expression in the hypoblast precedes that of *c-gsc*. On the other hand, *c-gsc* is an earlier marker of primitive streak cells. The expression domains of the two genes transiently overlap in Hensen's node and anterior mesendoderm, whereas only *c-otx2* is expressed in neuroectodermal areas. After a transient appearance in the progressing anterior node, *c-otx2* transcripts become stably confined to anterior neuroectoderm and mesendoderm. This late expression domain appears to be rather specific and suggests a role of this gene in specifying anterior head structures including the rostral brain and pharyngeal endoderm. The transient coincidental expression of *c-gsc* and *c-otx2* could be interpreted as an indication of an anterior cell fate and/or of intrinsic inductive properties.

In conclusion, a strong expression of *c-otx2* in the chick embryo is first associated with cells of presumptive anterior mesendoderm from their early determination and subsequently extends to anterior neuroectoderm.

EVOLUTION AND DEVELOPMENT

Otx-2 clearly represents a promising tool to test comparative hypotheses, at least in different species of Vertebrates. A primary question is the identification of true homologous developing structures like epiblast, hypoblast and early mesoblast in different Vertebrates. Even more important is probably the question of early determination of anterior mesendoderm and its relation with presumptive anterior neuroectoderm. A third question regards the distinction early in development between anterior and posterior neuroectoderm and its genetic control. An interesting implication of the evolutionary conservation of *orthodenticle* and its related vertebrate genes involves the distinction between head and trunk in the animal embryo. The specification of the head as a structure distinct from trunk regions, *i. e.* cephalization, was previously thought to have occurred independently in the vertebrate and invertebrate evolutionary lineages. The sequence conservation and related expression patterns observed for the *Otx* and *Emx* gene families seem to contradict this idea, suggesting instead that anterior patterning was already established in a primitive ancestor of both flies and man. This might imply that head emerged only once in evolution.

In the chick embryo *Otx2* is expressed at different levels of abundance in every cell of the early blastoderm, belonging to epiblast, hypoblast or mesoblast. It is not known whether the same is true for the mouse and for *Xenopus*, but it is highly likely. Even if it is not yet known whether all these *Otx2* transcripts are translated, we might suppose that this gene codes for something related to a general property of all primary embryonic tissues, distinct from presumptive extraembryonic structures. These transcripts progressively disappear from posterior developing tissues and remain confined in the anteriormost regions of the body, including mesendoderm and neuroectoderm. Whether these tissues share some properties with early blastoderm cells or, alternatively, the *Otx2* gene products play two different roles in the two different stages of development remains to be assessed.

We might also hypothesize that *otd/Otx* genes are directly related to the specification of anterior regions of the body plan. *Otx2*, for example, is one of the earliest genes expressed in anterior neuroectoderm of the mouse, demarcating rostral brain regions even prior to any sign of headfold formation. As a consequence, early *Otx2* expression appears intimately linked to anterior specification and head formation. Its gene product contains a homeodomain of the *bicoid* class and is able to recognize a *bicoid* target sequence (Simeone et al., 1993). It is tempting to speculate that an *otd*-like gene already specified anteriority in early metazoans. From this ancestral gene derive *otd* in flies and vertebrate genes of the *Otx* family. The maternal *bicoid* gene might have subsequently evolved in the evolutionary lineage leading to flies and other long germ-band insects.

Acknowledgements

We wish to thank Laure Bally-Cuif and all components of our lab for comments and helpful suggestions. This work was supported by grants from EC BIOTECH Programme, Progetti Finalizzati CNR "Ingegneria Genetica" and "ACRO", the Telethon-Italia Programme and the Italian Association for Cancer Research (AIRC).

REFERENCES

Bally-Cuif, L., Gulisano, M., Broccoli, V., and Boncinelli, E., 1994, *c-otx2* is expressed in two different phases of gastrulation and is sensitive to retinoic acid treatment in chick embryo, *Mech. Dev.* in press.

Boncinelli, E., 1994, Early CNS development: *Distal-less* related genes and forebrain development, *Current Opinion in Neurobiology* 4:29-36.

Cho, K.W.Y., Blumberg, B., Steinbeisser, H., and De Robertis, E.M., 1991, Molecular nature of Spemann's organizer: the role of the *Xenopus* homeobox gene *goosecoid*, *Cell* 67:1111-1120.

Cohen S., and Jürgens G., 1991, *Drosophila* headlines, *Trends Genet.* 7:267-272.

Figdor M., and Stern C., 1993, Segmental organization of embryonic diencephalon, *Nature* 363:630-634.

Finkelstein R., and Boncinelli, E., 1994, From fly head to mammalian forebrain: the story of *otd* and *Otx*, *Trends Genet.* 10:310-315.

Finkelstein R., and Perrimon N., 1991, The molecular genetics of head development in *Drosophila melanogaster*, *Development* 112:899-912.

Izpisúa-Belmonte J.C., De Robertis, E.M., Storey, K.G., and Stern, C.D., 1993, The homeobox gene *goosecoid* and the origin of organizer cells in the early chick blastoderm, *Cell* 74:645-659.

Keller, R., Shih, J., and Sater, A., 1992, The cellular basis of the convergence and extension of the *Xenopus* neural plate, *Developmental Dynamics* 193, 199-217.

Kuhlenbeck H., 1973, The Central Nervous System Of Vertebrates, S. Karger, Basel.

McGinnis W., and Krumlauf R., 1992, Homebox genes and axial patterning, *Cell* 68:283-302.

Pannese, M., Polo, C., Andrezzoli, M. Vignali, R, Kablar, B., Barsacchi, G., and Boncinelli, E., 1994, The *Xenopus* homologue of Otx2 is a maternal homeobox gene that demarcates and specifies anterior body regions, *Development* in press.

Salinas P.C., and Nusse R., 1992, Regional expression of the *Wnt-3* gene in the developing mouse forebrain in relationship to diencephalic neuromeres, *Mech. Dev.* 39:151-160.

Simeone A., Gulisano M., Acampora D., Stornaiuolo A., Rambaldi M., and Boncinelli E., 1992b, Two vertebrate homeobox genes related to the *Drosophila empty spiracles* gene are expressed in the embryonic cerebral cortex, *EMBO J.* 11:2541-2550.

Simeone A., Acampora D, Gulisano M., Stornaiuolo A., and Boncinelli E., 1992a, Nested expression domains of four homoeobox genes in the developing rostral brain, *Nature* 358:687-690.

Simeone A., Acampora D., Mallamaci A., Stornaiuolo A., D'Apice M.R., Nigro V., and Boncinelli E., 1993, A vertebrate gene related to *orthodenticle* contains a homeodomain of the *bicoid* class and demarcates anterior neuroectoderm in the gastrulating mouse embryo, *EMBO J.* 12:2735-2747.

Simeone A., Acampora D., Pannese, M., D'Esposito, M., Stornaiuolo A., Gulisano, M., Mallamaci A., Kastury, K., Druck, T., Huebner, K., and Boncinelli E., 1994, Cloning and characterization of two members of the vertebrate *Dlx* gene family, *Proc. Natl. Acad. Sci. USA* 91:2250-2254.

Tessier-Lavigne M. 1992. Axon guidance by molecular gradients, *Current Opinion in Neurobiology* 2:60-65.

NEURONAL ORGANIZATION OF THE EMBRYONIC FORE- AND MIDBRAIN IN WILDTYPE AND MUTANT MICE

Grant S. Mastick and Stephen S. Easter, Jr.

Dept. of Biology
University of Michigan
Ann Arbor, MI 48109 USA

INTRODUCTION

The first signs of spatial organization appear very early in the developing neural tube. The primary sign of regionalization is the formation of morphological subdivisions termed neuromeres, which have been recognized by embryologists for over 100 years as a series of bulges in the neural tube, separated by interneuromeric constrictions (reviewed in Vaage, 1969; Puelles et al., 1987).   The spatial organization and temporal appearance of neuromeres are evolutionarily conserved features of vertebrate embryos.

Recent studies have shown that these early morphological signs of regionalization correlate with distinct cellular and molecular patterns in the neural tube. While much of this work has focused on the hindbrain, the fore and midbrain are likely to utilize many of the same organizing principles.   Neuromeres correspond to polyclonal domains of neuroepithelial cells, as interneuromeric borders act as substantial barriers to cell migration in both the chick hindbrain (Fraser et al., 1990; Birgbauer and Fraser, 1994), and presumptive diencephalon (Figdor and Stern, 1993). Intercellular communication between neuroepithelial cells is limited across rhombomeric borders (Martinez et al., 1992). These cellular compartments are also distinct at the molecular level: regionalized expression patterns of putative regulatory genes are bounded by interneuromeric borders, implying that spatially organized codes could determine subsequent neuronal patterns in the brain (Hunt et al., 1991; Puelles and Rubenstein, 1993; Figdor and Stern, 1993).

The cellular and molecular compartmentalization of the neural tube generates specialized regulatory environments, which then could direct patterns of neuronal development.   Patterns of neuronal cell bodies and axon tracts should then reflect the underlying organization of the neural tube.   This prediction has been confirmed in the hindbrain. Motoneurons in the chick hindbrain are based in pairs of rhombomeres, so that r2 and r3 motoneurons project to gV, r4 and r5 to gVII, etc. (Lumsden and Keynes, 1989). Reticular neurons are also organized by rhombomeres, where several different classes of neurons are present in most rhombomeres, but with the exact combination varying in each rhombomere as "variations on a theme" (Metcalfe et al., 1986; Hanneman et al., 1988; Clarke and Lumsden, 1993). While late appearing patterns of axon tracts correlate with the borders of clonal domains in the chick diencephalon (Figdor and Stern, 1993), early patterns of neurons in the fore and midbrain remain unknown.

In a recent series of reports, we have undertaken a description of the fore and midbrain (Easter et al., 1993; Mastick and Easter, submitted). We have described the beginnings of brain development to test several specific predictions. 1. Clusters of early neuron cell bodies will be bounded by neuromeres. 2. position determines neuronal identity, so that specific neuron types will be spatially segregated, not intermixed. 3. Axonal projection patterns will be consistent with the neuromeric organization of the fore and midbrain.

INITIAL TRACT FORMATION IN THE MOUSE EMBRYO

Neuromeres

The neural tube of mouse embryos at E10 consists of a number of morphologically recognizable neuromeric subdivisions (Figure 1A). The neuromeres provide a series of landmarks, for which we have adopted nomenclature from chick embryos, which have a similar appearance (Puelles et al., 1987). Three neuromeres initially form as a series of bulges: prosencephalon, mesencephalon, and rhombencephalon, roughly corresponding to the adult structures of the forebrain, midbrain, and hindbrain. The simple neural tube then becomes more complex, bending and developing further subdivisions by E10. There is a ventral bend in the prosencephalon and mesencephalon to form the cephalic flexure. The primary neuromeres progressively become subdivided transversely in a segment-like fashion. This is most obvious in the rhombencephalon, with the formation of seven rhombomeres

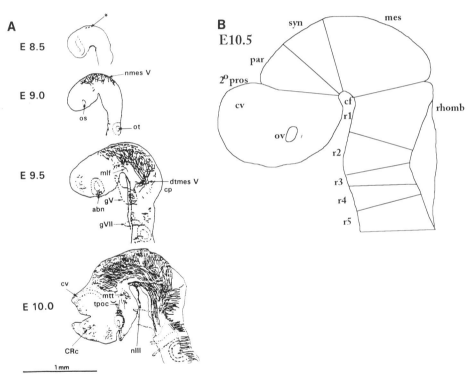

Figure 1. A. A schematic view of neuronal development in mouse embryos, labeled with an antibody to neuron-specific β-tubulin. The scale bar applies to all four ages. Abbreviations: abn, anterobasal nucleus; cp, cerebellar plate; cv, cerebral vesicle; CRc, Cajal-Retzius cells; dtmesV, descending tract of the mesencephalic nucleus of the trigeminal nerve; gV, ganglion of the trigeminal nerve; gVII, ganglion of the facial nerve; mlf, medial longitudinal fasciculus; mtt, mammillotegmental tract; nmesV, mesencephalic nucleus of the trigeminal nerve; nIII, oculomotor nerve; os, optic stalk; ot, otocyst; tpoc, tract of the postoptic commissure. (Reprinted from Easter, et al., 1993). B. Map of neuromeric organization at E10, with the visible transverse constrictions between neuromeres in unlabelled embryos indicated. The neuromeric nomenclature is that of Puelles, et al. (1987). Abbreviations: 2° pros, secondary prosencephalon; cf, cephalic flexure; mes, mesencephalon; par, parencephalon; rhomb, rhombencephalon; r1, rhombomere 1; syn, synencephalon.

(r1-r7). The anterior neural tube develops similar subdivisions. The caudal prosencephalon divides in two to form the parencephalon and synencephalon (together the presumptive diencephalon), with the remaining rostral tube termed the secondary prosencephalon, which develops two pairs of evaginations, the cerebral and optic vesicles. The mesencephalon remains morphologically undivided, although chick embryos do develop a minor mid-mesencephalon constriction (Vaage, 1969), suggesting the existence of two mesomeres.

The Earliest Neurons

The first neurons are also differentiating during these early neural tube stages, and have recently been described (Easter et al., 1993). The number of neurons in the prosencephalon and mesencephalon increases rapidly from their first appearance at E8.5, and they quickly form a small number of initial tracts over the course of the next two days (summarized in Figure 1A). This study used an antibody (TuJ1) against a neuron-specific b-tubulin (Moody et al., 1987; Lee et al., 1990b, a). This antibody labels neuron cell bodies before axons sprout, as well their axons, giving a powerful method to identify the earliest neurons and their axon projection patterns in whole mounts.

The first neurons in the embryo appear in a dorsal cluster near the synencephalon-mesencephalon junction on E8.5. By E9, the first axons are produced, and extend caudally across the surface of the mesencephalon, with a large committee of pioneers funneling together as they enter the rhombencephalon to form the first axon tract, the descending tract of the mesencephalic nucleus of the trigeminal nerve (dtmesV). By E9.5, the number of immunoreactive cells increases across the dorsal-ventral extent of the mesencephalon and synencephalon, and begins to obscure the axonal organization, but a few more discrete axon tracts are visible. In the ventral region of the cephalic flexure, a ventral tract called the medial longitudinal fasciculus is formed by axons projecting caudally through the mesencephalon into the ventral rhombencephalon. A bundle of axons exits the ventral neural tube at the cephalic flexure, and forms the oculomotor nerve, nIII. Neuronal differentiation in the secondary prosencephalon lags behind that in the rest of the brain, but two groups of neurons appear on E9.5, and form identifiable axon tracts by E10. The tract of the postoptic commissure (tpoc) originates from a cluster of neuron cell bodies at the base of the optic stalk, projecting caudally toward the parencephalon. Just rostral to the cephalic flexure, a ventral cluster of neurons project axons caudally, paralleling the tpoc and forming the mammillotegmental tract (mtt). (Note that our convention for describing the embryonic rostral-caudal and dorsal ventral coordinates assumes that the axis is bent by the cephalic flexure. Thus the neural tube coordinates depend on position relative to the flexure: for example, the tpoc axons have a caudal trajectory that projects upward, then curving to the right, and eventually downward, paralleling the cephalic flexure.)

The clarity of the antibody labeling becomes limited because of its non-selectivity, becoming less informative as the number of neuron cell bodies and axons increases with developmental stage. While the earliest stages are impressively simple, by the time the first axon tracts are apparent, E9.5, there is dense immunolabelling that obscures the location of the cell bodies that contribute axons to particular pathways. By E10.5, a continuum of immunoreactivity covers much of the surface of the mesencephalon and synencephalon, camouflaging the location of the initial tracts with respect to each other, as well as the appearance of any additional axon tracts. Two particular aspects of the neuronal organization remain unclear: the location of particular cell body types in relation to the neuromeric boundaries, and the spatial arrangement of types of cell bodies and axons with respect to each other .

THE ORGANIZATION OF NEURONAL CELL BODIES AND INITIAL TRACTS

Mapping of Cell Bodies to Neuromeres

Our approach was to extend the previous study in two directions: to more precisely determine the spatial organization of the initial neuron populations and their axon pathways generated during the first day of neuronal development (E8.5-9.5), and to identify and map any additional neuron types and new axon pathways that are generated during the second day (to E10.5). We have used the fluorescent lipophilic axon tracers diI and diO (Honig and Hume, 1986; Godemont et al., 1987) to selectively label individual axon tracts and the neuronal cell bodies that contribute axons to them (Mastick and Easter, submitted). The technique is to insert a crystal of dye into an intact fixed embryo, using a fine tungsten needle under a dissecting scope, into a hole made in the wall of the neural tube. All of the axonal membranes that pass through this point become fluorescently labeled, in both anterograde and retrograde directions. By selecting appropriate label sites in staged embryos, this technique allows precise spatial and temporal control of neuronal labelling, without requiring neuron-type specific antibodies or molecular probes. An example of this approach is shown in Figure 2A. A crystal of diI is inserted into the caudal dorsal mesencephalon of an E10 embryo, labelling the dtmesV axons projecting in a curving path into the rhombencephalon, as well as labelling the axons back to their cell bodies. To provide landmarks, small diO label sites were made to mark the rostral and caudal borders of the synencephalon. All of the labelled cells were found in the dorsal half of the embryo, with the majority spread in a wide zone across the dorsal mesencephalon. However, labelled somata were also found in the dorsal synencephalon. There was no apparent break in the population at the synencephalon-mesencephalon border, and no dtmesV somata were labelled rostral to the synencephalon. These results are consistent with mapping of dtmesV somata in more advanced chick embryos, where dtmesV somata are found in the posterior commissure, located in the posterior synencephalon (von Bartheld and Bothwell, 1993).

Similar labelling strategies were also used to map the locations of the two other neuron types in the synencephalon and mesencephalon. In the ventral half of the neural tube, the mlf nucleus also straddles the synencephalon-mesencephalon border, with about half of the cell bodies in each neuromere, although restricted to the rostral half of the mesencephalon. nIII somata are located in a tight nucleus in ventral rostral mesencephalon, adjacent to the floor plate, closely abutting but not crossing over into the synencephalon. These mapping experiments demonstrate that two tracts, dtmesV and mlf, originate from populations of neuron cell bodies that straddle the synencephalon-mesencephalon boundary, while nIII somata are limited to the mesencephalon. The neurons in the synencephalon-

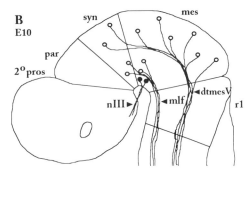

Figure 2. The neuromeric location of the neuronal cell bodies at E10. A. A fixed E10 embryo, with a diI crystal (*) inserted into the dorsal caudal wall of the mesencephalon (m). Smaller crystals of diO were placed at the rostral and caudal boundaries (arrowheads) of the synencephalon (s). The large caudally-directed bundle of dtmesV axons is indicated. Scale bar=?. B. A schematic diagram summarizing the locations of dtmesV, mlf, and nIII cell bodies in the E10 embryo, with respect to the neuromeric boundaries, based on double labeling strategies as in A. Open circles, dtmesV and mlf cell bodies; filled circles, nIII cell bodies. Modified from Mastick and Easter, submitted.

mesencephalon region appear to have an organization reminiscent of the rhombencephalon, where motoneuron populations straddle pairs of rhombomeres (Lumsden and Keynes, 1989). The neuromeric organization is neuron-type specific, however, as nIII neurons respect a border that the other two do not.

We have made several assumptions in this analysis. First, we have defined neuronal type as those neurons that project their axons into a particular pathway. We thus may be grouping together unrelated neurons from adjacent neuromeres that could be distinguished using other criteria. Secondly, we are mapping the locations of neuronal cell bodies after they have extended axons considerable distances. We thus do not know the locations of their cell bodies at the time that their identities were determined, and there are likely to be additional neuron types present at E10.5 that are just starting to project axons. Thirdly, we have used exclusively the visible landmarks for mapping, and indeed identify them as functional boundaries on the neuronal level, although additional subdivisions may also play a role. Indeed the restriction of the mlf and nIII cell bodies to the rostral half of the mesencephalon implies a cryptic boundary that divides this region in two.

Mapping of Cell Bodies With Respect to Each Other

Even at these early stages of the development of the mesencephalon and synencephalon, there are clearly more neuron types than neuromeres. As the number of different neuron types increases, the spatial organization of the neuronal cell bodies must become more complex. How are patterns of neurons generated? A likely scenario is that positional information in the neural tube specifies neuronal identity, similar to neuroblast specification in the *Drosophila* central nervous system by unique combinations of expression of regulatory genes (Chu-LaGraff and Doe, 1993), or motoneuron specification by a rhombomeric code of overlapping Hox gene expression (Hunt et al., 1991). The hypothesis, then, is that position determines identity, leading to the prediction that the early neuron types should be spatially segregated.

In the early mesencephalon and synencephalon, there are only a few neuron types, but they are closely associated. For example, from the neuromeric mapping of the mlf and nIII somata, it is apparent that these two groups could potentially intermix in the ventral rostral quadrant of the mesencephalon (Figure 2B). To determine whether these neuron types overlap, the two groups of axons were specifically labelled, mlf with diI, nIII with diO, and the locations of the cell bodies that give rise to the axonal populations were compared (Mastick and Easter, submitted). The two populations are closely associated, but clearly spatially segregated, with the mlf cell bodies located dorsal to the nIII nucleus. There may be a low level of errors on the cellular level, as 1 to 5 mlf somata were intermixed with the nIII somata. A similar double labelling strategy of dtmesV and mlf neurons showed that these two population are segregated dorsal-ventrally, in both the mesencephalon and synencephalon. Although our assay for cell body position depends on axonal projection, and is after the time of identity specification, our results are consistent with the hypothesis that position determines identity.

We have discovered an example of intermixing of neuronal types in the dorsal mesencephalon, however, where a second neuronal population is added to the dtmesV neurons (Mastick and Easter, submitted; summarized in Figure 3). The first neurons in the dorsal rostral mesencephalon clearly project their axons into the dtmesV pathway in immunolabelled E9 wholemounts (Easter et al., 1993), and retrograde labelling of this tract with diI labels these cell bodies at E10 (Mastick and Easter, submitted; Figure 2). However, with a label site in the dorsal rostral mesencephalon on E10, two axon populations are labelled: the caudally-directed dtmesV axons, and a second group of axons that project ventrally. These axons cross the ventral midline by E10.5, and turn caudally into the contralateral mlf, identifying them as tectobulbar neurons (Kroger and Schwarz, 1990). To determine if the tectobulbar cell bodies were intermixed with the dtmesV, we made two label sites, one in ventral mesencephalon, the other in caudal mesencephalon. In these double labelled embryos, we observed three zones of cell bodies: a dorsal zone with only dtmesV cell bodies, a lateral zone with only tectobulbar cell bodies, and an intermediate zone with extensive intermixing of the two types. Thus, in the dorsal mesencephalon, there is the initial appearance of dtmesV neurons at E8.5, then a later addition of tectobulbar cell bodies by E10. The addition of the tectobulbar neurons suggests that there may be

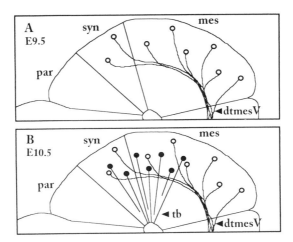

Figure 3. A schematic diagram showing that the dorsal mesencephalon and synencephalon contain a mixed population of neuron cell bodies by E10.5. A. E9.5, when diI label sites anterogradely label only caudally projecting dtmesV axons. Open circles, dtmesV cell bodies. B. E10.5, when a second population of neurons, tectobulbar (tb), appears in the dorsal mesencephalon and synencephalon. tb cell bodies (filled circles) are intermixed with those of dtmesV, and project axons ventrally across the floor plate. Modified from Mastick and Easter, submitted.

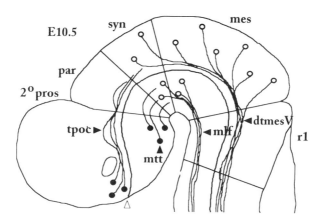

Figure 4. A schematic diagram of tpoc and mtt axons projecting into mlf and dtmesV pathways, based on pairwise double labelling of the tracts at E10.5. The presumed lateral midline separating the two longitudinal axon systems is indicated by an open arrowhead. Filled circles, cell bodies that project axons into the tpoc and mtt; open circles, dtmesV and mlf cell bodies. Modified from Mastick and Easter, submitted.

successive waves of neurogenesis in the dorsal mesencephalon, which could provide a mechanism for generating neuronal diversity in a particular region, by dynamic patterns of gene expression determining neuronal identity. Birthdating studies in the mesencephalon showed that dtmesV neurons are the earliest neurons produced in the midbrain, with birthdates confined to a single day, E8.5, while tectal birthdays begin a day later, and last for several days (Taber Pierce, 1973). In terms of axon guidance, our results also imply that, because of the two orthogonal axon trajectories, the dorsal mesencephalon contains at least two different axonal guidance cues, which are read independently by the two neuron types.

How Are Longitudinal Axon Tracts Organized?

A clear dorsal-ventral organization of the neural tube is apparent in the adult spinal cord, where a ventricular groove, the sulcus limitans, provides both a morphological division between the dorsal and ventral regions, forming the alar and basal plates, as well as a functional division between sensory (dorsal) and motor (ventral) neurons. The sulcus limitans does not continue rostrally into the brain proper, suggesting that the organization of the rostral end of the neural tube may be fundamentally different than that of the spinal cord. We have recently found that the early longitudinal axon tracts of the mouse embryo provide novel evidence that a longitudinal boundary, the functional equivalent of the sulcus limitans, extends rostrally into the forebrain (Mastick and Easter, submitted).

Immunolabelled whole mounts suggested that the first tracts to form are longitudinal, and consisted of caudally directed axons, but by E10, the density of the label obscures their organization (Easter et al., 1993). To determine the organization of the initial axon tracts, we used axon tracing to map the early longitudinal tracts, and have found them to be organized into two axon systems, one dorsal and one ventral (summarized in Figure 4). In mapping dtmesV and mlf somata, it was clear not only that these two populations of cell bodies are spatially segregated in the synencephalon and mesencephalon, but that their axons are segregated, as well. These results suggested that the lateral neural tube, which does not have any distinctive morphological feature, contains a border between these two types of somata and axons. We then examined the trajectories of the two earliest populations of axons from the secondary prosencephalon, the tpoc and mtt, on E10.5. Double labelling of these two tracts showed that they projected caudally, curving around the cephalic flexure in parallel, with tpoc axons remaining dorsal to the more ventral mtt. Additional labelling experiments confirmed that the tpoc pioneer axons enter the synencephalon at the midlateral level, and continue to extend caudally into the dtmesV pathway. Likewise, the mtt axons projected caudally into the mlf pathway.

These results suggest that the four earliest longitudinal tracts all consist of caudally directed axons, and that they form two parallel axon systems. We speculate that the division between these axon systems is an important organizing structure: a compartmental boundary that separates dorsal and ventral compartments of each neuromere. The tpoc thus appears to follow the dorsal edge of this boundary, and joins the dtmesV in following this boundary in the synencephalon and mesencephalon. The lateral midline is likely to be a barrier for dtmesV axon projections, as they tend to project ventrally first, then caudally along this barrier (Easter et al., 1993). The tpoc and dtmesV axons also likely share guidance cues.

Based on the tpoc axon trajectory, the longitudinal boundary extends into the forebrain, and terminates near the base of the optic stalk. If this longitudinal boundary does indicate the alar-basal boundary, then patterns of gene expression in the forebrain should mark one side or the other, with elongated borders. In fact, elongated boundaries of the expression patterns of the *Dlx*, *Wnt-3*, and *Gbx-2* genes in the basal forebrain have been used in a recent neuromeric model of the forebrain to place the alar-basal boundary at this location (Puelles and Rubenstein, 1993). According to this model, the alar-basal longitudinal boundary meets the dorsal and ventral midlines just ventral to the base of the optic stalk, and this location is the rostral end of the neural tube. In support of this assertion, genes that are expressed in the floor plate, such as Sonic hedgehog and HNF3β, have domains of expression that extend through the basal forebrain up to the base of the optic stalk (Echelard et al., 1993). A further implication of this model is that he optic and cerebral vesicles (presumptive telencephalon) are dorsal structures, and the basal forebrain ventral.

NEURONAL DEFECTS IN *Wnt-1*$^{-/-}$ MUTANT EMBRYOS

The genetic program that regulates the neuronal organization of the fore and midbrains is unknown. The ability to target mutations to cloned genes in mice has facilitated a genetic approach to the development of the brain, and increasing numbers of mutant mouse strains are becoming available. Mutant analysis of brain development has been largely limited to histological examination of the brains of embryos at advanced stages, with little information available about primary defects, that is, defects in neuronal organization at the early times that these genes are acting. From our perspective, the patterns of neurons described above offer very useful sets of landmarks in early embryos to interpret defects in mutant mice. In addition, patterns of axonogenesis in the altered neuromeric environment expected in mutant mice could provide insights into the mechanisms of axon guidance.

As a first step toward a genetic analysis of early neuronal development, we examined the neuronal development of *Wnt-1*$^{-/-}$ embryos, in collaboration with Dr. Andrew P. McMahon (Harvard). The *Wnt-1* gene is one of the best characterized candidates for directing the subdivision of the neural tube. *Wnt-1* is a member of a family of related genes that are expressed early in vertebrate embryogenesis (McMahon, 1992; Nusse and Varmus, 1992). Like its *Drosophila* homologue, *wingless* (*wg*), the mouse gene *Wnt-1* encodes a secreted glycoprotein that may play a role in directing cell fates. The primary role of *Wnt-1* is in the development of the brain, as it is transcribed early in the neural tube (Wilkinson et al., 1987; Parr et al., 1993). *Wnt-1* transcripts are initially detected on E8.0, before neural tube closure, throughout the dorsal-ventral extent of the presumptive mesencephalon except the ventral midline, but by E9.0-E9.5, transcripts become restricted to a dorsal strip in the mesencephalon and a ventrally-extending narrow ring just rostral to the mesencephalon/r1 border (Wilkinson et al., 1987; McMahon et al., 1992). An additional dorsal domain of expression also extends rostrally from the mesencephalon into the syn- and parencephalon.

Several lines of evidence indicate that a large region of the brain, including the mesencephalon and r1, is dependent on *Wnt-1* function for normal development. *Wnt-1* mutants generated by gene targeting result in embryos with large deletions in the brain (McMahon and Bradley, 1990; Thomas and Capecchi, 1990). Homozygous *Wnt-1*$^{-/-}$ mice lack the cerebellum and most or all of the midbrain on E14.5 and E16.5 (McMahon and Bradley, 1990; but see Thomas and Capecchi, 1990). The first defect is evident soon after neural tube closure. On E9.5, the brain of *Wnt-1*$^{-/-}$ embryos is smaller than wildtype (McMahon et al., 1992), principally because the mesencephalon and r1 appear to be severely reduced in size. This impression is supported by altered expression of the *engrailed* genes *En-1* and *En-2*, which are lost in the mutant at early stages (McMahon et al., 1992). In *Wnt-1*$^{-/-}$ embryos, the expression of *En-1* and *En-2* is initiated normally in both the mesencephalon and r1. However, *En* expression is reduced by E8.5 to a residual domain spanning the dorsal-ventral extent of r1, and by E9.5, this *En* expression also disappears. This suggests that deletion of the mesencephalon and r1 has juxtaposed the synencephalon and r2, the regions rostral and caudal to the intervening wildtype *En* domain. This interpretation is supported by altered expression of other genes (*Wnt-7b*, *Wnt-5a*) at E9.5 which also suggest deletions (McMahon et al., 1992).

In summary, the loss of *Wnt-1* function results in the early disruption of brain morphogenesis, and in the subsequent deletion of large regions of the brain. In the previous reports, the extent of the early disruptions could not be assessed very accurately from the whole mounted embryos, as they were not stained in a way that revealed neurons. To define more precisely the effects of the lesion on the initial tract formation of the brain, we examined early neuronal development in *Wnt-1*$^{-/-}$ embryos, using immunocytochemistry and fluorescent tracers to label early neurons and their axons (Mastick, et al., submitted). Homozygous mutant embryos have a severely reduced brain size, while wildtype and heterozygous littermates appear normal (McMahon et al., 1992). The brain has obviously been shortened along the rostral-caudal axis. The prosencephalon, including the eye and cerebral vesicles, is rotated dorsally, and moved closer to the rhombencephalon, which is marked by the branchial arches and otic vesicle. The interneuromeric constriction that marks the boundary between the mesencephalon and r1 is missing, and the curved dorsal-rostral surface of the brain now is adjacent to the thin roof of the rhombencephalon. Without additional landmarks, it is difficult to identify the brain regions that remain in unlabelled mutant embryos.

Neuron-specific immunolabelling provides the advantage of defining brain defects by groups of early neuronal landmarks, and we were able to place limits on the brain deletion, which includes the mesencephalon and r1. The ventral structures missing are the most clearly defined, as a deletion of the neuron-sparse ventral r1 marks the caudal boundary, and the loss of nIII places the rostral boundary at or anterior to the synencephalon-mesencephalon border. Dorsal defects are also present, although the boundaries are less clear. nIV, and the adjoining regions of the mesencephalon and r1 that form the interneuromeric constriction are deleted. Most or all of the cerebellar plate is also deleted, which is consistent with previous observations of the complete loss of *En* expression in r1 by E9.5 (McMahon et al., 1992), and the failure of the cerebellum to develop (McMahon and Bradley, 1990; Thomas and Capecchi, 1990). The rostral-dorsal boundary is in the dorsal synencephalon or mesencephalon, a region lacking landmarks.

Despite the disappearance of their neighboring neuromeres, the spared regions adjacent to the deletion, r2 and the synencephalon, appear to continue their program of neuronal development without delay or alteration. For example, synencephalic neurons in appropriate locations are produced and project axons caudally with the normal timing to pioneer the dtmesV and mlf pathways. We expected to find axon pathfinding errors caused by the absence of the mesencephalon and r1. Axons extending caudally from the synencephalon would encounter r2 upon crossing the caudal boundary of the synencephalon, in effect bypassing the normally intervening mesencephalon and r1. However, the only defect was an apparent reduction in the number of axons in both of the tracts, which is most likely caused by the elimination of most of their somata of origin, rather than by axon pathfinding errors. First, the rostral border of the deletion, implied by the loss of nIII, extends to the synencephalon-mesencephalon border. The deletion thus includes most dtmesV neurons, and half of the mlf neurons, as described earlier. This would have the simple effect of reducing the number of axons present in the tracts. Secondly, by three different methods, we found that the formation of the dtmesV and mlf occurs without errors. In immunolabelled whole mounts, tracts (with fewer than normal axons) were in the correct location within the neural tube. Immunolabelling in transverse sections at what is

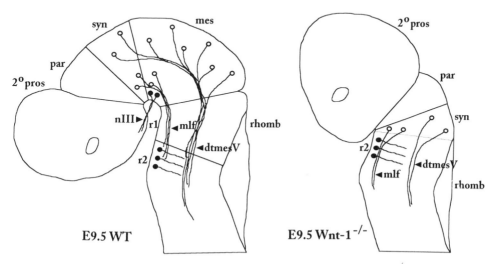

Figure 5. A schematic diagram of the neuronal organization of wildtype and *Wnt-1⁻/⁻* embryos at E9.5, labelled with an antibody to neuron-specific β-tubulin. The precocious neuronal labelling in ventral r2 is indicated. In the mutant, a dotted line indicates the novel boundary between synencephalon and r2, although no visible interneuromeric constriction is present. (Modified from Mastick, et al., submitted.)

now the rostral end of the rhombencephalon, r2, labelled caudally-directed axons in the normal locations of dtmesV and mlf. Also, by directly labelling synencephalic neurons with diI, no aberrant axonal projections were observed. We have no way to evaluate the possibility that mesencephalic dtmesV and mlf neurons are present, but unable to extend axons, as we do not yet have specific markers for these neuron types.

Although the navigational cues used by dtmesV and mlf axons are unknown, the regulative ability of these axons to compensate for their altered environment is remarkable. It is possible that some mesencephalic or r1-derived neural tube is retained in the mutant, possibly providing a less abrupt transition zone, but it is clear that most of these regions are deleted. The faithful tract formation in the absence of the mesencephalon and r1 implies that any navigational cues that lie in the region between the synencephalon and r2 are dispensable for these axons.

Wnt-1 action is likely to be mediated in part by *En-1*, as targeted mutations in the *En-1* gene have recently been shown to also cause a large mesencephalon and r1 deletion in homozygous mutant embryos, resulting in similar neuronal defects (Wurst et al., 1994). In older embryos than those examined here (E10.5), neurofilament immunolabelling shows a loss of both nIII and nIV. However, *En-1* function cannot fully account for *Wnt-1* action: the *En-1* brain deletion does appear to be less severe than in *Wnt-1* mutants, and some of the En-2 expression is retained. Interestingly, the *wingless* (*wg*) gene, a closely related *Drosophila* member of the *Wnt* family, functions to direct neuronal development as well as regulate expression of the engrailed gene (Chu-LaGraff and Doe, 1993). Genetic analysis of mouse homologues of other molecular players in the *wg* signalling pathway (Joyner et al., 1991; Millen et al., 1994; Siegfried et al., 1994; Noordermeer et al., 1994) may reveal how *Wnt-1* acts in the development of the embryonic mouse brain.

CONCLUSIONS

The embryonic mouse brain is highly organized at the neuronal level, and our descriptions of pattern formation of neurons confirm the importance of the molecular and cellular compartments that have recently been discovered. As more genes are discovered and mutated, we anticipate that early patterns of neuron cell bodies and initial tracts will be crucial for interpreting the role of regulatory genes in directing this complex developmental process.

ACKNOWLEDGMENTS

This work was supported by postdoctoral traineeships to GSM from NSF (9014275) and NIH (5T32HD07274 and 1F32 NS09701-01), and a research grant from NIH (EY00168) to SSE.

REFERENCES

Birgbauer, E. and Fraser, S. E. (1994). Violation of cell lineage restriction compartments in the chick hindbrain. Development 120, 1347-1356.

Chu-LaGraff, Q. and Doe, C. Q. (1993). Neuroblast specification and formation regulated by *wingless* in the *Drosophila* CNS. Science 261, 1594-1597.

Clarke, J. D. W. and Lumsden, A. (1993). Segmental repetition of neuronal phenotype sets in the chick embryo hindbrain. Development 118, 151-162.

Easter, S. S., Jr., Ross, L. S. and Frankfurter, A. (1993). Initial tract formation in the mouse brain. J. Neurosci. 13, 285-299.

Echelard, Y., Epstein, D. J., St-Jacques, B., Shen, L., Mohler, J., McMahon, J. A. and McMahon, A. P. (1993). *Sonic Hedgehog*, a member of a family of putative signaling molecules, is implicated in the regulation of CNS polarity. Cell 75, 1417-1430.

Figdor, M. C. and Stern, C. D. (1993). Segmental organization of embryonic diencephalon. Nature 363, 630-634.

Fraser, S., Keynes, R. and Lumsden, A. (1990). Segments in the chick embryo hindbrain are defined by cell lineage restrictions. Nature 344, 431-435.

Godemont, P., Vanselow, J., Thanos, S. and Bonhoeffer, F. (1987). A study in developing visual systems with a new method of staining neurones and their processes in fixed tissue. Development 101, 697-713.

Hanneman, E., Trevarrow, B., Metcalfe, W. K., Kimmel, C. B. and Westerfield, M. (1988). Segmental pattern of development of the hindbrain and spinal cord of the zebrafish embryo. Development 103, 49-58.

Honig, M. G. and Hume, R. I. (1986). Fluorescent carbocyanine dyes allow living neurons of identified origin to be studied in long-term cultures. J. Cell. Biol. 103, 171-187.

Hunt, P., Gulisano, M., Cook, M., Sham, M.-H., Faiella, A., Wilkinson, D., Boncinelli, E. and Krumlauf, R. (1991). A distinct *Hox* code for the branchial region of the vertebrate head. Nature 353, 861-864.

Joyner, A. L., Herrup, K., Auerbach, A., Davis, C. A. and Rossant, J. (1991). Subtle cerebellar phenotype in mice homozygous for a targeted deletion of the *En-2* homeobox. Science 251, 1239-1243.

Kroger, S. and Schwarz, U. (1990). The avian tectobulbar tract: development, explant culture, and effects of antibodies on the pattern of neurite outgrowth. J. Neurosci. 10, 3118-3134.

Lee, M. K., Rebhun, L. I. and Frankfurter, A. (1990a). Posttranslational modification of class III beta-tubulin. Proc Natl Acad Sci USA 87, 7195-7199.

Lee, M. K., Tuttle, J. B., Rebhun, L. I., Cleveland, D. W. and Frankfurter, A. (1990b). The expression and postranslational modification of a neuron-specific beta tubulin isotype during chick embryogenesis. Cell. Motil. Cytoskel. 17, 118-132.

Lumsden, A. and Keynes, R. (1989). Segmental patterns of neuronal development in the chick hindbrain. Nature 337, 424-428.

Martinez, S., Geijo, E., Sanchez-Vives, M. V., Puelles, L. and Gallego, R. (1992). Reduced junctional permeability at interrhombomeric boundaries. Development 116, 1069-1076.

Mastick, G. S. and Easter, S. S. J. (Submitted). Initial organization of neurons and axon tracts in the embryonic mouse fore- and midbrain.

Mastick, G.S. and Easter, S.S. Jr. (Submitted). Neuronal markers define the embryonic brain deletion of *Wnt-1*[-/-] mutant mice.

McMahon, A. P. (1992). The *Wnt* family of developmental regulators. Trends in Genetics 8, 236-242.

McMahon, A. P. and Bradley, A. (1990). The *Wnt-1* (*int-1*) proto-oncogene is required for development of a large region of the mouse brain. Cell 62, 1073-1085.

McMahon, A. P., Joyner, A. L., Bradley, A. and McMahon, J. A. (1992). The midbrain-hindbrain phenotype of *Wnt-1*[-]/*Wnt-1*[-] mice results from stepwise deletion of *engrailed*-expressing cells by 9.5 days postcoitum. Cell 69, 581-595.

Metcalfe, W. K., Mendelson, B. and Kimmel, C. B. (1986). Segmental homologies among reticulospinal neurons in the hindbrain of the zebrafish larva. J Comp Neurol 251, 147-159.

Millen, K. J., Wurst, W., Herrup, K. and Joyner, A. L. (1994). Abnormal embryonic cerebellar development and patterning of postnatal foliation in two mouse Engrailed-2 mutants. Development in press, .

Moody, S. A., Quigg, M. S. and Frankfurter, A. (1987). Development of the peripheral trigeminal system in the chick revealed by an isotype-specific anti-beta-tubulin monoclonal antibody. J Comp Neurol 279, 567-580.

Noordermeer, J., Klingensmith, J., Perrimon, N. and Nusse, R. (1994). *dishevelled* and *armadillo* act in the *Wingless* signalling pathway in *Drosophila*. Nature 367, 80-83.

Nusse, R. and Varmus, H. E. (1992). *Wnt* genes. Cell 69, 1073-1087.

Parr, B. A., Shea, M. J., Vassileva, G. and McMahon, A. P. (1993). Mouse *Wnt* genes exhibit discrete domains of expression in the early embryonic CNS and limb buds. Development 119, 247-261.

Puelles, L. and Rubenstein, J. L. R. (1993). Expression patterns of homeobox and other putative regulatory genes in the embryonic mouse forebrain suggest a neuromeric organization. Trends. Neurosci. 16, 472-479.

Puelles, L., Amat, J. A. and Martinez-de-la-Torre, M. (1987). Segment-related, mosaic neurogenetic pattern in the forebrain and mesencephalon of early chick embryos: I. Topography of AChE-positive neuroblasts up to stage HH18. J Comp Neurol 266, 247-268.

Siegfried, E., Wilder, E. L. and Perrimon, N. (1994). Components of *wingless* signalling in *Drosophila*. Nature 367, 76-80.

Taber Pierce, E. (1973). Time of origin of neurons in the brain stem of the mouse. Prog Brain Res 40, 53-65.

Thomas, K. R. and Capecchi, M. R. (1990). Targeted disruption of the murine *int-1* proto-oncogene resulting in severe abnormalities in midbrain and cerebellar development. Nature 346, 847-850.

Vaage, S. (1969). The segmentation of the primitive neural tube in chick embryos. Ergeb. Anat. Entwicklungsgesch. 41, 1-88.

von Bartheld, C. S. and Bothwell, M. (1993). Development of the mesencephalic nucleus of the trigeminal nerve in chick embryos: target innervation, neurotrophin receptors, and cell death. J Comp Neurol 328, 185-202.

Wilkinson, D. G., Bailes, J. A. and McMahon, A. P. (1987). Expression of the proto-oncogene *int-1* is restricted to specific neural cells in the developing mouse embryo. Cell 50, 79-88.

Wurst, W., Auerbach, A. B. and Joyner, A. L. (1994). Multiple developmental defects in *Engrailed-1* mutant mice: an early mid-hindbrain deletion and pattern defects in forelimbs and sternum. Development 120, 2065-2075.

EARLY EVENTS IN ESTABLISHMENT OF THE VERTEBRATE HEART

Thomas A. Drysdale, Kristin D. Patterson, Wendy V. Gerber, and Paul A. Krieg

Center for Developmental Biology
Department of Zoology
University of Texas at Austin
Austin, Texas 78712
USA

INTRODUCTION

In order to understand how the heart is formed in the embryo, it is essential to understand the cellular movements and interactions that are needed for heart formation. This is the domain of classical embryology studies. It is also important to understand the molecules that mediate these interactions and direct cellular differentiation. This is the domain of molecular biology. In isolation, neither of these fields of study will provide a complete picture of how a heart is formed. Recently, a combination of these approaches has greatly increased the general understanding of embryonic heart development. We wish to present an overview of heart embryology and also to show how current molecular results are enhancing our understanding of heart formation.

WHAT DO WE MEAN BY THE HEART?

The heart is often referred to as a single entity. In embryological studies, this hides a great deal of complexity. When initially formed, the heart appears to be a simple tube but in fact it already consists of three layers: the endocardium, myocardium, and pericardium (Fig 1A). The endocardium is the inner lining of the heart that contacts the blood pumped by the heart. This can be thought of as an extension of the endothelium, the interior lining of blood vessels (Coffin and Poole, 1988). The endocardium and endothelium appear to be functionally equivalent, although it remains to be seen if their origins are identical. They certainly express many of the same genes, which is consistent with their similar function. The myocardium is the musculature of the heart and is responsible for pumping of blood. Most often when the heart is referred to in an embryological sense, it refers to the myocardium, since the majority of markers for the heart are myocardial markers. This is also true in older studies, because the presence of beating tissue was the usual assay for the presence of heart and the myocardium is the only tissue that will contract. The pericardium is a thin sheath that surrounds the heart.

There are certainly many differences between the endocardium and myocardium, as would be expected between two such functionally distinct tissues. However, do these layers have a common origin or do they arise from distinct lineages? Of course, at the level of the fertilized egg they must share a common origin, but do they diverge from a common progenitor just before they are committed to one fate or the other? Studies in chick and zebrafish have attempted to answer this question. In chick, single cells express markers that are later restricted to either endocardial or myocardial cells (Linask and Lash, 1993) suggesting that they share a common origin. Expression of myocardial markers in the endothelium has also been reported (Tokuyasu and Maher, 1987; Han et al., 1992). In contrast to these results that suggest a common origin, attempts at fate mapping these two tissues find that they may arise from distinct lineages. Fate mapping in chick suggests that the myocardium comes from a unipotential stem cell (Mikawa et al., 1992). In zebrafish, injection of lineage tracers at the 256-512 cell stage has shown that, although some regions will only give rise to myocardial and not endocardial precursors, other regions have the potential to form both at this stage (Stainier et al., 1993). Essentially, the question of endocardial origins has not been resolved. However, the recent identification of genes likely to play a role in the determination of the endocardium and myocardium should prove useful in establishing the origins of these tissues. At present, no information appears to be available concerning the determination of the pericardium.

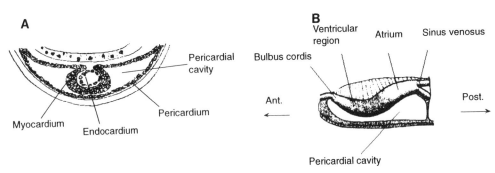

Figure 1. (A) The fusion of the lateral plate mesoderm results in three layers of tissue in the heart: the myocardium, endocardium, and pericardium. This process can be seen in this diagram of a cross section through the developing heart region of a frog embryo. Two layers of lateral plate mesoderm split, forming the pericardium and myocardium. The resulting space between the layers becomes the pericardial cavity. The endocardium is found trapped between the myocardium as it fuses at the midline. (B) Once the primitive heart tube is formed, it is subdivided into different regions as seen in this diagram of a frog heart. Note that the ventricular region is anterior and the atrium is posterior. Subsequent morphogenetic movement will reverse this orientation (adapted from Nelsen, 1953).

In addition to the various layers of the heart, several regions can be distinguished along the anterior-posterior axis (Fig. 1B). The most familiar regions are the atrium and ventricle, but the heart tube is also regionalized into the bulbis cordis, which goes on to form the conus ateriosus, and the sinus venosus. At this stage of development, the simple heart tube is similar in all vertebrates, suggesting that early patterning events are conserved and that it is the subsequent folding and cardiac morphogenesis that creates the differences seen between the various vertebrate classes. The advent of molecular markers is starting to provide some clues as to the determination of the various subdivisions of the vertebrate heart. For example, there are molecular differences between the atrial myocardium and ventricular myocardium. A chick myosin heavy chain sequence specific for the atrium has been identified (Yutzey et al., 1994) and a myosin light chain, specific to the ventricle, is known in mouse (O'Brien et al., 1993). Differences occur very early in the development of the heart indicating that regional characteristics are established soon after the heart is specified. These regional differences can be thought of as anterior-posterior patterning because when the heart is a simple tube, the ventricle is in the anterior part of the tube and the atrium is posterior. Subsequent folding of the heart reverses this orientation as the embryo ages. Consistent with other examples of embryonic anterior-posterior patterning, retinoic acid appears to be able to alter the anterior-posterior patterning of the heart (Stainier and Fishman, 1992; Yutzey et al., 1994). It is not known if the anterior-posterior patterning seen in the myocardium is reflected in the underlying endocardial or overlying pericardial tissue.

EMBRYONIC EVENTS IN HEART FORMATION

Mesoderm Induction and the Heart

The heart is of mesodermal origin. The fate maps of Keller (1976) show that, in *Xenopus*, the heart originates as two bilaterally symmetric patches of deep mesoderm. The heart primordium lies about 30° to 45° lateral to the dorsal midline, adjacent to the organizer (Fig. 2A). An understanding of how mesoderm is formed and subsequently patterned is therefore, a fundamental first step in understanding the development of the heart. Fortunately, studies of the amphibian embryo have made much progress in this area.

Formation of mesoderm requires an inductive signal from the underlying endoderm (Nieuwkoop, 1969). In *Xenopus*, the newly induced mesoderm forms a ring of tissue at the embryonic equator. On the dorsal side of the embryo the mesoderm forms the Spemann's organizer which goes on to form the notochord and prechordal plate. As one moves from dorsal to ventral on the fate map (Fig 2A) the types of mesoderm change. Other derivatives of mesodermal tissue include blood, kidney, and somites. These tissues are arranged in a graded character around the induced mesodermal ring. As can be seen from the mesodermal fate map (Fig. 2A) this only approximates a simple dorsal-ventral gradient, as distinct anterior-posterior differences are also present. In *Xenopus*, a model has been proposed to explain the emergence of the mesodermal pattern (Dale and Slack, 1987). This three signal model suggests that two signals emanate from the inducing endoderm. On the most dorsal side of the embryo, one signal induces the organizer region. The rest of the mesoderm is induced to form an extreme ventral character by a second signal. To account for the intermediate mesoderm, a third signal is hypothesized to be sent from the organizer. This third signal dorsalizes the ventral mesoderm to give the intermediate types of mesoderm. A number of molecules have now been identified that can either directly induce mesoderm or can act to modulate mesodermal patterning. In general, the three signal model holds up well as a method of generating a graded range of mesoderm, although accumulated results from a large number of experiments suggest that the actual number of molecules (or signals) is much larger than the three originally proposed (Moon and Christian, 1992; Smith et al., 1993; Ku and Melton, 1993).

The three signal model fits rather well with the experimental evidence on heart determination. Embryological studies of *Xenopus* show that heart specification occurs during gastrulation (Sater and Jacobson, 1989) and that signals required for specification emanate from the dorsal lip of the blastopore (Sater and Jacobson, 1990b). This is exactly how the three signal model suggests that heart induction should occur. Noggin is a secreted factor that has many of the properties expected for a dorsalizing agent made in the organizer (Smith et al., 1993). However, at present there is no evidence that it can induce heart tissue.

It is intriguing to note that a heart is always present in *Xenopus* embryos that have been dorsalized by lithium (Kao et al., 1986; Drysdale et al., 1994). This raises the possibility that the heart may actually be part of the organizer region, like the notochord and prechordal plate. In this case, induction of heart would be viewed as patterning within the organizer. A number of transcription factors have now been implicated in organizer patterning (Weinstein et al., 1994) but none appear to be expressed in the heart. Alternatively, the presence of heart tissue in lithium-treated embryos may reflect incomplete dorsalization and that the dose of lithium required for complete dorsalization is lethal.

How the Heart Gets Where it is Going

The events of mesoderm induction are followed by the movements of gastrulation. During gastrulation, the two patches of heart tissue are at the leading edge of the involuting

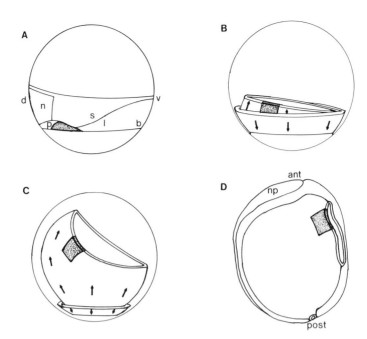

Figure 2. Fate map of the heart and its subsequent movements during gastrulation. (A). A fate map of the mesodermal mantle showing the approximate position of the heart primordium (stippled). There is a range of mesoderm formed in the embryo and the pattern of the fate map corresponds to the dorsal-ventral axis. Notochord (n) and prechordal plate (p) will form from mesoderm on the dorsal (d) side. Intermediate mesoderm includes the somites (s) and lateral plate mesoderm (l). The heart is derived from the most dorsal lateral plate mesoderm. On the ventral (v) side, blood (b) will form. In the other figures, the dorsal/ventral axis is maintained. During gastrulation, the mesoderm ring involutes (B; see arrows) bringing the mesoderm further into the embryo. The heart (stippled) is at the leading edge of the involuting mesoderm. Note that the heart primordia are paired and that the diagram only shows one side. (C) Gastrulation movements draw all of the mesoderm into the interior. Note that the dorsal side involutes further than the ventral and the heart is on the dorsal side. (D) Once gastrulation is complete, all of the mesoderm has involuted and the heart is now at the anterior (ant) end closer to the ventral side of the embryo. At this stage a neural plate (np) has formed. The two heart primordia then fuse at the ventral midline and subsequent movements make a heart tube and eventually a mature heart. These figures are based on the fate maps of Keller (1976) and Gerhart and Keller (1986).

mesoderm and come to lie on either side of the head (Fig. 2D). The patches of tissue then move laterally along the body wall until they fuse at the ventral midline. The cement gland, a structure used by the frog embryo to stick to substrates, is easily visible after neurulation and serves as an excellent guide as to the location of the heart. The heart fuses caudal to the cement gland. The area indicated as heart in Fig. 2 is not specific about the location of the precursors of the endocardium, myocardium, or pericardium. Fate maps that distinguish between these three layers has been attempted in chick and zebrafish, (see What Do We Mean by the Heart), but the low resolution of these maps has prevented a clear understanding of their origins. In classical terms, the region fated to form heart is part of the lateral plate mesoderm. Lateral plate mesoderm splits into two layers: the splanchnic and somatic mesoderm. In the cardiac region, the somatic portion, which is closest to the body wall, gives rise to the pericardium, while the internal, splanchnic portion goes on to form the myocardium. The two splanchnic layers fuse at the midline forming a tube. During the fusion process, they trap some tissue between them and the trapped cells come to lie at the interior of the tube. These trapped cells are the endocardium. How the endocardium ends up in this position is not understood. When other areas of lateral plate mesoderm split, the cavity that is created between the two layers is the coelom. In the heart, the coelom that is created by the splitting of the mesoderm is referred to as the pericardial cavity.

The process of splitting and fusion of the heart mesoderm has been visualized to some degree in chick (DeHaan, 1963), mouse (DeRuiter et al., 1992), and zebrafish (Stainier et al., 1993). A similar study has not been done with *Xenopus* embryos. The same movements may be occurring in *Xenopus* but they are likely to be difficult to observe. At the time this process occurs, the area through which the heart is moving is very congested and the bilateral tubes of heart mesoderm, seen in other organisms, may be quite compressed. Another difficulty is that the known differentiation markers are first expressed when the process of fusion is already well underway (Logan and Mohun, 1993; Chambers et al., 1994; Drysdale et al., 1994). The discovery of *XNkx-2.5* (Tonissen et al., 1994; this report), a heart marker expressed prior to fusion of the two cardiac primordia, should overcome this difficulty and enable a comparison between heart migration in *Xenopus* and other vertebrates.

How is the Heart Determined?

We know that a general range of mesoderm is established during mesoderm induction. The next step is the specification of cell types within that mesoderm. This aspect of mesoderm induction is not well understood. There appears to be some delay between mesoderm induction, which ends immediately before gastrulation (Jones and Woodland, 1987) and the actual specification of the heart. In *Xenopus*, heart is not specified until gastrulation movements are complete. During gastrulation, a signal from the organizer region tells the adjacent mesoderm to form heart tissue. After gastrulation, the presumptive heart region can be explanted and it will go on to form heart. In addition to the embryological evidence for this delay (Sater and Jacobson, 1989), treating embryos with retinoic acid early in gastrulation will block formation of heart (Sive et al., 1990; Drysdale et al., 1994). This suggests that there is at least a short delay from the end of mesoderm induction until the time at which heart is determined.

There are currently only two methods that can produce heart from tissue that would not normally make heart. These methods provide some clues as to how the heart may normally be formed. Sater and Jacobson (1990b) were able to show that the presence of an organizer region was able to induce adjacent, lateral mesoderm to form heart. This suggests that a heart-inducing signal emanates from the organizer region. Interpretation of this experiment is somewhat complicated because the lateral mesoderm region used in the experiments was normally fated to become heart. It is true that, if excised early in gastrulation and cultured in isolation, lateral mesoderm does not form heart, but this does not exclude the possibility that the presumptive heart region is somehow predisposed to react to the inductive signal coming from the organizer region. Alternatively, a greater range of ventral mesoderm may be competent to form heart in response to signals from the organizer region. An experimental strategy outlined below addresses this possibility.

A second method of forming heart tissue is to explant and culture organizer tissue (the dorsal lip of the blastopore at stage 10.5) in the presence of suramin. Explants of organizer tissue normally form notochord and prechordal plate mesoderm as well as anterior neural tissue, presumably through the induction of uncommitted ectoderm in the explant. Heart is not normally observed in organizer explants. When treated with suramin, notochord and anterior neural tissue are no longer present. Instead, many of the explants appear to form large amounts of heart tissue (Grunz, 1992; 1993). These studies relied on the presence of beating tissue as an assay for cardiac differentiation. We have repeated this experiment using expression of cardiac troponin I as a sensitive assay and find that most organizers treated with suramin will go on to form large amounts of heart tissue (Fig. 3). Suramin is known to interfere with the activity of certain growth factors, especially members of the FGF and TGF superfamilies (Coffey et al., 1987) but the mechanism by which suramin alters the fate of the organizer region is not understood.

MOLECULAR BIOLOGY OF HEART DEVELOPMENT

The above descriptions of heart development in frog can be used to define various events in the development of the heart. Molecular changes are now being defined that correspond to these embryological events. We will now briefly discuss some recent

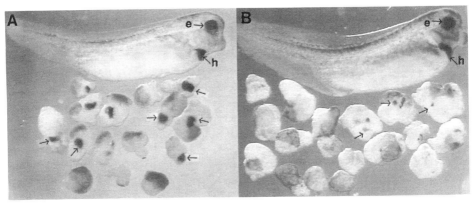

Figure 3. Dorsal blastopore lips were excised early in gastrulation and cultured in NAM/2 (Peng, 1991) in the presence of suramin (A) (150µM for 3 hours and then returned to NAM/2), or in NAM/2 alone (B). These were subsequently examined for the presence of cardiac troponin I transcripts by whole mount *in situ* hybridization (Harland, 1991). Expression of cardiac troponin I is limited to the heart (h) as seen in the whole embryos. The eye (e) appears dark due to natural pigmentation. When cultered in suramin (A), the dorsal lip explants failed to elongate and have regions that express high levels of cardiac troponin I (arrows) indicating the presence of myocardial tissue. Explants cultured in NAM/2 alone (B) only rarely contained myocardial tissue (arrows) probably due to including some presumptive heart tissue in the original explant.

studies that provide a description of heart development at the cellular and molecular level. In the section on growth factors, we will briefly summarize the current understanding of the cellular events involved in mesoderm induction and explain how these studies may guide us in thinking about heart induction. We know that heart is specified by the end of gastrulation (Sater and Jacobson, 1989), after the end of mesoderm induction (Jones and Woodland, 1987). It is not until considerably later, during the tadpole stage, that the heart begins to express cardiac-specific differentiation markers. What is happening during this delay? Very little is known about the genes expressed in specified, but undifferentiated heart tissue. We have recently isolated a frog homeobox gene, *XNkx-2.5*, that is expressed in myocardial precursor cells, and which is likely to be involved in the regulation of cardiac development (Tonissen et al., 1994). Some details of *XNkx-2.5* expression and its potential role in heart development will be discussed below.

Growth Factors and the Heart

From molecular studies, it is becoming clear that peptide growth factors are key elements in mesoderm induction. Some members of the transforming growth factor-β (TGFβ) superfamily, and members of the fibroblast growth factor (FGF) family have been shown to induce mesoderm from ectoderm using the animal cap explant as a responding tissue (reviewed in Smith, 1989). Using the same assay, activin, a member of the TGFβ superfamily, can induce mesoderm in a dose dependent manner, the lowest doses inducing more ventral type mesoderm and the highest doses inducing more dorsal tissue such as notochord (Green et al., 1992). The ability of a single molecule to induce a range of mesoderm, depending on concentration, appears to negate the need for a three signal model. However, an endogenous gradient of activin has yet to be demonstrated and FGF signaling is required for a response to activin signaling (Cornell and Kimelman, 1994; LaBonne and Whitman, 1994). It is possible that a precise concentration of activin could directly specify a mesodermal region to form heart. Because the concentration required to generate a specific type of mesoderm appears to be correlated with its position along the dorsal ventral axis, one could predict that a concentration lying between that required for notochord and somite specification might yield heart tissue. Using experiments by Green et al. (1992) as a guide, a concentration of 0.5 ng/ml might be predicted to specify heart mesoderm. The ability of activin to specifically induce heart was not studied until recently because heart-specific molecular markers, a necessity for these experiments, were not

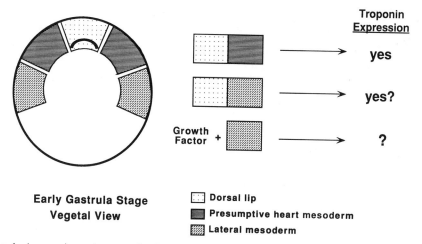

Early Gastrula Stage
Vegetal View

Troponin
Expression

yes

yes?

?

Growth Factor +

⬚ **Dorsal lip**
▤ **Presumptive heart mesoderm**
▨ **Lateral mesoderm**

Figure 4. An experimental strategy for finding growth factors that can induce heart tissue. The dorsal blastopore lip is considered to be the source of the natural inducer. The lateral mesoderm does not form heart when cultured in isolation but cardiac tissue may be formed when it is cultured in contact with the blastopore lip. This tissue may therefore serve as a substrate for detection of growth factors that play a role in cardiac induction. Details are contained in the text.

available. Molecular markers, specific for *Xenopus* heart tissue, have recently been cloned and it appears that activin is able to induce heart markers in *Xenopus* animal pole explants (Logan and Mohun, 1993). However, the levels of activin required to induce heart are 10 ng/ml, about 20 times that required for expression of notochord and prechordal plate gene expression. In addition, several explants had to aggregate in order for heart-specific gene expression to occur. These observations may be partially explained because of the use of explants rather than disassociated cells but the discrepancy from predicted values is still large. Alternatively, high levels of activin may be required because they are activating another signal transduction pathway. Indeed, it has been suggested that all mesoderm induction by activin may be working by activating the Vg1 receptor (Schulte-Merker et al., 1994). Vg1 is a member of the TGFβ family that has properties that make it an excellent candidate for being an endogenous mesoderm inducer, however, there is currently little evidence that the Vg1 precursor is cleaved to generate the active Vg1 in the embryo (Vize and Thomsen, 1994).

In axolotl, a salamander, there is evidence that specific peptide growth factors may be involved in the earliest stages of heart differentiation. Explants of prospective cardiac mesoderm were cultured with various known growth factors and then assayed for cardiac differentiation by the presence of beating tissue. When cultured alone, these explants rarely formed heart, but treatment of explants with PDGF-BB or TGFβ1 greatly increased the frequency of heart differentiation (Muslin and Williams, 1991). This work suggests that PDGF and TGFβ can at least mimic the role of natural inducer molecules and it is possible that these growth factors actually play the inducing role in the embryo. It must be remembered that the explants already had the potential to form heart. This makes it difficult to assign these factors an inductive role. PDGF-BB or TGFβ1 may simply promote the differentiation of heart tissue that has already been specified by a different factor. In *Xenopus*, receptors for FGF and PDGF-Rα are detectable at gastrulation (Freisel and Brown, 1992; Tannahill et al, 1992; Jones et al, 1993) giving these growth factors the potential to have a role in heart development, however, their distribution appears to be more widespread than would be expected if this was their primary role in early development.

We are carrying out experiments designed to identify growth factors, or indeed other molecules, involved in the early specification of heart. Although there is a growing body of evidence suggesting that growth factors play a role in heart development, many assays have been frustrated by the lack of unambiguous molecular markers for heart tissue. Previously, experiments have had to rely on beating tissue as an assay. This assay is time-consuming and may not work if small amounts of tissue are induced. To address this problem we have isolated the *Xenopus* homologue of the cardiac troponin I gene for use as a molecular marker for *Xenopus* myocardial tissue. Cardiac troponin I transcripts are heart specific at all stages of development (Drysdale et al., 1994). Two other markers have also been isolated, the *Xenopus* homologues of the myosin heavy chain α-isoform (Logan and Mohun, 1993) and the cardiac muscle-specific myosin light-chain (Chambers et al., 1994). The availability of these markers and the ability to easily isolate various types of mesoderm should enable us to identify growth factors involved in the initial induction of heart tissue. The general strategy for these experiments is to treat uncommitted ventral mesoderm with purified growth factors and assay for expression of cardiac troponin I (Fig. 4). Once a molecule capable of inducing cardiac troponin I expression is identified, we will determine if the potential factor is temporally and spatially expressed in a pattern consistent with heart specification.

Other questions can be addressed with this assay. As discussed earlier, most studies of heart development have focused on the myocardium. This is also true in our proposed growth factor studies as cardiac troponin I is a myocardial-specific marker. However, the same assay can also be used to detect growth factors involved in endocardial development or to determine if a single growth factor is capable of inducing both myocardium and endocardium. Recent cloning of a molecular marker for endocardial tissue, GATA-4 (Kelley et al., 1993), will facilitate these studies.

XNkx-2.5 and Heart Specification

To facilitate the study of the earliest events involved in specification and differentiation of the vertebrate heart, we have cloned a homeodomain-coding sequence whose early expression is limited to the cardiac muscle precursor cells (Tonissen et al., 1994). The *Xenopus* homeobox gene, *XNkx-2.5*, belongs to the NK class of homeodomain-containing proteins and appears to be related to the *Drosophila* gene *tinman* (Bodmer, 1993). It is also related to two independently isolated mouse homeobox sequences called *Nkx-2.5* (Lints et. al., 1993) and *csx* (Komuro and Izumo, 1993). At the amino acid level, the homeodomain of *XNkx-2.5* is 95% and 67% identical to that of mouse *Nkx-2.5* (*csx*) and *Drosophila tinman*, respectively (Tonissen et al., 1994). Outside of the homeodomain, *XNkx-2.5* and *tinman* appear to be completely divergent except for a conserved decapeptide near the N-terminus (Fig. 5). Not surprisingly, *XNkx-2.5* shows much greater similarity to mouse *Nkx-2.5*. The two proteins are 62% identical at the amino acid level and in addition to the conserved decapeptide and homeodomain, they also share a conserved segment downstream of the homeodomain (Tonissen et al., 1994; Fig. 5) In the mouse, *Nkx-2.5* is expressed very early during development in precardiac cells and continues to be expressed in the myocardium and the pharyngeal floor adjacent to the heart (Lints et al., 1993). In the early *Drosophila* embryo, *tinman* is expressed in mesodermal cells that will form the dorsal vessel, the insect equivalent of the heart, and other visceral musculature. Flies mutant for *tinman* fail to develop any visceral or cardiac tissue (Bodmer, 1993). The similarity in sequence and in expression pattern suggests that the role of *Nkx-2.5* and *tinman* proteins in heart development has been conserved over great evolutionary distances.

We are currently carrying out a detailed examination of *XNkx-2.5* expression in the frog embryo using *in situ* hybridization. *XNkx-2.5* is expressed in the presumptive heart region well before overt heart differentiation (Tonissen et al., 1994). *XNkx-2.5* transcripts first appear in two patches just ventral to the neural folds. As development proceeds, the staining pattern is consistent with the heart morphological movements described above. From post-neurula until tailbud stages, the extent of *XNkx-2.5* expression becomes continually more restricted until it is expressed only in differentiated heart tissue. This might indicate that the spatial pattern of *XNkx-2.5* expression coincides with the *Xenopus* embryonic heart field (Tonissen et al., 1994). Alternatively, it may detect migrating presumptive heart tissue that was not detected in the experiments of Sater and Jacobson (1990a). In later tadpoles, *XNkx-2.5* transcripts are detected in the foregut and pharynx in addition to the heart.

To further characterize expression of *XNkx-2.5*, RNA blot analysis was carried out using probes spanning the entire coding region of *XNkx-2.5* (Fig 6A). Two prominent

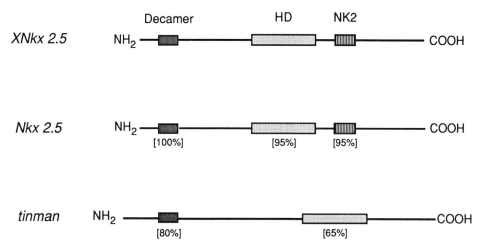

Figure 5. A comparison of the XNkx-2.5, Nkx-2.5, and tinman proteins. Three regions of high similarity exist between the XNkx 2.5 and Nkx 2.5 sequence. The homeodomain (HD), an NK2 domain (NK2), and a novel decamer sequence near the amino terminus. The tinman protein exhibits similarity in the conserved decamer but there is no recognizable NK2 domain. The numbers in brackets represent the amino acid identity with the XNkx 2.5 sequence

bands representing transcripts of 1.9 and 2.7 kb in length were visible at all stages from 15-26. These transcripts are present in reduced amounts later in development as well, although this is not apparent from the exposure shown. There is also a faint, but reproducible, band that is greater than 8 kb. The two smaller transcripts are present in approximately equal amounts and the ratio of the two species remains constant through early development. These results are in contrast to those observed in mouse where only a single 1.6 kb transcript in embryonic and adult tissues is reported (Lints et al., 1993; Komuro and Izumo, 1993). Since the clone we have isolated corresponds to the smallest transcript, our sequence does not provide any clues as to the nature of the larger transcripts. The first obvious possibility is that the transcripts differ at the site of poly-A addition. This possibility was supported by detection of a cDNA clone that contained sequences downstream of the poly-A addition site in the original *XNkx-2.5* clone. However, when a 1kb fragment of the extended 3' untranslated region (UTR) was used to probe the Northern blot, only the >8 kb band hybridized (data not shown), indicating that the two prominent transcripts are not due to use of alternative poly A addition sites. Another possibility is that the upper band is from a related but different gene. However, probing with the first 700 bp of the 3' UTR detects both the 1.9 and 2.7 kb transcripts at high stringency (data not shown), suggesting that these transcripts originate from the same gene. It is also possible that the two transcripts differ because of differential splicing patterns within the coding region. We have amplified sections of the coding region using PCR primer pairs and there is no evidence for the use of alternate exons within the coding region (data not shown). We are currently investigating the possibility that the differences between the two mRNAs lie at the 5' end, possibly involving different 5' exons and even different promoters.

Our RNA blot analysis, as well as RNase protection data (unpublished) first detect *XNkx-2.5* transcripts at mid-gastrulation. This is significant because this is the time at which heart tissue is thought to be first specified (see above). The onset of *XNkx-2.5* transcription coincides well with the time of heart specification and may represent an immediate early response to heart inductive signals. If this is the case, *XNkx-2.5* should be expressed in presumptive heart tissue even when protein synthesis is inhibited. This is currently being tested by culturing explants of presumptive heart plus dorsal lip in the presence of the protein synthesis inhibitor cycloheximide (Cascio and Gurdon, 1987) and subsequently assaying for *XNkx-2.5* expression.

Is *XNkx-2.5* always restricted to the heart? *XNkx-2.5* transcripts are detectable by RNase protection assays before a clear *in situ* pattern is visible. To determine whether the

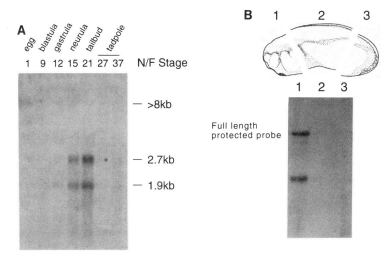

Figure 6. Expression of *XNkx-2.5* in the embryo. (A) An RNA blot, containing approximately 10µg of poly-A+ RNA per lane, was hybridized with random primed *XNkx-2.5* probes. Transcript size was determined by comparison with RNA size markers. Lanes are labeled with numbers corresponding to *Xenopus* embryonic stages (Nieuwkoop and Faber, 1994). (B) Post-neurula embryos were dissected into anterior (1), middle (2) and posterior (3) thirds. RNA from these portions were analyzed by RNAse protection for the presence of *XNkx-2.5* transcripts. At this stage, and earlier (data not shown), *XNkx-2.5* transcripts are localized to the anterior, heart-forming region of the embryo.

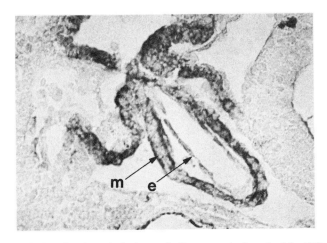

Figure 7. Longitudinal section through the heart of a *Xenopus* tadpole stained for *XNkx-2.5* transcripts by whole mount *in situ* hybridization. All detectable expression was limited to the myocardium (m). There was no expression of *XNkx 2.5* in the endocardium (e).

earliest expression of *XNkx-2.5* is localized to the heart forming region of the embryo, we have performed RNase protection analysis using RNA from dissected embryos. Our experiments show that *XNkx-2.5* RNA is localized to the heart region of the embryo, even during the early stages of expression (Fig. 6B and unpublished). This is consistent with the expression pattern of mouse *Nkx-2.5*. In contrast, *Drosophila tinman* is expressed broadly in early mesoderm before becoming restricted to pre-cardiac mesoderm (Bodmer, 1993).

Is *XNkx-2.5* expression restricted to the myocardium? To pinpoint the cells in which *XNkx-2.5* is expressed, we sectioned embryos stained by whole-mount *in situ* hybridization. In the tailbud stage embryo shown, the heart has already formed into an inner, endocardial tube surrounded by an outer, myocardial tube (Fig. 7). It is quite apparent that *XNkx-2.5* transcripts are localized exclusively to the myocardium. This is also true in the mouse (Lints et al., 1993). Since *XNkx-2.5* transcripts are also expressed in the presumptive heart of the early embryo before cardiac differentiation occurs, it appears to function as an excellent marker for the developing myocardium and will allow us to visualize the morphogenesis of this tissue in *Xenopus*.

Another *Xenopus* gene whose expression is limited to the heart at early stages is *GATA-4* (Kelley et al., 1993). It is also expressed during gastrulation and is a member of the GATA family of DNA binding proteins suggesting an important role in heart specification. Interestingly, expression of this gene is limited to the endocardium. As described above, the origin of the endocardium is still not well understood. Since both of these genes are expressed at the time of heart specification, the presence of both *GATA-4* and *XNkx-2.5* transcripts in the same cell at these early stages, might indicate a common origin for the two tissues. Aside from the technical difficulty of such an analysis, one caveat to interpreting any result is the finding that the probable mouse homologue of *GATA-4* is expressed in both endocardium and myocardium (Heikinheimo et al., 1994).

An important remaining question concerns the function of *XNkx-2.5* in the development of the heart. Our current hypothesis relies solely on the nature of the protein. Homeodomain-containing proteins are well-established as transcriptional regulators (McGinnis and Krumlauf, 1992). Some of the best characterized homeodomain-containing proteins, such as those included in the *Drosophila* HOM-C and the vertebrate Hox clusters, function in axis formation and limb development by providing positional information in the embryo (McGinnis and Krumlauf, 1992). *XNkx-2.5* is a member of the NK-class of homeodomain-containing proteins. These proteins were first characterized in *Drosophila*, although many have been isolated from other organisms including planarians, *C. elegans*, leech, *Xenopus*, rat, and mouse (Kim and Nirenberg, 1989; Garcia-Fernandez et al., 1993; Okkema and Fire, 1994; Guazzi et al., 1990). Transcription from many of these genes seems to be restricted to individual cell-types, suggesting they may function in the specification or differentiation of tissues rather than providing positional information in the embryo.

Given the expression pattern of *XNkx-2.5*, we propose that it is functionally involved in the early development of the *Xenopus* heart. Our expression profiles show that *XNkx-2.5* transcripts are first detected during gastrulation, peak at late neurula stages and decrease to about 20% of peak levels by early tailbud stages, when the expression of myocardial differentiation markers begins. This profile is not exactly what we would expect if *XNkx-2.5* is a direct transcriptional regulator of cardiac differentiation genes (although we must remember that we are observing the RNA profile and that the protein profile could be different). Perhaps it is responsible for keeping the prospective heart tissue in an undifferentiated state until an appropriate environment for heart formation exists. This hypothesis is possible but has to be reconciled with its continued expression in differentiated heart tissue and adult heart. Clarification of the role of *XNkx-2.5* is likely to become simpler once target genes regulated by *XNkx-2.5* have been identified.

XNkx-2.5 gives us a molecular correlate of heart specification as defined in the earlier recombination experiments of Sater and Jacobson (1989; 1990b). Nieuwkoop and Faber (1994) describe the differentiation of the heart starting at stage 27/28. This observation also has molecular correlates. The three heart-specific molecular markers: myosin heavy chain α-isoform (Logan and Mohun, 1993), cardiac troponin I (Drysdale et al., 1994), and the cardiac muscle-specific myosin light-chain (Chambers et al., 1994) are all first expressed at this time. The first expression of another muscle gene, cardiac α-actin, occurs in the somites, but the onset of expression in the heart coincides with expression of the other differentiation markers (Logan and Mohun, 1993). It appears that,

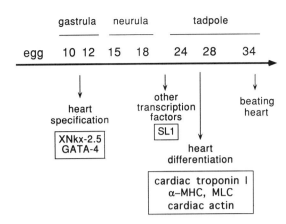

Figure 8. A time-line of *Xenopus* cardiac development. The middle arrow represents the Nieuwkoop and Faber (1994) stages of *Xenopus* development. Heart specification occurs during gastrulation (stage 11-12) . This corresponds very closely to the first expression of *XNkx-2.5* and *GATA-4* suggesting that these two genes are early responses to induction of heart tissue. Cytological heart differentiation occurs at stage 28 (Nieuwkoop and Faber, 1994) and this corresponds to onset of expression of all known heart differentiation products in *Xenopus*. Differentiation precedes actual beating of the heart by over half a day. There is a considerable delay between heart specification and differentiation. A requirement for the expression of other transcription factors, such as SL1, could be the reason for this delay.

in the frog, all of the known myocardial differentiation markers are first expressed at the same time. This simultaneous expression of differentiation markers contrasts with results in chick and mouse (reviewed in Lyons, 1994) where sequential activation of various differentiation products appears to be required for proper heart function.

We have now established a time line of heart development with points representing both embryological events and expression of molecular markers (Fig. 8). A remarkable feature of this time line is the long delay between specification and differentiation. At the time of specification, as defined by embryological means, molecular studies reveal the onset of expression of two heart-specific transcription factors, *XNkx-2.5* and *GATA-4*. The time of differentiation is indicated both by morphological observations and expression of differentiation markers. How can we account for the delay between these two events? Other tissues in the *Xenopus* embryo, such as somite, have a much shorter lag time before differentiation begins and so this delay is not intrinsic to frog embryogenesis.

One reason for the delay may be that the expression of transcription factors such as *XNkx-2.5* and *GATA-4* is the first step in a long pathway and thus there is a requirement for additional factors before differentiation can occur. There are candidates for such factors. Members of the MEF2 family of helix-loop-helix transcription factors are expressed in the heart, albeit not exclusively, prior to heart differentiation in the mouse (Edmondson et al., 1994). A *Xenopus* homologue, SL1, is also expressed shortly before heart differentiation and ectopic SL1 expression alone can initiate some aspects of the heart differentiation program (Chambers et al., 1994). An intriguing possibility is that SL1, or another helix-loop-helix protein may act cooperatively with *XNkx-2.5* to activate gene expression (Grueneberg et al., 1992).

Other potential transcription factors involved in cardiac differentiation include other helix-loop-helix proteins (Litvin et al., 1993), evi-1 (Perkins et al., 1991), *hox 1.5*, (Chisaka and Capecchi, 1991), Mhox (Cserjesi et al., 1992), Xhox 7.1 (Su et al., 1991), N-myc (Moens et al., 1993) and the RXRα receptor (Sucov et al., 1994). Although these factors may play a role in cardiac development, none of them appear to be specific to the heart.

In summary, the embryological studies describing heart development are now being supplemented with cellular and molecular information. In time, the identification of transcription factors and growth factors will eventually allow us to explain the complex series of movements, inductive events and tissue specifications at the molecular level.

REFERENCES

Bodmer, R. 1993, The gene *tinman* is required for specification of the heart and visceral muscles in *Drosophila*. *Development* 118: 719-729.

Cascio, S. and Gurdon, J. B. 1987, The initiation of new gene transcription during *Xenopus* gastrulation requires immediately preceding protein synthesis. *Development* 100: 297-305.

Chambers, A. E., Logan, M., Kotecha, S, Towers, N., Sparrow, D., and Mohun, T. J. 1994, The RSRF/MEF protein SL1 regulates cardiac muscle-specific transcription of a myosin light-chain gene in *Xenopus* embryos. *Genes & Dev.* 8: 1324-1334.

Chisaka, O., and Capecchi, M. R. 1991, Regionally restricted developmental defects resulting from targeted disruption of the mouse homeobox gene *hox-1.5*. *Nature* 350: 473-479.

Cserjesi, P., Lilly, B., Bryson, L., Wang, Y., Sassoon, D. A., and Olson, E. N. 1992, Mhox: a mesodermally restricted homeodomain protein that binds an essential site in the muscle creatine kinase enhancer. *Development* 115: 1087-1101.

Coffey, R.J., Leof, E.B., Shipley, D.D. and Moses, H.L. 1987, Suramin inhibition of growth factor receptor binding and mitogenicity in AKR-2B cells. *J. Cell. Physiol.* 132: 143-148.

Coffin, J. D. and Poole, T. J. 1988, Embryonic vascular development: Immunohistochemical identification of the origin and subsequent morphogenesis of the major vessel primordia in quail embryos. *Development* 102: 735-748.

Cornell, R. A. and Kimelman, D. 1994, Activin-mediated mesoderm induction requires FGF. *Development* 120: 453-462.

Dale, L., and Slack, J. M. W. 1987, Regional specification within the mesoderm of early embryos of *Xenopus laevis*. *Development* 100: 279-295.

DeHaan, R. L. 1963, Migration patterns of the precardiac mesoderm in the early chick embryo. *Exp. Cell Res.* 29: 544-560.

DeRuiter, M. C., Poelmann, R. E., Vander Plas-de Vries, I., Mentink, M. M. T., and Gittenberger-deGroot, A. C. 1992, The development of the myocardium and endocardium in mouse embryos. *Anat. Embryol.* 185: 461-473.

Drysdale, T. A., Tonissen, K. F., Patterson, K. D., Crawford, M. J., and Krieg, P. A. 1994, Cardiac troponin I is a heart specific marker in the *Xenopus* embryo: Expression during abnormal heart morphogenesis *Dev. Biol.* in press.

Edmondson, D. G., Lyons, G. E., Martin, J. F., and Olson, E. N. 1994, *Mef2* gene expression marks the cardiac and skeletal muscle lineages during mouse embryogenesis. *Development* 120: 1251-1263.

Freisel, R., and Brown, S. A. 1992, Spatially restricted expression of fibroblast growth factor receptor-2. *Development* 116:1051-1058.

Garcia-Fernandez, J., Gaguna, J. and Salo, E. 1993, Genomic organization and expression of the planarian homeobox genes *Dth-1* and *Dth-2*. *Development* 118: 241-253.

Gerhart, J. C., and Keller, R. E. 1986, Region-specific cell activities in amphibian gastrulation. *Ann Rev. Cell Biol.* 2: 201-229.

Green, J. B. A., New, H. V. and Smith, J. C. 1992, Responses of embryonic *Xenopus* cells to activin and FGF are seperated by multiple dose thresholds and correspond to distinct axes of the mesoderm. *Cell* 71: 731-739.

Grueneberg, D. A., Natesan, S., Alexandre, C., and Gilman, M. Z. 1992, Human and *Drosophila* homeodomain proteins that enhance the DNA-binding activity of the serum response factor. *Science* 257: 1089-1095.

Grunz, H. 1992, Suramin changes the fate of Spemann's organizer and prevents neural induction in *Xenopus laevis*. *Mech. Dev.* 38: 133-142.

Grunz, H. 1993, Dorsalization of Spemann's organizer takes place during gastrulation in *Xenopus laevis* embryos. *Dev. Growth Diff.* 35: 25-32.

Guazzi, S., Price, M. De Felice, M., Damante, G., Mattei, M-G., and Di Lauro, R. 1990, Thyroid nuclear factor 1 (TTF-1) contains a homeodomain and displays a novel DNA binding specificity. *EMBO J.* 9: 3631-3639.

Han, Y., Dennis, J. E., Cohen-Gould, L., Bader, D. M. and Fischman, D. A. 1992, Expression of sarcomeric myosin in the presumptive myocardium of chicken embryos occurs within six hours of myocyte commitment. *Dev. Dynam.* 193, 257-265.

Harland, R. M. 1991, *In situ* hybridization: An improved whole-mount method for *Xenopus* embryos. *Methods in Cell Biology* **36**: 685-695.

Heikinheimo, M., Scandrett, J. M., and Wilson, D. B. 1994, Localization of transcription factor GATA-4 to regions of the mouse embryo involved in cardiac development. *Dev. Biol.* 164: 361-373.

Jones, S. D., Ho, L., Smith, J. C., Yordan, C., Stiles, C. D., Mercola, M. 1993, The *Xenopus* platelet-derived growth factor-α receptor: cDNA cloning and demonstration that mesoderm induction establishes the lineage-specific pattern of ligand and recptor gene expression. *Dev .Gen* .14: 185-193.

Jones, E. A. and Woodland, H. R. 1987, The development of animal caps in *Xenopus*: a measure of the start of animal cap competence to form mesoderm. *Development* 101: 557-563.

Kao, K. R., Masui, Y., and Elinson, R. P. 1986, Lithium-induced respecification of pattern in *Xenopus laevis* embryos. *Nature* 322: 371-373.

Keller, R. E. 1976, Vital dye mapping of the gastrula and neurula of *Xenopus laevis*. II. Prospective areas and morphogenetic movements of the deep layer. *Dev. Biol.* 51: 118-137.

Kelley, C., Blumberg, H., Zon, L. I., and Evans, T. 1993, *GATA-4* is a novel transcription factor expressed in endocardium of the developing heart. *Development* 118: 817-827.

Kim, Y. and Nirenberg, M. 1989, *Drosophila* NK-homeobox genes. *Proc. Nat. Acad. Sci.* 86: 7716-7720.

Komuro, I and Izumo, S. 1993, *Csx*: A murine homeobox-containing gene specifically expressed in the developing heart. *Proc. Nat. Acad. Sci.* 90: 8145-8149.

Ku, M., and Melton, D. A. 1993, *Xwnt-11*: a maternally expressed *Xenopus wnt* gene. *Development* 119: 1161-1173.

LaBonne, C. and Whitman, M. 1994, Mesoderm induction by activin requires FGF-mediated intracellular signals. *Development* 120: 463-472.

Linask, K. K., and Lash, J. W. 1993, Early heart development: Dynamics of endocardial cell sorting suggests a common origin with cardiomyocytes. *Dev. Dynam.* 195: 62-69.

Lints, T. J., Parson, L. M., Hartley, L., Lyons, I., and Harvey, R. P. 1993, *Nkx 2.5*: A novel murine homeobox gene expressed in early heart progenitor cells and their myogenic descendants. *Development* 119: 419-431.

Litvin, J., Montgomery, M. O., Goldhamer, D. J., Emerson Jr., C. P., and Bader, D. M. 1993, Identification of DNA-Binding protein(s) in the developing heart. *Dev. Biol.* 156: 409-417.

Logan, M and Mohun, T. 1993, Induction of cardiac muscle differentiation in isolated animal pole explants of *Xenopus laevis* embryos. *Development* 118: 865-875.

Lyons, G. E. 1994, *In situ* analysis of cardiac muscle gene program during embryogenesis. *Trends in Card. Med.* 4: 70-77.

McGinnis, W. and Krumlauf, R. 1992, Homeobox genes and axial patterning. *Cell* 69: 283-302.

Mikawa, T., Borisov, A., Brown, A. M. C., and Fishman, D. A. 1992, Clonal analysis of cardiac morphogenesis in the chicken embryo using a replication defective retrovirus: I. Formation of the ventricular myocardium. *Dev. Dynam.* 193: 11-23.

Moens, C. B., Stanton, B. R., Parada, L. F., and Rossant, J. (1993) Defects in heart and lung development in compound heterozygotes for two different targeted mutations at the *N-myc* locus. *Development* 119: 485-499.

Moon, R. T., and Christian, J. L. 1992, Competence modifiers synergize with growth factors during mesoderm induction and patterning in *Xenopus*. *Cell* 71: 709-712.

Muslin, A. J. and Williams, L.T. 1991, Well-defined growth factors promote cardiac development in axolotl mesodermal explants. *Development* 112: 1095-1101.

Nelsen, O. E.,1953, Comparative Embryology of the Vertebrates. The Blakiston Company, New York.

Nieuwkoop, P. D. 1969, The formation of mesoderm in urodelean amphibians. I. Induction by the endoderm. *Roux's Arch. Dev. Biol.* 162: 341-373.

Nieuwkoop, P. D., and Faber, J. 1994, Normal Table of Xenopus laevis (Daudin) Garland Publishing, New York.

O'Brien, T. X., Lee, K. J., and Chien, K. R. 1993, Positional specification of ventricular myosin light chain 2 expression in the primitive murine heart tube. *Proc. Natl. Acad. Sci.* 90: 5157-6161.

Okkema, P. G. and Fire, A. 1994, The *Caenorhabditis elegans* NK-2 class homeoprotein CEH-22 is involved in combinatorial activation of gene expression in pharyngeal muscle. *Development* 120: 2175-2186.

Peng, H. B. 1991, Solutions and Protocols. *Methods in Cell Biology* 36: 657-662.

Perkins, A. S., Mercer, J. A., Jenkins, N. A., and Copeland, N. G. 1991, Patterns of Evi-1 expression in embryonic and adult tissues suggest that Evi-1 plays an important regulatory role in mouse development. *Development* 111: 479-487.

Sater, A. K. and Jacobson, A. G. 1989, The specification of heart mesoderm occurs during gastrulation in *Xenopus laevis*. *Development* 105: 821-830.

Sater, A. K. and Jacobson, A. G. 1990a, The restriction of the heart morphogenetic field in *Xenopus laevis*. *Dev. Biol.* 140: 328-336.

Sater, A. K. and Jacobson, A. G. 1990b, The role of the dorsal lip in the induction of heart mesoderm in *Xenopus laevis*. *Development* 108: 461-470.

Schulte-Merker, S., Smith, J. C., and Dale, L. 1994, Effects of truncated activin and FGF receptors and of follistatin on the inducing activities of Vg1 and activin: Does activin play a role in mesoderm induction? *EMBO J.* 13: 3533-3541.

Sive, H. L., Draper, B. W., Harland, R. M., and Weintraub, H. 1990, Identification of a retinoic acid-sensitive period during primary axis formation in *Xenopus laevis*. *Genes & Dev.* 4: 932-942.

Smith, J. C. 1989, Mesoderm induction and mesoderm inducing factors in early amphibian development. *Development* 105: 665-667.

Smith, J.C. 1993. Mesoderm-inducing factors in early vertebrate development. *EMBO J.* 12: 4463-4470.

Stainier, D. Y. R., and Fishman, M. C. 1992 Patterning the zebrafish heart tube: Acquisition of anteroposterior polarity. *Dev. Biol.* 153: 91-101.

Stainier, D. Y. R., Lee, R. K., and Fishman, M. C. 1993, Cardiovascular development in the zebrafish I. Myocardial fate map and heart tube formation. *Development* 119: 31-40.

Su, M-W., Suzuki, H.R., Solursh, M. and Ramirez, F. 1991, Progressively restricted expression of a new homeobox-containing gene during *Xenopus laevis* embryogenesis. *Development* 111: 1179-1187.

Sucov, H. M., Dyson, E., Gumeringer, C. L., Price, J., and Chien, K. R. 1994, RXRα mutant mice establish a genetic basis for vitamin A signaling in heart morphogenesis. *Genes & Dev.* 8: 1007-1018.

Tannahill D., Isaacs H.V., Close M. J., Peters G., Slack J. M. 1992, Developmental expression of the *Xenopus int-2* (FGF-3) gene: Activation by mesodermal and neural induction. *Development* 115: 695-702.

Tokuyasa, K. T. and Maher, P. A. 1987. Immuno cytochemical studies of cardiac myofibrillogenesis in early chick embryos. I. Presence of immunofluorescence titin spots in premyofibril stages. *J. Cell Biol.* 105: 2781-2793.

Tonissen, K. F., Drysdale, T.D., Lints, T.J., Harvey, R.P. and Krieg, P.A. 1994, *XNkx-2.5*, a *Xenopus* gene related to *Nkx-2.5* and *tinman*: Evidence for a conserved role in cardiac development. *Dev. Biol.* 162: 325-328.

Vize, P. D. and Thomsen, G. H. 1994, Vg1 and regional specification in vertebrates; a new role for an old molecule. *TIG* 10: 371-376.

Weinstein, D. C., Ruiz i Altaba, A., Chen, W. S., Hoodless, P., Prezioso, V. R., Jessel, T. M., and Darnell, J. E. 1994, The winged-helix transcription factor *HNF-3β* is required for notochord development in the mouse embryo. *Cell* 78: 575-588.

Yutzey, K. E., Rhee, J. T., and Bader, D. 1994, Expression of the atrial-specific myosin heavy chain AMHC1 and the establishment of the anteroposterior polarity in the developing chicken heart. *Development* 120: 871-883.

ECTOPIC EXPRESSION OF *Wnt-1* INDUCES ABNORMALITIES IN GROWTH AND SKELETAL PATTERNING OF THE LIMBS

József Zákány[1], Marianne Lemaistre[2] and Denis Duboule[1]

[1]Department of Zoology, University of Geneva, Sciences III, Quai Ernest Ansermet 30, 1211 Geneva 4, Switzerland

[2]European Molecular Biology Laboratory, Meyerhofstrasse 1, Postfach 10.2209, 6900 Heidelberg 1, Germany

INTRODUCTION

Wnt genes are members of a family of vertebrate genes related to the *Drosophila* gene wingless (*wg*, Nusse and Warmus, 1992; McMahon, 1992). They encode secreted molecules (Papkoff and Schryver, 1990; Jue et al., 1992) which are thought to be of importance in patterning and growth control during ontogenesis (McMahon and Bradley, 1990; Thomas et al., 1992; Thomas and Capecchi, 1990; Dickinson et al., 1994). Several such genes are transcribed in localised domains during limb budding and morphogenesis (McMahon, 1992; Gavin et al., 1990; Dealy et al., 1993). We report here further observations on a congenital limb malformation in a mouse transgenic line that ectopically expresses *Wnt-1* in limb buds (Zakany and Duboule, 1993).

RESULTS

Hemizygote animals from this family showed strong skeletal malformations affecting exclusively the four limbs (Fig.1). In the forelimbs, the radius and distal epicondyle of the

Organization of the Early Vertebrate Embryo
Edited by N. Zagris *et al.*, Plenum Press, New York, 1995

humerus were absent. The bony mass was reduced distally and incomplete sets of carpals, metacarpals and phalanges were observed (Fig.1b,c). In hindlimbs, loss of skeletal element primarily affects the autopod; digit I, its metatarsal and tarsal supporting bones were lost as well as digit II. In addition, distal phalanges were missing (Fig.1c,h). In both fore and hindlimbs, proximal-distal fusions were often observed. Distal truncations of tibia, fibula and ulna occured, with fusion between fibula and talus, resulting in a remodelling of the ankle joint (Fig.1c,f).

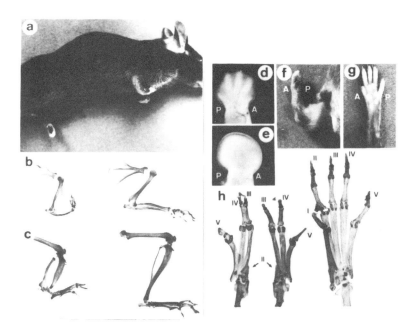

Figure 1. Transgenic *Wnt-1* animals. a) An adult male is shown to illustrate the reduction of the limbs to stumps. The paws are rotated to 90° and the overall sizes of the limbs are reduced. b) Forelimb and c) hindlimb skeletons of a transgenic mouse (left) and its non-transgenic sibling (right). Skeletal elements are shorter. d, e) E13 fetal hindlimbs from either normal (d) or transgenic fetus (e) are shown. Note the larger dimensions of the transgenic footplate. f, g) Morphologies of hindlimbs of abnormal (f) and normal (g) littermates. Note the lack of digit separation and the reduced size of the limb as compared to the control. h) Two abnormal hind paws (left) are shown together with that of a normal littermate (right). In the transgenic paws, digit I as well as the corresponding metatarsal and tarsal bones are missing. The calcaneum articulates directly to the third cuneiforme. The third metatarsal, cuboideum, metatarsal IV and metatarsal V are present, although shorter than usual. The small plantar paired bones on the distal epiphysis of metatarsal III and IV are fused and assume different, abnormal shapes. A, anterior; P, posterior; s, scapula; h, humerus; u, ulna; r, radius; fe, femur; fi, fibula; t, tarsals; ta, talus; ti, tibia; p, phalange; mt, metatarsal. Reprinted with permission from Nature (Zakany and Duboule, 1993). Copyright (1993) MacMillan Magazine Limited.

In order to determine if this dominant phenotype was linked to the ectopic expression of the *Wnt*-1 gene, hemizygote transgenic fetuses were analysed for the distribution of *Wnt-1* transcripts by in situ hybridization (Fig.2 and Zakany and Duboule, 1993). In transgenic embryos, a strong ectopic expression was detected in the limbs (Fig.2). Ectopic *Wnt-1* mRNAs was first observed in day 10 (E10) fetal limbs, distally. In E11 and E12 fetuses, the transgene was highly expressed at the distal periphery of the developing limbs, in a continuous sub-apical domain (Fig.2b). The amount of ectopic transcripts in limbs decreased by E13. Ectopic expression was also detected in the E13 dorsal thalamus and RNAse protection analysis revealed that *Wnt-1* transcripts in limbs did not initiate at an endogenous promoter but rather, in the transgene promoter as protected fragments of the expected sizes were obtained (Zakany and Duboule, 1993). As the ectopic *Wnt-1* expression domain correlated with the localisations of the phenotypic alterations (see below), we conclude that this dominant phenotype is the consequence of ectopic expression of the *Wnt-1* transgene rather than that of insertional mutagenesis, which occurs, as a dominant trait, with extreme low frequency (DeLoia and Solter, 1990; Meisler, 1992).

We analysed the expression of two genes known to be expressed during limb development, in hemizygous mice. *Hoxd-13* is a *Hox* gene involved in the control of patterns in the mesenchymal compartment, probably by regulating local growth rates. *Msx-1* is expressed in both the ectoderm and mesenchyme and was thus postulated to be of importance in epithelial-mesenchymal interactions. In control limbs, the expression domain of the *Msx-1* gene is strongest along the distal periphery of the autopod. Expression is also well detectable between the skeletal condensations. The most anterior metatarsal condensation did not individualize yet in the hindlimb, which is reflected in the absence of the fourth inter-metatarsal expression domain. In transgenic limbs, the expression domain involves the entire distal periphery of the autopods and appears to be abnormally broad, anteriorly. The inter-metacarpal and inter-metatarsal domains of expression are missing, except for a minor focus between the prospective metacarpal IV and III (fl), which however, is separated from the peripheral one. The expanded peripheral *Msx-1* expression domain matches the ectopic *Wnt-1* domain (Fig 2c,d).

In control limbs, *Hoxd-13* expression is strongest in the mantle of the autopodal condensations while a residual signal is seen in the inter-metacarpal and inter-metatarsal domains (Fig.2e). The expression domains extend into the incipient anterior-most condensations in both limbs. In transgenic limbs, *Hoxd-13* expression is detected throughout the posterior three-fourth of the autopod, in both limbs (Fig.2f). The lack of differential distribution reflects the lack of full grown condensations. The anterior fourth of the autopod, which corresponds to the broadening of *Wnt-1* and *Msx-1* domains, remains *Hoxd-13* negative, in contrast to the control limbs. This region would normally give rise to the condensations II and I. These condensations and their derivatives are severely defective in the forelimb and totally absent in the hindlimb of transgenic mice.

Even though hemizygous mice were severely impared in their capacities to mate, mice homozygous for the transgene were obtained at the predicted mendelian ratio. Such specimen showed a reinforced phenotype with limbs heavily truncated in their most distal parts. A further reduction in the length of the tibia was observed (Fig.3) and skeletal elements which were still present in hemizygous animals were absent, as examplified by the metatarsal II and III (Fig.3). Consequently, homozygous limbs were shorter and smaller than hemizygous

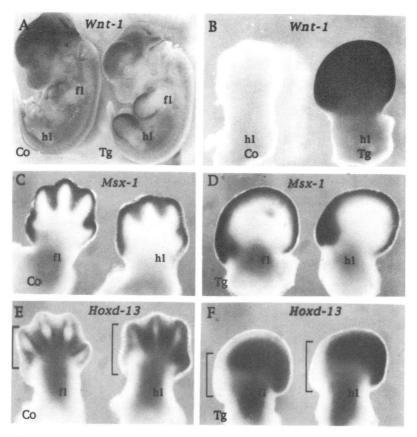

Figure 2. Expression pattern of *Wnt-1*, *Msx-1* and *Hoxd-13* gene products in control and *Wnt-1* transgenic limbs. A. A control and a transgenic fetuses, hybridized with the *Wnt-1* probe. Staining along the roofplate of the central nervous system, and the edge of the IVth ventricle shows the expression of the endogenous *Wnt-1* gene in both fetuses. Staining in limbs of the transgenic fetus is due to ectopic expression of the *Wnt-1* transgene. B. Hindlimbs of control and transgenic fetuses, hybridized with *Wnt-1* specific probe. A strong distal staining is seen in the mesoderm of the transgenic limb. C, D and E, F; Expression pattern of the *Msx-1* and Hoxd-13 genes, respectively. Co, control; fl, forelimb; hl, hindlimb; Tg, transgenic.

ones. This strong alteration of the appendicular skeleton, however, did not affect these animals in their feeding behaviour so that the survival was comparable to that of normal or hemizygous littermates. In several instances, homozygous females were fertile and gave rise to litters of transgenic *Wnt-1* hemizygous mice.

The vertebrate limb develops in a proximo-distal sequence of growth and patterning (e.g. Tabin, 1991). The growth of the stucture is achieved within a distal, subectodermal area. Cells in this 'progress zone' are maintained in a highly mitotic and undifferentiated state under the influence of the apical ectodermal ridge (AER, Saunders, 1977) which is, in turn, stimulated by a factor(s) produced by the underlying mesenchyme cells. While the limb grows, cells continually leave this zone (Summerbell et al., 1973) and condense following a precise pattern of successive bifurcation processes (Shubin and Alberch, 1986). Early condensations are composed of quiescent cells and a mantle of proliferating cells (Herken, 1980) that overlap with the 'progress' zone at the tip of the condensations. The ectopically

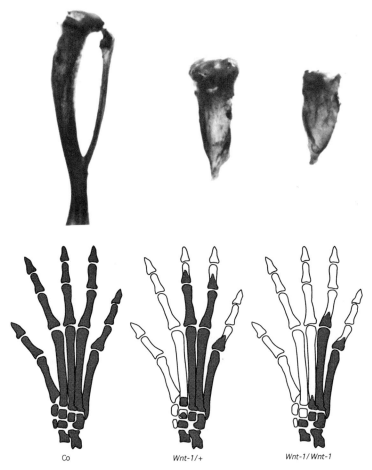

Figure 3. Upper panels: Reduction in the length of the tibia in hemizygous (center) or homozygous (right) animals. The bottom part shows a summary of the overall reduction and truncation of transgenic hemizygous or homozygous hindlimbs. Only those bones indicated in grey are present. Controls are on the left.

expressed *Wnt*-1 RNA accumulates in these distal areas. The presence of the *Wnt*-1 product could thus result in increased numbers of proliferating cells which do not undergo chondrogenic condensation, and could produce the observed phenotype. The analysis of hemizygote limbs at various stages of fetal development revealed a substantial enlargement of the E13 autopod along the dorsal-ventral and anterior-posterior axes (Fig.1d, e), suggesting defective growth control.

In normal E13.5 hindlimbs, the five metatarsal blastemas are distinct and extend to the marginal zone of the footplate with sparser areas and indentations between the tips. These condensations are composed of elongated cells oriented perpendicular to the long axis of the rudiment that are themselves not engaged in DNA synthesis. In transgenic limbs, cells inside the remaining condensations were elongated but BUdR incorporation revealed abnormally sustained DNA synthesis. However, most metatarsal condensations were absent and replaced by a mass of mitotic cells expressing *Wnt-1* transcripts (Zakany and Duboule, 1993).

CONCLUSIONS

In summary, ectopic *Wnt-1* specific RNA accumulation first predates and then colocalizes with the accumulation of non-condensing, proliferating mesenchymal cells. Condensations are observed in more proximal or central parts of the limbs, where *Wnt-1* expression is either weaker or not seen . This spatial and temporal correlation suggests a direct effect of the WNT-1 protein on the retardation of mesenchymal condensation and the accumulation of proliferating mesenchymal cells. The cell masses that are unable to contribute to chondrogenic condensations are apparently arrested at a previous stage of development, as suggested by the primitive, non evolving patterns of both the *Msx-1* and *Hoxd-13* gene transcripts. These alterations in limb development may result from an interference between the ectopic WNT-1 protein and other members of the family of WNT proteins which are normally present. For example, the ectopically produced protein could, through endogenous WNT receptor molecules, prevent distal cells from condensing by maintaining them in a large, abnormal 'progress zone'. The deletions of skeletal elements observed in the adults may reflect the inability of these proliferating cells to condense and undergo chondrogenesis. These results further suggests that members of the *Wnt* gene family may be mediating epithelial-mesenchymal interactions that control limb growth and may thus act at the interface between growth and patterning.

ACKNOWLEDGMENTS

We would like to thank Dr. J. Papkoff for the *Wnt*-1 cDNA clone and Dr. B. Robert for the *Msx-1* probe. J. Z. is on leave from the Institute of Genetics of the Biological Research Center of the Hungarian Academy of Sciences, Szeged, Hungary. This work was supported by funds from the Swiss National Fund, the Canton de Genève, the Claraz foundation and the HFSPO.

REFERENCES

Dealy, C. N., Roth, A., Ferrari, D., Brown, A. M. C., Kosher, R. A., 1993, *Wnt-5a* and *Wnt-7a* are expressed in the developing chick limb bud in a manner suggesting roles in pattern formation along the proximodistal and dorsoventral axes. *Mech. Devel.* 43:175.

DeLoia, J. A., Solter, D., 1990, A transgene insertional mutation at an imprinted locus in the mouse genome. *Development, Supp.* 73.

Dickinson, M.E., Krumlauf, R., and McMahon, A.P., 1993, Evidence for a mitogenic activity of *Wnt-1* in the mammalian central nervous system. *Development* 120:1453.

Gavin, B., McMahon, J. A., and McMahon, A. P., 1990, Expression of multiple novel *Wnt-1/int-1*-related genes during fetal and adult mouse development. *Genes Dev.* 4:2319.

Herken, R., 1980, Proliferation behaviour and surface coat in the limb bud blastema of mouse embryos. An autoradiographic study. *in. Teratology of the Limbs*, Merker, H.-J., Nau, H. and Neubert, D. eds. De Gruyter and Co. Berlin.

Jue, S.F., Bradley, R.S., Rudnicki, J. A., Varmus, H.E., and Brown, A.M.C., 1992, The mouse *Wnt-1* gene can act via a paracrin mechanism in transformation of mammary epithelial cells. *Mol. Cell Biol.* 12:321.

McMahon, A. P., 1992, The *Wnt* family of developmental regulators. *Trends Genet.* 8:263

McMahon, A. P. , Bradley, A., 1990, The *Wnt-1 (int-1)* proto-oncogene is required for development of a large region of the mouse brain. *Cell* 62:1073.

Meisler, M.H., 1992, Insertional mutation of "classical" and novel genes in transgenic mice. *Trends Genet.* 8:341.

Nusse, R., and Varmus, H.E., 1992, *Wnt* genes. *Cell* 69:1073.

Papkoff, J. and Schryver, B., 1990, Secreted *int-1* protein is associated with the cell surface. *Mol. Cell. Biol.* 10:2723.

Saunders, J.W., 1977, The experimental analysis of chick limb bud development. *in:Vertebrate Limb and Somite Morphogenesis* eds. Ede, D.A., J.R. Hinchliffe, and Balls, M. Cambridge University Press.

Shubin, N. H., and Aberch, P. A., 1986, Morphogenetic approach to the origin and basic organization of the tetrapod limb. *Evol. Biol..* 20:319.

Summerbell, D., Lewis, J.H., and Wolpert, L., 1973, Positional information in chick limb morphogenesis. *Nature*, 244:492.

Tabin C., 1991, Retinoids, homeoboxes and growth factors: Toward molecular models for limb development. *Cell* 66:199.

Thomas, K. R. and Capecchi, M. R., 1990, Targeted disruption of the murine *int-1* proto-oncogene resulting in severe abnormalities in midbrain and cerebellar development. *Nature* 346:847.

Thomas, K. R. , Musci, T.S., Neumann, P. E., and Capecchi, M. R., 1991, *Swaying* is a mutant allele of the proto-oncogene *Wnt-1*. *Cell* 67:969.

Zákány, J. and Duboule, D., 1993, Expression of the *Wnt-1* gene in developing limbs correlates with abnormalities in growth and skelettal patterning. *Nature* 362:546.

DIFFERENTIATION OF THE SPLANCHNOPLEURE IN THE GENITAL AREA: MORPHOLOGICAL AND EXPERIMENTAL INVESTIGATIONS ON AVIAN EMBRYOS

Stefanie Himmelmann and Heinz Jürgen Jacob

Ruhr-University Bochum
Department of Anatomy and Embryology
Universitätsstraße 150
D-44780 Bochum, Germany

INTRODUCTION

In amniotes development of the gonad is initiated by formation of the genital ridge in the embryo proper, and by the arrival of primordial germ cells (PGCs) from extra-gonadal sites at the genital ridge during the early stages of embryonic development.

In avian embryos the PGCs are of epiblastic origin (Eyal-Giladi et al., 1981). Previous to the formation of the primitive streak the PGCs translocate into the hypoblast. At primitive streak stages, they separate from the endoderm (Clawson et al., 1969) and gather anteriorly between the area pellucida and area opaca in a crescent-shaped region called the germinal crescent (Swift, 1914). Simultaneously with the development of the area vasculosa, the PGCs enter the vascular system. Singh et al. (1967) observed a maximum number of PGCs in the vascular system at stages 13HH to 15HH according to Hamburger et al. (1951). In the genital area of the embryo, the PGCs leave the blood vessels of the splanchnic mesoderm and finally migrate into the genital ridges (Fujimoto et al., 1989). In vitro- and in vivo-experiments suggest that the PGCs are chemotactically attracted by the genital ridges (Dubois, 1968; Kuwana, 1986) and that this chemotaxis is not species-specific (Rogulska, 1971; Nakamura et al., 1991).

In vertebrates, the formation of the early gonadal primordia is the same in all embryos, whether destined to become male or female (indifferent stage). In chick and quail embryos the development of the gonads begins between days 2 and 3 of incubation (Hamilton, 1952). The gonadal primordia arise from a strip of the coelomic epithelium of the splanchnopleure, called the germinal epithelium, which can be distinguished from the adjacent splanchnopleure by its thickness (Ukeshina et al., 1987). Morphological investigations show that the germinal epithelium appears at the 20 somite level and extends posteriorly to the somite 26 or 27 (Hamilton, 1952). Rodemer et al. (1986), however, experimentally determined the prospective gonad forming areas extending from the somite level 17 to 21. During the indifferent stage of the gonadal primordia, primary sexual cords arise as proliferations of the germinal epithelium (Swift, 1915) and carry numerous PGCs with themselves into the underlying mesenchyme. These are destined to form the seminiferous tubules in the male and the medullary cords in the female. The sex of the embryo can be determined definitely beginning at day 6 and 7 of incubation (Hamilton, 1952): morphologically by the relative sizes of the two gonads[1] and histologically by the sex-specific differentiation of the cortex and medulla.

Till this day, two aspects of gonadal development are obscure: 1. when does the splanchnopleure of the prospective gonad forming area become determined to develop into the genital ridge and 2. which factors control the sex-specific differentiation of the gonadal primordia. These questions can be answered by means of heterotopic grafting: the splanchnopleure of the prospective gonad forming areas is excised from quail embryos at various stages of development, and it is grafted into the coelom of chick hosts. The intracoelomically developed fragments are examined histologically. The fate of the grafted quail tissues can be followed using the Feulgen reaction (Le Douarin et al., 1969) or by staining with Weigert's iron hematoxylin (Himmelmann et al., 1994). The genetic sex of the donor embryos is identificated using Gasc's method (1973) and is compared with the sex of the intracoelomically developed gonadal fragments. Moreover, we are interested in the attraction of PGCs by the grafts (by splanchnic epithelium). Quail and chick germ cells (GCs) are distinguished by an immunohistological procedure (Pardanaud et al., 1987) in the gonadal fragments.

MATERIALS AND METHODS

Interspecific grafting experiments between quail and chick embryos are carried out because it is possible to distinguish between cells of these two species (quail-chick-marker-technique according to LeDouarin et al., 1969).

Quail embryos in stages from 13HH to 20HH of development are used as donors. A blood sample is taken and treated according to Gasc's method (1973). The genetic sex of the quail embryos is identified from nuclear metaphase stages of Giemsa stained blood smears.

The donor embryos are freed from their yolk and handled in phosphate buffered saline. The splanchnopleure or splanchnic coelomic epithelium with underlying mesenchyme is ex-

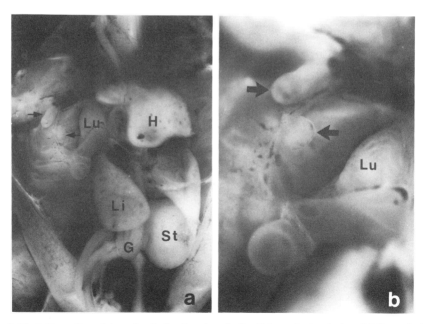

Figure 1. Host situs with an intracoelomic developed graft after 7 days of reincubation. (a) General view. 8x. (b) Detail. 22x. Arrows: graft; G: gut; H: heart; Li: liver; Lu: lung; St: stomach.

---

[1] In the female chick and quail embryos the right gonadal primordium gets reduced.

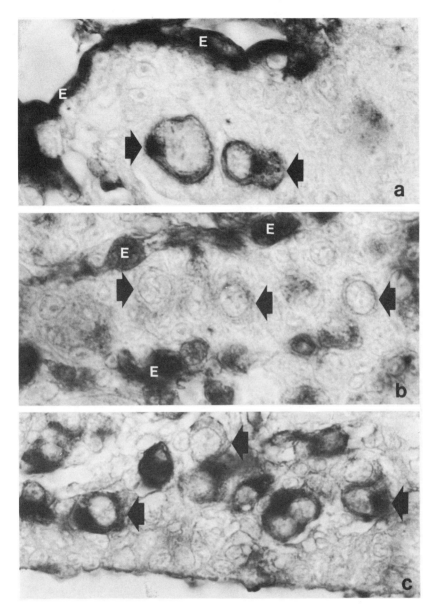

Figure 2. Stage and sex specific appearance of quail GCs after immunostaining with QH1. (a) Detail of an indifferent gonadal primordium at day 4 of incubation. GCs and endothelial cells are obvious QH1-positive. 1875x. (b) Detail of a male gonadal primordium at day 10 of incubation. GCs show no reaction with QH1 whereas endothelial cells are still stained. 1875x. (c) Detail of the cortex of a female gonadal primordium at day 12 of incubation. GCs are QH1-positive. 1875x. Arrows: GCs; E: endothelial cells.

cised from various levels of the prospective gonad forming areas[1]. The length of the grafts varies from 300 μm to 700 μm. Sometimes the mesodermal tissues remain associated with endoderm or ectoderm.

To test the origin and vitality of the excised tissues, some explants are prepared as semithin sections. The great majority of the excised tissues are grafted into the intraembryonic coelom of chick embryos incubated about 2,5 to 3 days. Hamburger's method (1960) of intracoelomic grafting is used.

After 7 days of reincubation the chick hosts are dissected (Figure 1). Intracoelomically developed quail fragments are fixed in Serra's fluid and stained according to the Feulgen reaction. Alternatively Weigert's iron hematoxylin is used. Both stainings permit distinction of quail and chick cells (Himmelmann et al., 1994): In quail cells the nuclei normally possess a large quantity of nucleolus-associated heterochromatin whereas chick nuclei contain dispersed granules of heterochromatin.

In PGCs, unfortunately, the quail nuclear marker is less obvious (Tachinante, 1974). Hence, an immunohistological technique (Himmelmann et al., 1994) for demonstrating quail PGCs in previously Feulgen stained tissues is applied to selected sections of intracoelomically developed fragments. We use the monoclonal antibody QH1 which specifically binds to quail endothelial cells as well as to quail PGCs (Pardanaud et al., 1987). In gonads of normal quail embryos, we notice stage and sex specific differences in QH1's affinity for GCs (Figure 2): At day 10 of incubation the GCs in the ovary are still QH1-positive whereas the GCs in the testis show no reaction with QH1. The quantitative analysis of GCs in intracoelomically developed gonadal fragments is restricted to the female fragments. In chimeric combinations of quail and chick tissues the origin of GC can be determined with certainty because QH1 has no affinity for chick cells.

RESULTS

In this study, 117 intracoelomic graftings were carried out. 82 host embryos survived 7 days of reincubation and reached stages 35HH to 37HH revealing normal development. A total of 26 grafts were isolated and evaluated histologically. The results are summarized in Table 1: By means of semithin sections (results not shown) origin and vitality of the excised quail tissues are tested. PGCs, however, are noticed in the explants obtained from donor embryos at stage 15 to 16HH (27 pairs of somites) or older stages. Hence, the grafts are subdivided into two groups: group I without PGCs and group II with PGCs from the quail donor. In group I the excised mesodermal tissues necessarily remain associated with endoderm or ectoderm to obtain intracoelomic development. The gonad forming potency of the grafted tissues is dependent on the age of the donor embryos. In quail embryos with at least 27 pairs of somites (stage 15 - 16HH), the splanchnic epithelium with underlying mesenchyme is able to develop into a gonad.

The sex of the intracoelomically developed gonadal fragments corresponds to the genetic sex of the donor embryo. The host embryo does not affect sex determination and formation of the gonadal grafts. There is no relation between the sex of the host gonads and the sex of the intracoelomically developed gonadal fragments (Table 2). Thus we notice both identical and different sex combinations.

The histological picture of the intracoelomically developed gonadal fragments corresponds to normal quail gonads at day 10 of incubation: The male gonadal fragments (Figure 3 a+b) show a characteristic proliferation of the medulla and regression of the cortex. The testes are covered with the thin germinative epithelium. The cortex is flattenend and has condensed itself into an encapsulating layer (tunica albuginea). Within the medulla, the sexual cords contain the GCs and are separated by areas of mesenchymal cells (interstitium). In contrast to the testes, the ovaries (Figure 3 c+d) display intense proliferation resulting in a voluminous cortex, whereas the medulla becomes more loosely organized. Clusters of GCs separated by cortical cords of somatic cells are visible in the cortex, and distended medullary cords (rete ovarii) are present in the medulla.

The intracoelomically developed gonadal fragments always contain GCs. By means of immunostaining with QH1 the origin of the GCs is analysed in 4 female gonadal fragments

---

[1] According to Rodemer et al. (1986) and Hamilton (1952) the prospective gonad forming area is found in the region from somite level 17 to 26.

Table 1. Summary of the intracoelomically developed grafts.
According to semithin sections the grafts are subdivided into two groups: group I without PGCs and group II with PGCs of the quail donor. The number refers to the isolated and histologically examined grafts.

| group | excised tissues (before reincubation) | | intracoelomic grafts |
	number	grafted tissues	developed tissues
I	**2** (1x stage 13HH, 1x stage 14HH)	splanchnopleure and entoderm	gut
	1 (1x stage 14–15HH)	intermediate cell mass with Wolffian duct, splanchnopleure and entoderm	mesonephros, Wolffian duct and gut
	3 (1x stage 13HH, 2x stage 14HH)	ectoderm, somatopleure, intermediate cell mass with Wolffian duct and splanchnopleure	integument (pulps of feathers), costal cartilage, suprarenal gland, mesonephros and Wolffian duct
	5 (2x stage 13HH, 1x stage 14HH, 1x stage 15HH, 1x stage 15–16HH)	ectoderm, somatopleure, intermediate cell mass with Wolffian duct, splanchnopleure and entoderm	integument (pulps of feathers), costal cartilage, mesonephros, Wolffian duct, gonad (at stage 15–16HH) and gut
II	**3** (1x stage 16HH, 2x stage 17HH)	ectoderm and coelomic epithelium with underlying mesenchyme	integument (pulps of feathers), costal cartilage, müllerian duct, mesonephros and gonad
	10 (1x stage 16HH, 1x stage 17HH, 2x stage 18HH, 3x stage 19HH, 3x stage 20HH)	coelomic epithelium with underlying mesenchyme	mesonephros and gonad
	2 (2x stage 16HH)	coelomic epithelium with underlying mesenchyme and entoderm	mesonephros, gonad and gut

(figure 4). The results are summarized in table 3: Most of the GCs originate from the quail donor. Along with the grafted tissues, PGCs were carried into the coelom of the host. These PGCs increase by mitosis. Additionally we find chick GCs which must have been migrating from the host into the graft.

Table 2. Sex combinations between gonads of host and intracoelomic developed gonadal fragments.
"m" stands for male sex, and "f" stands for female sex. The number ($\Sigma = 13$) refers to the combinations discovered in the grafting experiments.

| combination | sex | | number |
	gonad of host	intracoelomic graft (gonad of donor)	$\Sigma = 13$
equal sex	m	m	4
	f	f	4
different sex	m	f	2
	f	m	3

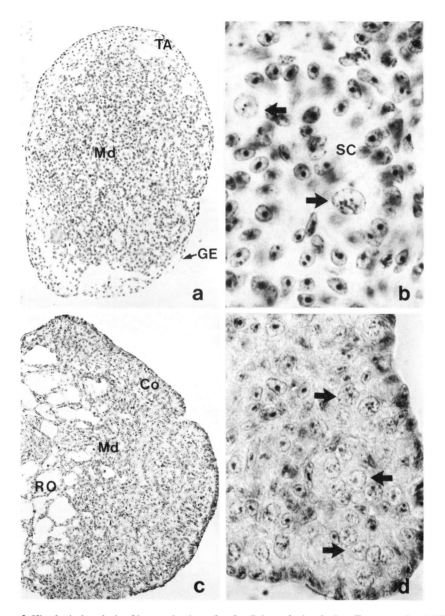

Figure 3. Histological analysis of intracoelomic grafts after 7 days of reincubation. Two examples. (a) Feulgen stained transverse section through the fragment of a male gonadal primordium. 268x. (b) Detail of a: GCs in the sexual cords of the medulla. 1542x. (c) Transverse section through the fragment of a female gonadal primordium stained with Weigert's iron hematoxylin. 233x. (d) Detail of c: clusters of GCs in the cortex. 1200x. Arrows: GCs; Co: cortex; GE: germinative epithelium; Md: Medulla; RO: rete ovarii; SC: sexual cords; TA: tunica albuginea.

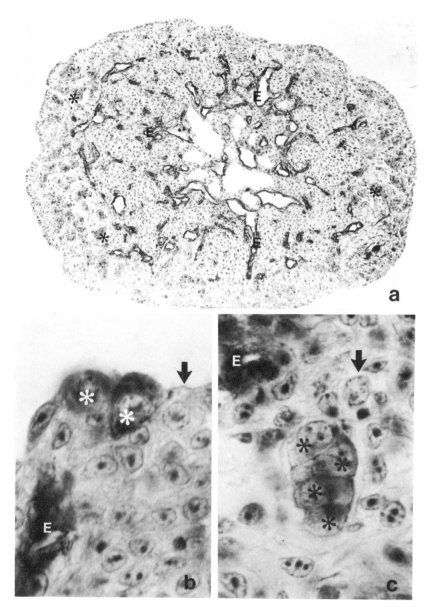

Figure 4. Immunohistological analysis of GCs in intracoelomic developed gonadal primordia. Previously Feulgen stained transverse section immunostained with QH1 and restored by Weigert's iron hematoxylin. (a) Fragment of a female gonadal primordium. 252x. (b + c) Details of a: QH1-positive GCs originate from the quail donor. From the host chick GCs migrate into the graft. 2083x. Arrows: chick GCs; asterisks: quail GCs; E: quail endothelial cells.

Table 3. Quantitative analysis of GCs in intracoelomic developed gonadal fragments.
The number of serial sections (8μm thick) corresponds to the length of the gonadal grafts. The number of evaluated sections is put in brackets. The sum (Σ) of PGCs exclusively refers to the evaluated sections. The mean value (x) stands for the number of PGCs per section.

excised tissue		intracoelomic graft (gonadal tissue)					
stage (of donor)	size (region according to somites)	length (number of sections)		PGCs			
				quail		chick	
				Σ	$\bar{x}$	Σ	$\bar{x}$
17HH	20–25	124	(18)	1945	108	17	1
18HH	20–25	141	(19)	5120	270	158	8
19HH	17–25	92	(11)	1116	102	175	16
20HH	17–20	90	(6)	268	45	11	2

DISCUSSION

The gonadal primordium arises from a strip of thickened coelomic epithelium of the splanchnopleure, and it evidences a high capacity for self-differentiation (Greenwood, 1924; Haffen, 1977; Haffen et al., 1977).

In the present study gonadal development is investigated by means of the quail-chick-marker-technique (Le Douarin et al., 1969). We are interested in the following questions: 1. when is the splanchnopleure of the prospective gonad forming area determined to form the gonadal primordia in avian embryos, 2. are the PGCs attracted by the splanchnic epithelium or by tissues originated from the splanchnopleure, and 3. which factors control the sex-specific differentiation of the gonadal primordia.

Heterotopic grafting experiments are carried out. Quail embryos from stages 13HH to 20HH are used as donors. In the region from somite level 17 to 26 (prospective gonad forming areas), splanchnopleure or coelomic epithelium with underlying mesenchyme is excised. The tissues are grafted into the intraembryonic coelom of chick hosts which were already incubated for 2,5 to 3 days.

Origin and vitality of the excised quail tissues is tested by means of semithin sections. Moreover, we only notice PGCs in the grafts obtained from donors of stage 15 - 16HH (27 pairs of somites) or older stages. Morphological investigations in chick embryos correspond to our observation: According to Meyer (1964) and Ukeshima et al. (1987) PGCs begin to settle in the coelomic epithelium at stage 16HH of development.

The prospective gonad forming potency of the grafts is dependent on the age of the donor embryo. Beginning at stage 15-16HH (27 pairs of somites) of quail embryos, the splanchnic epithelium combined with underlying mesenchyme is able to develop into a complete gonad. At the same time, the presence of PGCs is detected in the splanchnopleure (semithin sections). Hence, we suppose that PGCs exert a positive influence on survival and differentiation of the splanchnopleure at a heterotopic site. We exclude that PGCs produce an inductive effect because sterile gonadal primordia still possess the capacity for sex-specific self-differentiation (McCarrey et al., 1978) independent of the presence of PGCs.

The intracoelomically developed gonadal fragments always contain GCs. By means of immunostaining with QH1 (Pardanaud et al., 1987) the origin of GCs can be analysed quantitatively. Most of the GCs originate from the quail donor. Along with the grafted tissues PGCs have been carried into the coelom of the host. Their number increases strongly by mitosis. Additionally we find chick GCs which must have been actively migrating from the host embryo into the graft. Migration may depent on chemotaxis as supposed by in-vitro (Dubois, 1968; Kuwana et al., 1986) and in-vivo experiments (Rogulska, 1969).

The sex-specific differentiation of intracoelomically developed gonadal fragments corresponds to the donor sex which has been genetically identified by nuclear metaphase stages of Giemsa stained blood smears (Gasc, 1973). Haffen (1977) gets similar results in vitro. In our experiments the sex of the host embryo does not affect the differentiation of the graft. We suppose that sexual hormones are not produced during the reincubation period. Haffen et al. (1977) are able to change the sex specific features of the gonads in avian embryos by injection of male and female sexual hormones respectively.

ACKNOWLEDGEMENTS

The authors are grateful to Dr. I. Flamme for his help with the immunohistological procedures. We thank H. Hake, U. Ritenberg and M. Köhn for excellent technical assistance and R. Kumpernatz for typing the manuscript.

REFERENCES

Clawson, R.C. and Domm, L.V., 1969, Origin and early migration of primordial germ cells in the chick embryon: a study of the stages definitive primitive streak through eight somites, Am. J. Anat. 125: 87- 112

Dubois, R., 1968, La colonisation des ébauches gonadiques par les cellules germinales de l'embryon de poulet, en culture in vitro, J. Embryol. exp. Morph. 20: 189-213

Eyal-Giladi, H. and Kochav, S., 1981, Avian primordial germ cells are of epiblastic origin, J. Embryol. exp. Morph. 65: 139-147

Fujimoto, T., Ukeshima, A., Miyayama, Y., Kuwana, T., Yoshinaga, K. and Nakamura, M., 1989, The primordial germ cells in Amniotes: their migration in vivo and behaviors in vitro, in: Motta, P.M. (ed.), "Developments in ultrastructure of reproduction: a celebrative symposium: The "Opere Omnia" of Marcello Malpighii; proceedings of the VIII. International Symposium on Morphological Sciences", held in Rome, July 10-15, 1988, New York: 13-21

Gasc, J.M., 1973, Sur les résultats d'identification précoce du sexe génotypique des embryons d'oiseaux, C. R. Acad. Sci. Paris 277: 1925-1928

Greenwood, A.W., 1925, Gonad grafts in the embryonic chicks and their relation to sexual differentiation, Brit. J. Exp. Biol. 2: 165-187

Haffen, K., 1977, Sexual differentiation of the ovary, in: Zuckermann, L. and Weir, B.J. (eds), "The Ovary I", Academic Press, New York, San Francisco, London: 69-112

Haffen, K. and Wolff, E., 1977, Natural and experimental modification of ovarian development, in: Zuckermann, L. and Weir, B.J. (eds.), "The Ovary I", Academic Press, New York, San Francisco, London: 393-446

Hamburger, V., 1960, "A manual of experimental embryology", Univ. of Chicago Press, Chicago

Hamburger, V. and Hamilton, H.L., 1951, A series of normal stages in the development of the chick embryo, J. Morph. 88: 49-92

Hamilton, H.L., 1952, "Lillie's development of the chick", Holt, Rinehart and Winston, New York

Himmelmann, S., Hake, H. and Jacob, H.J., 1994, Restoring and immunohistological examination of Feulgen stained avian embryonic tissues using iron hematoxylin and endothelial cell specific antibody, Biotech. Histochem. 69: 55-59

Kuwana, T., Maeda-Suga, H. and Fujimoto, T., 1986, Attraction of chick primordial germ cells by gonadal anlage in vitro, Anat. Rec. 215: 403-406

Le Douarin, N. and Barq, G., 1969, Sur l'utilisation des cellules de la cailles japonaise comme "marqueurs biologiques" en embryologie expérimentale, C. R. Acad. Sci. Paris 269: 1543-1546

McCarrey, J.R. and Abbott, U.K., 1978, Chick gonad differentiation following excision of primordial germ cells, Dev. Biol. 66: 256-265

Meyer, D.B, 1964, The migration of the primordial germ cells in the chick embryo, Dev. Biol. 10: 154- 190

Nakamura, M., Maeda, H. and Fujimoto, T., 1991, Behavior of chick primordial germ cells injected into the blood stream of quail embryos, Okajimas Folia Anat. Jpn. 67: 473-478

Pardanaud, L., Buck, C. and Dieterlen-Lièvre, F., 1987, Early germ cell segregation and distribution in the quail blastodisc, Cell Differ. 22: 47-60

Rodemer, E.S., Ihmer, A. and Wartenberg, H., 1986, Gonadal development of the chick embryo following microsurgically caused agenesis of the mesonephros and using interspecific quail-chick chimaeras, J. Embryol. exp. Morph. 98: 269-285

Rogulska, T., 1969, Migration of the chick primordial germ cells from the intracoelomically transplanted germinal crescent into the genital ridge, Experientia 25: 631-632

Rogulska, T. Ozdzenski, W. and Komar, A., 1971, Behaviour of mouse primordial germ cells in the chick embryo, J. Embryol. exp. Morph. 25: 155-164

Singh, R.P. and Meyer, D.B., 1967, Primordial germ cells in blood smears from chick embryos, Science 156: 1503-1504

Swift, C.H., 1914, Origin and early history of the primordial germ cells in the chick, Am. J. Anat. 15: 483-516

Swift, C.H., 1915, Origin of the definitive sex cells in the female chick and their relation to the primordial germ cells, Am. J. Anat. 18: 441-470

Tachinante, F., 1974, Sur les echanges interspecifiques des cellules germinales entre le poulet et la caille, en culture organotypique et en greffes coelomiques, C. R. Acad. Sci. Paris 278: 1895-1898

Ukeshima, A., Kudo, M. and Fujimoto, T., 1987, Relationship between genital ridge formation and settlement site of primordial germ cells in the chick embryos, Anat. Rec. 219: 311-314

Participants in the NATO - ASI "Organization of the early vertebrate embryo".
(Photograph supplied by L. Chatelin)

INDEX